CORPORATE AVIATION MANAGEMENT

Second Edition

KENNETH J. KOVACH

KENDALL/HUNT PUBLISHING COMPANY
4050 Westmark Drive   Dubuque, Iowa 52002

Previously entitled *Corporate and Business Aviation.*

Cover photo: Gulfstream IV-SP (Special Performance) jet. Courtesy of Gulfstream.

Copyright © 1994, 1998 by Kendall/Hunt Publishing Company

Library of Congress Catalog Card Number: 98-66415.

ISBN 0-7872-4482-1

All rights reserved. No part of this publication may be reproduced, stored in a retrieval system, or transmitted, in any form or by any means, electronic, mechanical, photocopying, recording, or otherwise, without the prior written permission of the copyright owner.

Printed in the United States of America
10 9 8 7 6 5 4 3 2 1

About the Author

Dr. Ken J. Kovach instructs for various schools, including Embry-Riddle Aeronautical University (Associate Professor & Outstanding European Faculty Member 1994 & 1996), University of Maryland, and other universities (over 320 total courses taught). Prior to his teaching positions, he worked for 23 years in the United States Air Force, performing duties in logistics, transportation, plans, operations, patient care, and administration. His duties involved the European Airborne Command and Control Post (EC-135), management in Joint Command and Joint Chief of Staff activities, aerial delivery, air terminal and Logair, air freight, operations and plans, air passenger and cargo movements, railroad operations, hospital care, vehicle maintenance & operations, SATO (schedule airlines) activities, traffic management, and VIP passenger transportation.

Initially, Ken spent almost five years as an enlisted medical technician specializing in psychiatric care. His qualifications through the Airman Education and Commissioning Program enabled him to earn his Air Force commission as a Transportation officer. In this field, he attended intensive training programs in traffic management, personnel management, logistics, air cargo, air passenger movements, strategic management and planning, and vehicle operations & maintenance. He flew on numerous combat missions in Vietnam while supervising aerial delivery, air cargo, and passenger air operations. Ken supervised the movement of the entire 1st Calvary division during a major combat relocation, earned the airman's and air medal, and was recognized as the Air Force's Outstanding Transportation Officer in 1969. In total, he earned over 30 DOD and Air Force awards during his career. Within Air force Logistics Command, Ken negotiated civilian contracts, represented management in labor disputes, and was directly involved in employee/employer settlements. Many times, he has been involved in major logistical activities, coordinating closely with Supply and Procurement officials in conducting important domestic and international air movements of personnel and material.

In transportation planning and management, Ken supervised the largest United States Air Force railroad operations; operated one of two DOD tire depots for aircraft tires; operated the northeast United States Logair activity; managed all aspects of aerial port activities; developed and supervised base mobility plans at RAF Mildenhall and Wright-Patterson AFB; managed a 1500 vehicle fleet and civilian contract operations; provided control and expertise relating to air/ground movement of cargo and passengers (70,000 tons/100,000 passengers annually); contracted and supervised commercial air cargo carriers; prepared industrial and operations' budgets; and recruited, trained, and supervised administrative staff.

Ken has conducted major research efforts in personnel management, training, labor resources, teaching, personnel and student assessment, and corporate aviation. His publications include this *Corporate Aviation Management* text, an NBAA program in aircraft selection, outfitting, and retrofitting, three independent study courses for Embry-Riddle, a research writing guide for University of Maryland, a Major Applied Research Project, and various educational writings.

Dr. Kovach received his doctorate from Nova University (EdD-Higher Education), Master from Wayne State University (MA-Guidance & Counseling), and Bachelor from the University of Tennessee (BS-Business). Ken is active with the Institute of Transport Administration, a founder member of the Institute of Air Transport, member of the American Statistical Association and American Counseling Association, advisor for corporate/business aviation, and committee chair for research and statistics' programs.

Ken and his wife, Sally, have two daughters, Katie and Christine. His interests are fitness, sports, research activities, and helping students learn.

Contents

List of Figures   ix
List of Tables   xiii
Preface   xv
Acknowledgments   xvii

1 History of Corporate and Business Aviation

Overview   1
Goals   1
History   1
    Background   1
    U.S. Aviation Structure   3
    General Aviation   4
    Early Days (1903–1940s)   6
    World War II   9
    Post-World War II
      (1950s–1960s)   11
    1970s   14
    1980s–1990s   15
Summary   18
Activities   19
Chapter Questions   19

2 Regulations and Associations

Overview   21
Goals   21
Regulation Development   21
Aviation Regulations   22
    Airline Deregulation   28
    Fringe Benefit Rule   30
    General Aviation Revitalization
      Act   30
IBAC and Regional Associations   33
    International Business Aviation
      Council   33
    National Business Aviation
      Association   34
        Who are NBAA's
          Members?   36
        Diversity of NBAA
          Membership   37
        NBAA Activities   39
    Canadian Business Aircraft
      Association   39
    European Business Aviation
      Association   44
    German Business Aircraft
      Association   45
    Business Aircraft Users
      Association, Ltd.   45
    Other Business Aircraft
      Associations   45
Summary   46
Activities   46
Chapter Questions   47

3 Value/Benefit of Using Business Aircraft

Overview   49
Goals   49
Introduction   49
Value/Benefits of Business Aviation   50
Disadvantages of Using a Business
    Aircraft   61
Overview of Benefits   63
Programs and Promotions   64
    Fortune 500 Study   64

v

No Plane-No Gain   66
　　　Travel$ense® Program   66
　Summary   67
　Activities   68
　Chapter Questions   68

4 Corporate/Business Aviation Decision

　Overview   69
　Goals   69
　Introduction   69
　Three Cs   70
　Transportation Choices   72
　Determining the Transportation Service   72
　Value of an Executive/Employee's Worth   73
　Travel Analysis   78
　　　Data Collection   78
　　　Spreadsheet   79
　　　Travel Pattern Map   80
　　　Travel Pattern Graph   81
　Travel$ense® Program   85
　Business Strategies   87
　Summary   88
　Activities   89
　Chapter Questions   89

5 Business Aviation Aircraft

　Overview   91
　Goals   91
　Introduction   91
　General Aviation   92
　　　Statistical Analysis   95
　　　NBAA Fleet   98
　Airports and Pilots   101
　Aircraft Manufacturers   104
　Aircraft Selection   106
　　　Project Management   106
　　　Aircraft Selection Processes   107

　Aircraft Costs   109
　Aircraft Selection Matrix   110
　Obtaining Aircraft   113
　　　Buying   113
　　　Leasing   114
　　　Chartering   114
　　　Time Sharing, Joint Ownership, and Interchange   115
　　　Fractional Ownership   116
　　　　　Business Jet Solutions   116
　　　　　Executive Jet Aviation   116
　Aircraft Examples   117
　Summary   117
　Activities   160
　Chapter Questions   160

6 Aviation Department Management

　Overview   161
　Goals   161
　Introduction   161
　Fixed-Base Operators   162
　Aircraft Management Companies   167
　Aviation Department   167
　　　Aviation Department Manager   167
　　　Organizational Structure   174
　　　Placement   177
　　　Aviation Personnel   177
　Recruitment Sources   180
　Selection Process   181
　Number of Pilots and Mechanics   183
　Salaries   184
　Other Department Personnel   185
　Budgeting   186
　Operations Manual   188
　Air Crew Expenses   191
　Air Crew Certificates and Ratings   191
　Other Flight Management Issues   192
　Summary   193
　Activities   194
　Chapter Questions   195

7 Flight Operations

Overview   197
Goals   197
Introduction   198
General Flight Topics   199
Transportation Requests   200
Scheduling/Dispatching   200
Crew Duty Limitations   203
Personal Appearance and Conduct   205
Use of Intoxicants or Drugs   207
Carrying Drugs and Weapons   207
Air Crew Expenses   207
Air Crew Certificates and Ratings   208
Aircraft Fueling   208
Flight Preparations   210
    Preflight Planning   210
    Minimum Navigation Performance Specifications   211
    Reduced Vertical Separation Minimum   211
    Documentation   212
    Communications   212
    Flight Crew Health and Fitness   212
    Flight Handlers   212
    ETP Computations   213
    Miscellaneous Topics   214
Enroute Operations   214
Cockpit Resource Management   215
Extended Range Operations   216
Postflight Actions   216
Other Flight Operations Topics   217
    Air Support of Billund   223
    Universal Weather & Aviation, Incorporated   223
Summary   227
Activities   231
Chapter Questions   232

8 Aircraft Maintenance

Overview   233
Goals   233
Introduction   234
Inspection Programs   234
Aviation Maintenance Department   235
Maintenance Issues   236
    Technicians   236
    Minimum Equipment List (MEL)   236
    ATA System Codes   238
Aircraft Maintenance Methods   238
    In-House Maintenance   241
    Contract Aircraft Maintenance   242
    Combination Aircraft Maintenance   242
    CAMP Systems Incorporated   243
    SaSIMS System   248
Summary   249
Activities   254
Chapter Questions   254

9 Safety and Security

Overview   255
Goals   255
Introduction   255
Security   257
Information/Physical Security   258
Aircraft Security   260
Safety   261
Aviation Safety Statistics   262
Accident Causes   263
Safety Programs   265
    ASRS Program   266
    NTSB Safety Programs   267
Accident Investigations   268
Miscellaneous   269

Summary   270
Activities   271
Chapter Questions   272

10 Current/Future Issues

Overview   273
Goals   273
Introduction   274
SMART OPS   276
    Safety/Security   276
    Management   277
    Aircraft   278
    Regulations   279
    Technology   280
    Operations   281
    Planning and Systems   281
Miscellaneous   282
Summary   282
Activities   283
Chapter Questions   283

Appendices

A. Glossary   285
B. U.S. Government Printing Offices   301
C. Aviation Internet Addresses   303
D. Federal Aviation Regulations   307
E. State Aviation Offices   313
F. Aviation Associations   317
G. Aviation Manufacturers   321
H. Company Operations Manual Extracts   325
I. Case Studies   355

References

Index

List of Figures

1-1 Business Aviation 2
1-2 Sikorsky 9
1-3 Gulfstream I Aircraft 12
2-1 FAA Regional Boundaries 23
2-2 Hub-and-Spoke Model 29
2-3 Annual New U.S. Manufactured General Aviation Unit Shipments/Billings 32
2-4 Annual New U.S. Manufactured General Aviation Aircraft Shipments by Type of Aircraft 33
2-5 NBAA Member Companies, 1979–1996 38
2-6 NBAA Members by State 39
2-7 CBAA Organization Chart 42
2-8 CBAA Canadian Membership by Province, Summer 1991 43
2-9 CBAA U.S. Membership by State, Summer 1991 43
2-10 EBAA Position Policy Papers 44
3-1 Airports Serving Metropolitan Atlanta 54
3-2 Travel Time and Convenience Business Aircraft Service Compared to Scheduled Airline Service 55
3-3 Ground Time Requirements 57
4-1 Spread Sheet 79
4-2 Travel Pattern Map 80
4-3 Travel Pattern Graph (Expando Incorporated) 81
4-4 Travel Pattern Graph 83
5-1 General Aviation Active Aircraft by Primary Use 93
5-2 General Aviation Hours Flown by Primary Use 94
5-3 Number of Active General Aviation Aircraft by Type and Primary Use—1995 (excluding Commuters) 95
5-4 General Aviation Hours Flown by Aircraft Type and Primary Use—1995 (in thousands of hours) 96
5-5 General Aviation Aircraft Shipments 97
5-6 U.S. Civil Aircraft Imports from the World—1992–1996 Units and Dollar Value ($millions) 98
5-7 Civil Helicopter Shipments 99
5-8 NBAA Member Aircraft by Weight and Type 100
5-9 U.S. Civil and Joint Use Airports, Heliports, Stolports, and Seaplane Bases on Record by Type of Ownership—1996 102
5-10 Estimated Active Pilots and Flight Instructors by FAA Region and State—1996 103
5-11 Estimated FAA Active Pilot Certificates Held by Category and Age Group of Holder—1996 104
5-12 Extract of B&CA's Planning and Purchasing Handbook 105
5-13 Aircraft Operating Costs Form 111
5-14 Aircraft Selection Matrix 112
5-15 FlexJet Benefits 118
5-16 FlexJet Options 119
5-17 FlexJet Program Details 120
5-18 NetJets Investment Summary 121
5-19 Executive Jet Financing and Leasing Program 122
5-20 The Citation S/II 123
5-21 The Citation V Ultra 124
5-22 The Hawker 1000 125
5-23 Super King Air B200 126
5-24 Beechjet 400A 126
5-25 Cessna Caravan 127
5-26 Caravan Models 128
5-27 SJ30-2 129

5-28	Falcon 900B 130	5-66	Range Capability from New York 153
5-29	Falcon 900B Interior 130		
5-30	Falcon 900B Cockpit 131	5-67	Range Capability from Los Angeles 153
5-31	Falcon 900B Specifications 131		
5-32	Falcon 900EX 132	5-68	How Do TiltRotor Aircraft Operate? 154
5-33	Falcon 900EX Interior 133		
5-34	Falcon 900EX Cockpit 134	5-69	TiltRotor Aircraft 155
5-35	Falcon 900EX Specifications 134	5-70	VIP Transporter 156
5-36	Falcon 2000 135	5-71	Sikorsky S-76C 157
5-37	Falcon 2000 Interior 135	5-72	S-76C Six/Eight Passenger Corporate Transport Cabin Configuration 158
5-38	Falcon 2000 Cockpit 136		
5-39	Falcon 2000 Specifications 136		
5-40	Challenger 601 3R 137	5-73	S-76C Cost of Operation—Corporate Transport Service, Fixed Costs 159
5-41	Challenger 601 3R Interior A 137		
5-42	Challenger 601 3R Interior B 137	5-74	S-76C Cost of Operation—Corporate Transport Service, Variable Costs 159
5-43	Challenger 604 138		
5-44	Challenger 604 Specifications 139		
5-45	Gulfstream Fleet 140	**6-1**	**Asheville Regional Airport 163**
5-46	Gulfstream's Service Center (200,000 sq ft at Savannah, Georgia) 141	6-2	Asheville Regional Airport 163
		6-3	Asheville Jet Center 164
5-47	Gulfstream IV-SP 142	6-4	Trevor King, Commercial Director at MAGEC Aviation Ltd, Luton, United Kingdom 165
5-48	Gulfstream IV-SP Cockpit 142		
5-49	Gulfstream V (First Flight, November 28, 1995) 143		
		6-5	Receptionists at MAGEC Aviation Ltd, Luton, UK. Maureen Durrant, Barbara Manner, Carol Meads 165
5-50	Gulfstream V 143		
5-51	Gulfstream V Specifications 144		
5-52	Gulfstream V Cabin 145	6-6	Reception Area at MAGEC Aviation Ltd, Luton UK. Henry Hewitt and Sally Kovach 165
5-53	Boeing Corporate Aircraft 146		
5-54	Boeing 737s 147		
5-55	Boeing 737 Family 147	6-7	MAGEC Aviation Ltd Hangar. Clean Spacious Inside Aircraft Parking Area for Customers 165
5-56	Boeing Total Orders 148		
5-57	Boeing Aircraft Design 148		
5-58	Boeing Engine 149	6-8	MAGEC Aviation Aircraft Parking Area, Luton Airport, United Kingdom. 166
5-59	Example Corporate Boeing Interior 149		
		6-9	MAGEC Aviation Aircrew Ground Transportation, Luton Airport, United Kingdom 166
5-60	Overall Size Comparison 150		
5-61	Corporate Cabin Comparison 150		
5-62	Interior Size Comparison 150	6-10	Dave Quinn, MAGEC Aviation Chef/Manager of Catering Activities 168
5-63	Boeing Corporate Jet Interior 151		
5-64	Boeing Corporate Jet Interior 151		
5-65	Boeing 737 Volume 152	6-11	Jet Aviation 169

6-12	Sample Corporate Flight Department Reporting Chart—Larger Flight Department 175	7-15	Weather Services/Fuel 231
6-13	Sample Corporate Flight Department Reporting Chart—Smaller Flight Department 175	**8-1**	**Pilot Decision Sequence When Operating without a MEL 237**
		8-2	Pilot Decision Sequence When Operating with a MEL 238
6-14	Hiring Model 182	8-3	ATA Specification 100 Manufacturer's Technical Data-Type of Numbering System Used 239
7-1	**Company Aircraft Travel Request Form and Passenger Manifest 201**	8-4	ATA Specification 100 Manufacturer's Technical Data-Requirements for Tab Dividers and Text Content in Standard Publications 240
7-2	MAGEC Aviation Flight Dispatch 202		
7-3	MAGEC Aviation Flight Dispatch 202		
7-4	MAGEC Aviation's Chief Pilot, Captain John Robinson 205		
7-5	United Technologies Hawker 1000 and Captain David Isaacson after successful cross-country flight (Hartford to San Diego) 206	8-5	CSI Link Authorized Manufacturers and Service Centers 244
		8-6	CAMP Computerized Aircraft Maintenance Program 245
		8-7	CAMP Systems Reports 246
7-6	MAGEC Aviation Aircraft Fueling Operations, Luton Airport, United Kingdom 209	8-8	CSI Aircraft Maintenance Management Services Available 248
		8-9	SaSiMS 249
7-7	NBAA International Operations Checklis 218	8-10	SaSiMS Staff 250
		8-11	SaSiMS Basic Aircraft Data 251
7-8	IBAC Geographic Regions 219	8-12	SaSiMS Maintenance 252
7-9	MentorPlus FliteStar Corporate Flight Planning Software 220	8-13	SaSiMS Service Bulletins/ Airworthiness Directives 253
7-10	Air Support of Billund 223	**9-1**	**Aircraft Accident Rates, 1984-1996 (per 100,000 Flight Hours) 263**
7-11	Air Support of Billund 224		
7-12	FlightPak® Systems 228		
7-13	Pilot's Choice℠ 229		
7-14	Flight Operations/Ground Handling 230		

List of Figures **xi**

List of Tables

1-1	General Aviation by Primary Use (FAA) 5	4-5	Issues Affecting Business in the Millennium 88
1-2	General Aviation Aircraft by Type and Primary Use (GAMA) 5	5-1	NBAA Average Annual Flight Hours by Aircraft Type 101
2-1	**Federal Aviation Administration 24**	5-2	Project Management Steps 108
2-2	Regulatory Issues for Corporate Flight Managers (Examples) 25	5-3	Tactical Mission Profile 110
2-3	Other Federal Regulators 27	**6-1**	**Responses to the Question, Should an ADM Be a Pilot? 171**
2-4	Members of IBAC 35	6-2	Skill Areas for Aviation Managers 173
2-5	NBAA Publications 40	6-3	Corporate Executive Expectations for Flight Managers 173
3-1	**Benefits of Using Business Aircraft 51**	6-4	Domestic/International Differences for Consideration by Aviation Managers 176
3-2	Listing of Benefits of Using Business Aircraft (personal survey, 1995–1997) 52	6-5	Flight Management Items 190
3-3	Benefits of Business Aircraft According to the Passengers 53	6-6	Aviation Management Issues 192
3-4	Ground Transportation Elements 56	**7-1**	**Flight Operations Items 198**
3-5	Ground Times Comparison—Airline versus Corporate 56	7-2	Flight Issues Identified in Air Crew Survey 199
3-6	Cost/Time Comparison (commercial versus corporate air travel) 60	7-3	Scheduling Software Vendors 203
3-7	Disadvantages of Using Business Aircraft 61	7-4	Tangible and Intangible Benefits of Cockpit Resource Management 216
3-8	Listing of Disadvantages of Using Business Aircraft (personal survey, 1995–1997) 62	**8-1**	**In-House Aircraft Maintenance 241**
		8-2	Contract Aircraft Maintenance 242
3-9	Social, Economic, and Political Benefits of Using Business Aircraft 63	8-3	Features and Benefits of the CAMP Program 243
3-10	Aircraft Operators versus Non-operators (Fortune 500, 1984–1990) 65	**9-1**	**Suggestions to Avoid Terrorists' Acts 258**
		9-2	Security Awareness Questions 259
4-1	**Outline of PRC Aviation Valuation Methods 74**	9-3	Rotorcraft Accident Rate (100,000 flight hours) 264
4-2	Summary of Salary Multipliers for Determining Employee's Worth 75	9-4	Human-Factor Causes of Aircraft Accidents 264
4-3	MAF Value of Responsibility ($400,000,000 Corporate Sales) 77	9-5	Phase of Flight in which Helicopter Accidents Occurred (1982–1995) 266
4-4	Travel$ense® Default Settings 86	9-6	NTSB Accident/Incident Reporting Requirements 269
		10-1	**Items Affecting Corporate Aviation in the Future (May 1997 Survey Results) 275**

Preface

A new title *Corporate Aviation Management* reflects the purpose of this revised text which is to provide a reference and learning book for students studying corporate and business aviation and for professionals who are familiarizing themselves with an overview of the industry. It is meant to be a base upon which anyone can increase his or her knowledge about this important and vital segment of aviation. The National Business Aircraft Association and the General Aviation Manufacturers Association have taken a very active and praiseworthy approach to providing information. It is with the NBAA's support, especially John Olcott's, that this text is based.

Changes to This Edition

Chapter 1. The first chapter on history has been expanded to include more details, but remains a starting point upon which the reader can understand definitions and the structure of U.S. aviation.

Chapter 2. With a historical basis, this chapter now includes pertinent aviation regulations and important business aviation associations.

Chapter 3 (previously chapter 2). Chapter 3 has been updated with more information about the advantages and disadvantages of using business aircraft. Important to this presentation is the information collected while instructing this subject the past three years over the internet for Embry-Riddle Aeronautical University. Students, mostly pilots and aviation managers, provided challenging questions, informational reports, and valuable insight into many aspects of working in a corporate flight department.

Chapter 4 (previously chapter 3). Chapter 4 contains new information, creation of the 3Cs, and emphasis on accomplishing a travel analysis.

Chapter 5 (previously chapter 6). Chapter 5 allows for more details in aircraft selection through project management, or the team approach. Many aircraft are presented to provide those unfamiliar with the quality and availability of aircraft, to include Boeing's corporate airliners.

Chapter 6. The heart of this text is the material in chapter 6 for those who have accepted the industry's importance to management's productivity. Many, many actions are needed to manage and control a vital business tool, the business aircraft. Above all, it is strongly suggested that the value-added approach be emphasized by aviation managers. The need to develop an aviation department must be identified and supported by showing the value the operation brings to a firm and its people. One of the major documents needed for effective aviation management is a company's Operations Manual, and its importance is demonstrated by extracts of manuals in Appendix H.

Chapter 7. Major activities involving flight operations are presented in chapter 7, with the importance of international air operations being emphasized.

Chapter 8. Aircraft maintenance's contribution to business aviation continues to be promoted, and a review of various maintenance methods is accomplished. To provide examples of two system programs, CAMP Systems and SaSIMS are included.

Chapter 9. A new chapter on security and safety is presented at the request of R. Speas, now of R. Dixon Speas Associates, Inc., and numerous students. This chapter will be further developed as time goes on.

Chapter 10. This chapter includes a review of current and future issues made through the SMART Ops outline. The major business aviation concern continues to be designated by business aircraft operators as airport and airway access.

From Owen Lee's initial suggestion to develop a text in this subject to supplement Embry-Riddle's need for course material through to the many people who are actively working in the industry and are promoting its development, this revision has been written to provide readers more information and details.

Various appendices have been added or expanded, to include a listing of aviation internet addresses, an expanded glossary, and revised case studies. Overall, the text has grown in size, not to consider quantity but to address the important subjects desired by students and professionals during the last three years. Further, an instructor's manual will be available and will provide answers for all chapter questions, examples of completed case studies, and suggestions for research reports.

Acknowledgments

John Olcott, President of NBAA, is especially thanked for his continued support and full encouragement to promote the subject and his permission to use the well-developed NBAA material. NBAA professionals provided outstanding support in many areas, and people like David Almy, Jason Wolfe, Cassandra O'Bosco, and Jay Evans are thanked for their help. C. Dennis Wright, now IBAC Director General, is especially thanked for his information and support. Special thanks goes again to Trevor King, Commercial Director at MAGEC Aviation Ltd, for his time and effort to provide information and allow personal visits and discussions. Trevor remains a solid supporter to anyone interested in business aviation.

Henry Hewitt continues his refinishing operation from Luton airport, and is always supportive of my interest in this subject. Likewise, many, many students throughout the years have provided excellent suggestions and material, most notably David Isaacson, Ralph Abate, Barabara Batko, Dan Bilodeau, Fred Van Cleave, Kevin Dellinger, Dave English, Robert Feerst, Scott Filline, Keith Groen, Christian Kennedy, Jim Micko, James Mumma, Stephen Knudson, Jim Philbin, Brian Poliner, Martin Quinlan, James Scorza, John Toland, Jeff Troccolo, Mitch Vuernick, Edward Wright, Tim Wrobel, and undoubtedly others who have provided information and suggestions. Mike Sargent, Embry-Riddle instructor, is thanked for his detailed suggestions which he personally presented. To those who I unintentionally did not name, my apologies, but many thanks for your support.

Many professionals deserve special thanks for their support and information also. Christine Atherton and Dave Lotterer, ATA; Kim Beaumont & Jane Rovolis; Mike Brown, Raytheon; Ulf Cassel, Cassel Aero; Robert Fisher, HAI; Richard Gorman, FalconJet; David Harrington, FAA; Marinda Hochadel-Jolly, Universal Weather & Aviation; John H. House and Richard Gorman, FalconJet; Kent Jackson and Joseph Brennan; C. Jacobs, EBAA; Per Jensen and Jens Pisarski, Air Support; Brian Jessen, GAMA; Leo Knaapen, Bombardier; Gerald Kovach, Asheville Jet Center; Klaus Lammich, GBAA; Derek Leggett, BAUA; Brian Leonard, CAMP Sysytems; Brian Leutschaft, Mentor Plus; Elizabeth May, Business Jet Solution; Janette Prince, AOPA; Kevin Russel, Executive Jet; Mary Shaner, Boeing Commercial; R. Dixon Speas; Richard Stone, ISASI; Gulfstream; and many others. Officials of United Technologies are also thanked for their excellent inputs. Again, any name missing is unintentional and will be thanked in the next revision.

Not least, thanks goes to my wife, Sally, who read the material many times before you got to it. Her editing, although with an English twist, proved invaluable. Her patience and guidance were needed many times. The names in the case studies are fictional, and if there is any coincidence to actual names, it is unintentional. Also, every attempt was made to contact those who provided information or whose material I used, and if a source is not provided, my fullest apologies because I always wish to provide full credit to others who work so hard and develop valuable information. Corrective action will be taken as soon as known.

Suggestions and ideas are always welcomed. Please send them to me direct at Ken J. Kovach, PSC 37 Box 3414, APO AE 09459 for U.S. mail or 31 Hempfield Road, Littleport, England Cambs CB6 1NW for England. My e-mail address is INTERNET:74321.135 @compuserve.com, or just 74321,1325 for compuserve users. Enjoy!

History of Corporate and Business Aviation

CHAPTER 1

Overview

You are introduced to the definitions of corporate and business aviation, the U.S. aviation structure, a historical background of corporate and business aviation, and key terms applicable to the industry. Although the history of corporate and business aviation is indistinct, various industry operators and researchers have provided sufficient material to achieve a general understanding of its development through the years. Important regulations are referred to but will be more fully described in the next chapter. Review Appendix A for a glossary of terms.

Goals

- ✈ Define corporate and business aviation.
- ✈ Understand the U.S. aviation structure.
- ✈ Describe the early development of corporate and business aviation.
- ✈ Identify aviation rules applicable to corporate and business aviation.
- ✈ Understand historical aspects of general aviation especially in regards to corporate and business aviation.

History

Background

Exactly what is *corporate and business aviation?* Actually, there are two parts to the definition. *Corporate aviation* has been defined by the National Business Aviation Association (NBAA) as the use of an aircraft, owned or partially leased, by a company or business person for the transportation of people, cargo, mail, or combination in furtherance of a firm's business, and which is flown by a professional pilot who receives a direct salary or compensation for piloting. *Business aviation* is the use of an aircraft by a non-professional pilot. In other words, corporate and business aviation is "made up of companies and individuals who

Figure 1-1. Business Aviation. Photo courtesy of *Corporate Distinction, A Journal of British Aerospace.* Vol. 3, No. 2, 1992, page 6.

use aircraft as tools in the conduct of their business" (National Business Aircraft Association [NBAA], 1997, p. 2), and it is in this context that the material following has been presented. More specifically, when a professional pilot is paid to fly such an aircraft, it is called **corporate aviation.** When a business person flies him- or herself in conducting business, it is called business aviation. See Figure 1-1.

The Federal Aviation Administration presented these two activities within its *business transportation* and *executive transportation* definitions (1995, pp. K-2–K-3). Basically, business transportation applies when an aircraft is used by an individual in the course of non-flying business. When an aircraft is used, not for compensation, by a corporation or company for company business and professional pilots are paid to fly the aircraft, then this is executive aviation. The latter term has fallen in disfavor by industry users for the more acceptable designation of *corporate aviation.* This may have resulted because corporate operators did not want the public to believe that company aircraft were just for the top executives of a firm. The term *royal barge* was coined during the early years (prior to 1970) to designate that company aircraft were being used to comfortably transport only the top managers of a firm. However, many FAA officials still use the terms executive or corporate transportation and executive transportation to designate professional pilots flying company employees (Federal Aviation Administration [FAA], 1995; Office of Aviation Policy and Plans, 1994).

You won't find corporate aviation and business aviation defined in Federal Aviation Regulation (FAR) Part 1 or in other federal rules. You will, however, find various relevant definitions such as air carrier. For example, in FAR Part 241, air carrier is defined as "any citizen of the United States who undertakes, whether directly or indirectly or by lease or any other arrangement, to engage in air transportation" (Office of Federal Register, 1994, p. 219).

What this means is that aviation is considered by U.S. government officials to fall under commercial, military, or general aviation. An air carrier is within the commercial (sometimes called civil) category because a company is using an aircraft in order to make money. The FAA's assistant administrator for system safety (1996) defined air carrier as "any air operator under FAR Parts 121, 127, or 135" (1996, p. G-1). The FAA's Office of Aviation Safety Policy and Plans placed business transportation and executive/corporate transportation under its primary use category. While this may appear confusing, what FAA, General Aviation Manufacturer's Association (GAMA), Aircraft Owners and Pilots Association (AOPA), and other agencies appear to be doing is reporting aviation groups under various statistical headings, but corporate and business aviation falls under the grouping of general aviation in all reports. AOPA, for example, included commuters under general aviation. GAMA excluded commuters but included air taxi operators under active general aviation aircraft.

Military aviation is deemed understandable. What remains, then, is the category of general aviation. Before specifying the differences within this category, a review of the U.S. commercial aviation structure will be presented.

U.S. Aviation Structure

The surface transportation modes within the United States have been generally classified by their operating revenues. The Civil Aeronautics Board followed suit in 1981 by eliminating the many categories of aviation (e.g., trunk carriers, supplemental carriers, feeders, local service, etc.) and establishing four categories based on gross operating revenues. The FAA (1995) defined commercial air carriers as air carriers "certificated in accordance with FAR Part 121 or 127 to conduct scheduled services on specified routes . . . Majors [over $1 billion], Nationals [between $100 million and $1 billion], Large Regionals [between $20 million and $100 million], and Medium Regionals [less than $20 million]" (p. K-3). These are the categories generally referred to by aviation analysts and writers.

FAR Part 241, Section 04, indicates that the large certificated air carrier groupings are divided into three basic categories:

    a. Group I (0–$100 million),
    b. Group II ($100 million–$1 billion), and
    c. Group III (over $1 billion).

Group I is further divided into two subgroups:

    a. air carriers earning $20 million to $100 million and
    b. air carriers earning below $20 million.

These later subgroups relate to the large and medium regional air carriers. The FAA's Office of Airline Statistics updates these groupings and air carrier classification on an annual basis.

One category needs clarification. The small regionals are commercial operators (not air carriers) earning money for common carriage but operating smaller aircraft. These are classified as commuters. Their aircraft size is limited to 30 seats or less or a maximum payload of 7,500 pounds (3,402 kilograms) of cargo for intrastate movement and maximum seating capacity under 20 seats or a maximum payload of less than 6,000 pounds (2,722 kilograms) of cargo for interstate movement. Air taxi operators also earn money for air transportation services, but fall under FAR Part 298 and do not hold a certificate of public convenience and necessity (Section 401 certificate). They may use small aircraft or large aircraft after complying with applicable regulations; for example, FAR 121 or FAR 135 (Office of Federal Register, 1994). Air taxi operations are usually reported by the FAA, Airline Owners and Pilots Association (AOPA), and General Aviation Manufacturers Association (GAMA) along with general aviation operations, but air taxi operators fall outside the definition of general aviation.

General Aviation

The FAA (1995) defined general aviation as "all civil aviation activity except that of air carriers certificated in accordance with FAR Parts 121, 123, 127, and 135" (p. K-4). The NBAA (1997) simply reported "General aviation includes all aircraft not flown by airlines or the military" (p. 2). Herein lie corporate and business aviation operations. GAMA listed 11 general aviation aircraft categories while the FAA's primary use listing included 10 categories. The listings are identified in Tables 1-1 and 1-2. The important aspect here is that corporate and business aviation falls under the general aviation category within the United States aviation industry (don't be confused by the agencies using commuters and air taxis under general aviation operations because of statistical reporting). Various types of people and companies use aircraft for business, and the aircraft used differ from single-engine piston to large Boeing 747s. Some operators have only one aircraft while large companies may operate a large fleet of multi-engine, turbine-powered aircraft.

Today, cargo, mail, and people are flown on business aircraft to increase productivity and satisfy managerial demands. Not only are company officials flown to customer locations, but customers are even brought to the companies. For example, a producer of large equipment may find it easier to bring potential buyers to a plant location. Additionally, business aircraft are not only flown on demand, but many are operated on a scheduled basis. In England, for example, Ford Air Transport provides daily (morning and evening) air transportation services to the European continent for company employees.

Most corporations that operate aircraft use a two-person professional aircrew who primarily are responsible to fly the company aircraft. The majority of business aircraft are owned by individuals or companies, but arrangements such as fractional ownership, chartering, leasing, time sharing, interchange agreements, partnerships, and aircraft management contracts are used. Also, most of the business aircraft are not flown for hire, but operate under FAR Part 91. If monies are to be made using business aircraft, then "FAR 135 may apply which covers on-demand commercial operations" (NBAA, 1997, p. 2). FAR Part 125 applies when

4 *Chapter 1*

Table 1-1. General Aviation by Primary Use (FAA)

Aerial Application — Agriculture, health, forestry, cloud seeding, firefighting, insect control.
Aerial Observation — Aerial mapping/photography, survey, patrol, fish spotting, search and rescue, hunting, highway traffic advisory, sight-seeing (not FAR Part 135).
Air Taxi - FAR Part 135 passenger and cargo operations, excluding commuter air carrier.
Business Transportation — Individual use of an aircraft for business transportation.
Commuter Air Carrier — Performs, under FAR Part 135, at least five scheduled round trips per week or carries mail.
Executive/Corporate Transportation — Company flying with a professional crew.
Instructional — Flying under the supervision of a flight instructor (excludes proficiency flying).
Other — Experimentation, R&D, testing, government demonstrations, air shows, air racing.
Other Work Use — Construction work (not FAR Part 135), helicopter hoist, parachuting, aerial advertising, towing gliders.
Personal/Recreation — Flying for personal reasons (excludes business transportation).

Source: Office of Aviation Policy and Plans. (1994). *General aviation and air taxi activity survey—calendar year 1994*. Washington, DC: U.S. Government Printing Office. p. B-4.

Table 1-2. General Aviation Aircraft by Type and Primary Use (GAMA)

Corporate	Business
Personal	Instructional
Aerial Application	Aerial Observation
Sight-seeing	External Load
Other Work	Air Taxi
Other	

Source: General Aviation Manufacturers Association. (1996). *General aviation statistical databook*. Washington, DC: U.S. Government Printing Office. p. 10.

an aircraft has a "seating configuration of 20 or more passengers, or a maximum payload capacity of 6,000 pounds or more when common carriage is not involved" (Office of Federal Register, 1994, p. 486).

The major *benefit* of business aircraft, according to the NBAA, is flexibility. Company managers can control most aspects of employee travel plans, and changes can be made instantly when needed—hard to do while using a commercial airline. The major *purpose* of using business aircraft is to increase management's productivity.

1: History of Corporate and Business Aviation

Early Days (1903–1940s)

Corporate and business aviation probably began with the invention of the *flying machine*, the Wright brothers' patent of their 1903 invention. One could not argue that the Wright brothers used their invention for their business, even though it was mainly aircraft manufacturing at that time. They traveled to locations using their aircraft to do business. Historians, however, credit Lincoln Beachey in 1905 as using aviation to promote business. On September 16, 1905, Charles B. Knox hired Lincoln Beachey to fly an airship over Portland, Oregon, to display an advertisement for Knox Gelatin. This 18-year-old pilot was then credited as the first known corporate pilot (Whempner, 1982).

Whempner (1982) also emphasized the fact that nations developed through the improvement of commerce and trade. Transportation and communication systems are necessary for this to happen. During the early 1900s, various societies used their advantages of resources, specialization, and economies of scale. This was the beginning of globalization of world marketing. Communications were used to inform the buyers, and transportation was used to get the products to the customers. Transportation, then, provided for time and place utility. In other words, people and/or products moved to *where* they were required *when* they were required. Today's main speed advantage of air transportation was not the overwhelming advantage in the early days of aviation. Then, the railroads provided fast transportation over the distances traveled; however, rapid improvements in aviation's infrastructure and equipment slowly gained the advantage of fast transportation for the airlines.

This, then, was the stimulant for air transportation; that is, speed, time savings, or both for the shipper or passenger. Throughout the early development of the United States people moved westward, communities developed at trade and commerce centers, and the Industrial Revolution created the need to transport goods to distant places.

Early business leaders and managers realized the advantages of the surface modes of transportation for business and pleasure. They induced that the factor in common was that time was important. Add to this the enthusiasm and activity of the early aviators, and aircraft were soon in demand to provide quick and essential service. Company managers could not provide equipment support or customer service any faster than by using air transportation.

The Wright brothers were not successful immediately after their 1903 invention because people either didn't believe them or the need for air travel was not emphasized. In 1908, however, after the Wright brothers gained a U.S. Army contract for aircraft, people slowly became aware of the new transportation mode during air displays, public gatherings, and business contracts. Wells and Chadbourne (1987) reported that by 1911 several manufacturers were building general aviation aircraft, but only at a rate of about three or four per year. In fact, by 1914 when the First World War broke out, it was reported that fewer than 200 aircraft had been commercially produced since 1903. The Curtis Aeroplane and Motor Company was the largest of the early aircraft manufacturers.

These early aircraft manufacturers began to develop various aircraft. In 1910, Clyde Cessna began building airplanes (approximately 180,000 Cessnas have been built by 1997), and Glenn H. Curtis developed a seaplane in 1911. The Loughead (later renamed to Lockheed) brothers gave 600 air rides at $5 a person during early airshows. This income was enough to start the Lockheed Aircraft Corporation. Other names became associated with American aviation: John K. Northrop, Glenn L. Martin, Donald Douglas, and Walter Beech. Beech developed the Travel Air aircraft and promoted time savings by flying. His advertisements in his news publications identified the use of privately owned aircraft as a profitable business tool (this concept was not firmly established until the 1980s). Some early Travel Airs were used to move critical parts to harvesting equipment in the wheat valley. Besides the Travel Air, other aircraft names were becoming known: Stinson, Waco, Fairchild, and Moth. Business managers were attracted to the attention gained with the use of aircraft produced at the time. Hotson (1991) stated that the "clear leaders in U.S. business aviation were the oil companies . . . Phillips Petroleum had a Travel Air and Continental Oil had a Ryan and a Travel Air, Standard Oil and Texaco opted for Ford Trimotors while Shell . . . chose Fokker Trimotors" (p. 4).

Baran (1990) reported that the James Heddon Company used three open-cockpit de Havilland biplanes in 1920 to market its products. The Heddon company developed fishing lures at that time and used aircraft to beat its competitors to the customers. Even the Wright Aeronautical Corporation paid $19,500 for its Loening Air Yacht, a monoplane pusher with a semi-enclosed five-seat cabin, in the early 1920s. The Goodyear Tire & Rubber Company of Akron, Ohio, used a Fokker in 1929 for testing aircraft tires and brakes, and later came to use its aircraft for company business.

The early aircraft were not designed for passenger comfort—probably one reason executives in those days did not use their own company aircraft to travel. Company managers purchased early-developed aircraft for various reasons, including fast movement of critical parts, advertisement and marketing advantages, and using technology much as we use new technology today.

The First World War resulted in many aircraft improvements. It also resulted in the training and availability of many pilots. Whempner (1982) stated that "some of the surplus military aircraft found their way into the homes of civilians who thought the aircraft could be used to promote commerce" (p. 2). He related the story of the many Baby Ruth candy bars that were dropped by paper parachutes in 1931—an advertisement that created great interest in those days.

Hotson (1991) stated that by the late 1920s, an intercity airline structure developed in the United States, and increases in radio ranges and advanced instrumentation fostered the development of corporate flying during the Depression years. He declared that company officials who flew their aircraft through the troubled years did much to advance the corporate aviation concept.

The Ford Trimotor became a stalwart aircraft for Standard Oil and other large companies. It could carry up to 13 passengers almost 500 miles (805 kilometers) in greater comfort and style than could be given in other early aircraft.

Castro (1995) cited several early aircraft. The Travel Air 6000 was a six-seat commercial high-wing monoplane with a 220 hp engine realizing a 125-mph (109 knots) speed and almost 700 mile (1,126 kilometers) range. He described the Fairchild 71 with its 410 hp Wasp radial engine, 140-mph (122 knots) speed, and 17,000 foot (5,181 meters) altitude. He also reported that the "greatest hindrance to the growth of business aviation during the late 1920s was not the capabilities of the aircraft but the quality of ground-support services" (p. 308).

Aviation infrastructure had not yet been developed sufficiently enough to expand the use of business aviation beyond the main centers of business. Navigation instruments, aeronautical charting, and landing and other facilities were still needed. Wells and Chadbourne (1987) reported that in 1927, 34 non-aviation companies were using corporate aviation, and by 1930, 300 companies were using the Stinson, Waco, Travel Air, and Fokker aircraft.

One of the greatest stimulants for aviation was Charles Lindbergh's historic solo flight across the Atlantic on May 20, 1927. His New York to Paris flight put the spotlight on the utility and dependability of flying. Lindbergh's flight was a catalyst that sparked great interest in aviation. Airshows were held, and people began to become attracted to flying. They also saw that flying was safe and reliable. Lindbergh had sparked an interest in air transportation that has never diminished.

The oil and petroleum companies were the main users of corporate aircraft during the early days of aviation. In the film *Wings at Work (n.d.)*, Lockheed officials unfolded the story of early corporate/business aviation. Newspapers like the *Daily Oklahoman* used the Travel Air; Movie Tone used the Fokker aircraft; Socony Oil used the Bellanca; and Richfield Oil used the Vega. John A. Whitney bought a Sikorsky and made it the finest corporate aircraft of the day. Radio newspeople, polar explorers, and others successfully used aircraft as a business tool to get their jobs done. During the 1930s, employment rose because over 600 airports were built. Business people were not tied to the big cities; therefore, business expansion took place. One such event was presented by B. E. Batko (personal communication, February 1, 1997). She reported that the Detroit Metro Airport (DTW) could not have been developed if it was not for the DLAND (Development of Landing Areas for National Defense) Act prior to World War II. The Detroit airport went from one square mile to four square miles, and in 1997 was at 6,700 acres with an operating budget of $130.5 million.

Throughout the 1930s and 1940s, aircraft advances were further realized in terms of size, speed, comfort, and capacity. Fairchild could not find a suitable aircraft, so he built one. International airmail service began from Miami to Cuba. Clarence Chamberlain took off from the deck of an aircraft carrier. Amelia Earhart and Ruth Elder accomplished flying feats. New records were set for speed and distance. Beechcraft introduced the Model 17 in 1932, and by 1935, the Model 18 was in full production. The Model 18 was designed as luxury executive transport. It could get business people in comfort and style to their destinations quicker than a DC-3 airliner. The era of wonderful nonsense (1930s and 1940s) was a time when aviators gave airshows and barnstormed for thousands of spectators. Business people became familiar with aircraft and their fast, safe, and economical uses. They could use aircraft without worrying about airline schedules, layovers, connections, or other transport concerns. Business aircraft were used to deliver critical parts or perishable items, to deliver products to hard-to-reach areas, to promote the company, and to increase customers. For the most part, however, only managers of companies with large resources were forward thinking enough to develop the aircraft as a tool.

The U.S. Post Office was the major motivator for aircraft improvements and did such through airmail contracts and policies. For example, in 1918 the Post Office provided

Sikorsky
September 14, 1939. The first flight of the VS-300, with Igor at the controls.

Figure 1-2. Sikorsky. Photo courtesy of United Technologies Corporation.

$100,000 for airmail subsidies, and later through various airmail acts in 1925, 1930, and 1934, it awarded contracts by pieces of mail, then weight and volume, and finally, by size of aircraft, respectively. With larger aircraft, airline managers realized (in 1936) that their businesses would prosper by moving passengers rather than with airmail contracts. People began to realize that aircraft were useful and safe.

Technological advances came about also during this period. A turbojet engine was invented by Sir Frank Whittle in 1937 (Smith, 1992). Later, along with its derivative, the turboprop, more engine power resulted and aircraft speed was increased. Turboprop aircraft are today the favorite of corporate aviation department managers.

Not to be forgotten are *helicopters* for corporate and business aviation use. The autogyro of the 1930s was the first step in the development of helicopters and an absolutely vital contribution to corporate and business aviation. Igor I. Sikorsky began testing his VS300 on September 14, 1939 (see Figure 1-2), and on May 30, 1940, successfully flew the helicopter (Wartenberg, 1989). Other manufacturers, such as Bell, Hiller, Kaman, and Hughes, also developed helicopters for World War II.

In 1949, Southern California Edison used an early helicopter to haul material to hydroelectric generating plants in the mountains of California. In a conversation with R. W. Cloud, a student had reported that the use of the helicopter was an experiment to replace Jerry, Pete, Dan, and Jim—four pack mules used for the same purpose. The experiment was a success, the mules were put to a good rest, and it became clear to everyone involved that increased productivity, high quality inspections, and rapid movement of manpower were possible at costs even less than conventional means.

World War II

Most researchers and analysts credited the Second World War with being the turning point for corporate and business aviation. Aircraft being built then were larger (initially

1: History of Corporate and Business Aviation 9

motivated by provisions of the Airmail Act of 1934), more pilots were available, and there was a great military need for using air transportation. Prior to the war, Taylor built Piper Cubs and Ford the Trimotor. The war stopped civil aircraft production in its tracks. Every resource was diverted to the war effort, and while this may have seemed counterproductive at first, one major outcome of the war effort was the development of aviation's economy and structure for the postwar days. Military production resulted in a quarter of a million new aircraft, and 75 percent of the pilots received their first training in new Piper Cubs. New aircraft were pressurized and could be flown higher and faster. Airplane production was ceaseless until the war stopped. Corporate managers obtained large military aircraft and redesigned them into luxury transportation carriers; for example, the C-45 and C-47 were favorites to use. Additionally, the U.S. government provided a guarantee for an aircraft loan, resulting in banks becoming involved in aircraft financing and other agencies deciding to provide aircraft loans.

After the war, the availability of cheap war surplus airplanes and aircraft parts contributed to the modification of military aircraft and the reduction of new designs. The wartime C-45s, or Beech 18s, and the C-47s were modified by companies like Dallas Aero Service, Remmert-Werner, Pacific Airmotive, and others. Luxury was the theme of the day.

Important to corporate aviation's growth and development was the help provided by the U.S. government to corporations. Wartime requirements had necessitated industrial support. Military contracts had to be negotiated with corporate managers. Salespeople, engineers, and technicians had to travel extensively throughout the United States and even visit foreign countries. Civilian managers convinced government officials that aircraft were needed to provide for their critical transportation requirements. Some larger military aircraft were modified and redesigned to provide such corporate travel. Companies hired pilots, and aviation departments were formed to use government-provided aircraft. It was only natural to internalize these departments after the war ended into corporate flight departments.

Whempner (1982) provided an interesting story of travel during this time. A colonel, who was enroute in San Antonio traveling to Chicago to hear a special presentation, bumped off the person who was going to give the lecture. This was an example of the high priority given to military travelers and the need for civilian air transportation. Another example included a low-priority ticket holder from Honeywell. The president of the company was stuck at Chicago's O'Hare airport enroute to Minneapolis. Two of Honeywell's sales engineers who were traveling on the company's Northrup Delta 6 aircraft happened to be in the terminal restaurant. After realizing that his company's plane was available, Honeywell's president was soon traveling to Minneapolis. This was an example of the effect of military priority and the misunderstanding of the value of a company aircraft to top managers. Soon after, however, Honeywell purchased two C-47s, and the company has had a strong aviation department ever since.

Also during the war period, the U.S. government provided financial incentives to build airports. People soon learned that there was economic value in having an airport in their community. With the addition of new airports, new and improved navigation systems, a greater number of improved aircraft and trained pilots available, and the demand for expanded business operations, corporate and business aviation became firmly entrenched in American society. The prohibitive costs of air travel, the nonavailability of efficient aviation equipment and systems, and inadequate managerial attention to corporate/business aviation appeared to be significantly reduced. At this time, however, the aircraft manufacturers focused on pleasure flying rather than business flying. Several large corporations used the aircraft for more than just pleasure. The air transportation problems of civilians during the war made many corporate managers realize that a business aircraft could provide time savings and flexibility in traveling.

Post–World War II (1950s–1960s)

Castro (1995) reported that during the 1950s, aircraft manufacturers began looking toward the business flyer as a key market for growth and profit. Beech, Cessna, and Piper continued to provide aircraft for government and private flyers, but also produced aircraft with more seats, more comfort, increased speed, and greater range for the business flyer. According to Castro, even President Eisenhower used the Aero Commander for personal short-haul travel.

Aircraft manufacturers realized that mass production of aircraft was not the answer for their survival and growth. They had to produce aircraft for specific customers, and one group that seemed to be a lucrative market was the business flyers. Safety and capability were foremost in people's minds at the time. Aircraft not only had to be comfortable, they had to be fast and have the capacity and range to be useful. The commercial airlines were beginning to use the new jet airliners in the late 1950s, but the corporate operators lacked a specifically designed aircraft of such capability.

Wells and Chadbourne (1987) provided another interesting story of how people's attitudes had been influenced by the flying feats of William P. Odom. Odom flew 5,723 miles (4,972 nautical miles) nonstop in a Beech Bonanza late in 1949. Cross-country flying was shown to be very possible for the small aircraft flyer of the times. Also, after the war the Goodyear Tire & Rubber Company used its aircraft between Akron and Toronto to move company employees because it otherwise involved a two-day airline trip with an aircraft change in Cleveland, Ohio. Goodyear used its Grumman Widgeon (CF-BVN) for the trips and for company executives who had summer homes in the Muskoka lake country. With a wartime bargain, Goodyear bought a twin-engine Anson V training aircraft and installed seats. Later, a DC-3 was bought and used (Hotson, 1995).

Then the Korean War broke out in June 1950. Production of aircraft for the business flyer fell, but many positive spinoffs resulted from the war. Wells and Chadbourne (1987) reported that more omni radio stations were in operation; general aviation was assigned its own frequency, 122.8, titled unicom. Lear developed the autopilot, and avionics improved. Excellent advances in engine power through the turbocharger and turbine engine came about. U.S. government demands for new technology in aircraft fostered an increased demand for research and development by the aircraft manufacturers. Development of faster, larger, and more capable aircraft for the military provided many benefits for the corporate aircraft operator.

The British Armstrong Siddeley, the French Vesta, and the Turbomica Marbore II, better known as the Continental J-69, came on the scene. Competition among the aircraft manufacturers was fast and furious to not only survive but to develop a profitable customer base.

Figure 1-3. Gulfstream I Aircraft. Photo courtesy of Gulfstream Aerospace Corporation.

The U.S. Air Force further encouraged aircraft development by requesting a small jet twin-engine airplane seating four people. Lockheed built the Jet Star, North American produced the Sabreliner, Rockwell produced the Standard, and European aircraft manufacturers started civil aircraft production. An example of the latter's effort was the De Havilland DH-125 (later Hawker Siddeley [HS 125], British Aerospace [BAe 125], and Hawker 125). For a short time in the U.S., the Hawker was sold by Beech and was called a Beechcraft Hawker (BH-125) by some.

The first aircraft specifically designed for corporate aviation was Grumman's turboprop model 159, or G-I, built in May 1959 (Figure 1-3). It had a large, pressurized cabin and a longer range (2,000 statute miles or 1,738 nautical miles). From its revised flight manual in 1980, it can be seen that the G-I had a take-off weight of 36,000 pounds, Rolls Royce Dart Mark 529-8 type engines, and true airspeed indicated of 257 knots (295 mph) at 34,000 feet (10,363 meters) and 295 knots (339 mph) at 26,000 feet (7,924 meters). Initially, 19 customers purchased the aircraft, but at the end of its production in 1969, around 200 customers were satisfied with its capacity of 19 passengers (later modified to accommodate around 40 passengers), and reputation for safety and reliability ("Story of Gulfstream," n.d.).

Aircraft manufacturers competed with Grumman for customers by designing new corporate aircraft. Beech worked on the King Air and Lear the Lear Jet. The Gates Lear Jet was the first jet aircraft designed and produced specifically for corporate and business operators. Introduced in 1963, it was a beautiful aircraft with clean, aerodynamic lines, a sleek needle nose, small stubby wings, and two pure-jet engines. It could fly at over 550 miles per hour

12 *Chapter 1*

and had a rapid rate of climb to a high cruising altitude, which could put it above most bad weather. This aircraft was a replacement for the only general aviation jet aircraft that had originally been designed for the military, the North American Sabreliner and the Lockheed JetStar. Priced at about $490,000 and with low operating costs, the Lear Jet allowed many corporate managers the transition into the jet age, one of speed and convenience. One nickname for the Lear Jet was the executive mail tube! Not to be left at the post, Grumman developed the G-II twinjet in 1966, of which 202 aircraft were produced before the G-III came on the scene ("Story of Gulfstream," n.d.). Reviewing the Lockheed film *Wings at Work* provides for an excellent overview of the corporate and business aircraft manufacturing developments during this time.

Not to be forgotten because of its extreme importance to corporate and business aviation is the helicopter and its development. From Igor I. Sikorsky's earliest boyhood feat of building a model helicopter powered by rubber bands in 1901, through his attempts at building a flying machine during the early 1900s, and until his success with the VS-300, helicopter use was considered mainly for the military.

During the Second World War, the American use of the helicopter was minimal, and it wasn't until April 23–24, 1944, that the first American heliborne rescue came about. Afterwards, manufacturers (Sikorsky, Bell, Hiller, Kaman, and Hughes) improved their designs, and production and use increased. The main changes to the helicopter have come about in its components. Wooden rotor blades were changed to metal and then to composites, piston engines changed to turbine engines, retractable landing gears were used, improvements were made to communication equipment, and better flight control systems were developed. With the development of the jet engine for fixed-wing aircraft, the turboshaft engine improved the helicopter. Fuel consumption was as good or better than for the reciprocating engine and the weight was less; however, cost of the new engines was not on the plus side.

With the improvements made to helicopters, more were used by the military in the Korean War and by corporate managers, especially for their indispensable roles in serving remote locations and off-shore oil platforms. Types and numbers of helicopters, as well as fixed-wing aircraft, will be reviewed in Chapter 4.

Besides new aircraft for the corporate/business traveler, other factors stimulated an increase in business aviation. Viafora (1986) listed nine items as follows:

1. Major industrial shift to marketing business,
2. Corporate emphasis on research and development,
3. European common market and its industrial recovery plan,
4. Marshall Plan of funds and technology,
5. Vital communications to ease international tension,
6. Expanding business climate,
7. Transition to air transport (value proven during the war),
8. Increasing demand for air services, and
9. Corporate aviation scenario materializing.

1970s

At the start of the 1970s, the general aviation industry was being affected by the *environmentalists*. The concern for controlling noise became a major issue for the Federal Aviation Administration. In 1971, the Noise Control Act was passed. This act set down strict noise reinforcement policies. Aircraft manufacturers redesigned their engines and airframes to reduce noise to acceptable levels. President Johnson signed into law several acts which made people develop social awareness and become more considerate of each other.

The *royal barge* concept was de-emphasized because of the growing demand for productivity from company assets, the changing attitudes of company managers that aircraft were only for top executives, and the overall business use of company assets. Perhaps some top executives still have private use of company aircraft and are treated like royalty, but most, if not all, corporate top managers have realized that business aircraft are excellent company tools, and their use is provided for various company personnel and matters.

Other major impacts during the 1970s were two fuel crises—in 1973 and 1978. Oil shortages later forced jet fuel and avgas up to record price levels. Aviation flight department managers had to start reducing costs. One way was to sell off fuel-inefficient aircraft; another that managers resorted to was tankering. Tankering meant that extra fuel had to be carried because pilots could not obtain fuel at some airports or because the price of aviation fuel was just too high at some airports used. Besides people's concern for the environment and fuel, the American economy went into a recession during this period.

On October 24, 1978, Public Law 95-504, better known as the Airline Deregulation Act, was passed (actually this was an amendment to the Federal Aviation Act of 1958). Airline managers, pilots, and many others have been debating the pros and cons of airline deregulation ever since the act was signed into law. During the previous decade, the Civil Aeronautics Board (CAB) had regulated aviation more closely than any other transportation mode. The CAB's economic regulation was in the form of setting airline routes, fares, prices, and other aspects of the commercial airline operations. In 1977, Fred Smith of Federal Express lobbied with other air cargo managers to have President Carter sign the Air Cargo Deregulation Act of 1977 (PL 95-163). While the passage of this legislation had no effect on corporate and business aviation users to any extent, the passenger deregulation rules had a dramatic impact on corporate and business aviation operations.

For the commercial airline managers, the market forces determined the survival, profit, and growth of their airlines. Airline managers immediately had the opportunity to control their prices, routes, and use of the aircraft. High-cost route operations were eliminated, aircraft were scheduled on more money-making segments, and the use of the hub-and-spoke system greatly increased.

These changes affected the corporate and/or business traveler in different ways. First, when airline managers pulled their flight services out of small communities, the community involved had to be declared an essential air service point by the Department of Transportation or all flight services could be lost. If a small commuter or other air operator didn't provide air services for these communities, then corporate and business travelers either used another mode of transportation or used business aviation to get to their locations. The airlines' hub-

and-spoke system resulted in more passengers traveling to an airline's central airport operation to make onward connections. For business travelers, it has been estimated by subject analysts that about one-third of business travelers use the commercial airlines to make onward connections, and another one-third do business at a main airport without further travel, and the remaining one-third need to go to locations where the commercial carriers do not go. The effect of the airlines using their hub-and-spoke systems more meant that business people had to spend more time traveling. This, in itself, motivated many business managers to consider and obtain business aircraft. They just were no longer satisfied with the airlines' schedules and services.

Another development was the increasing use of corporate aircraft for international flight operations for corporate managers. Because of changing commercial airline systems, the globalization of world markets, new business concepts being applied (e.g., just-in-time systems), and other business activities developing during the latter part of the 1970s, corporate managers decided to use their aircraft for international flights. This had a major impact on corporate flight departments. Now everyone in the flight department had to become aware of new rules, policies, and procedures. Flight training was expanded, aircraft maintenance requirements extended to far-away operations, and flight management companies were needed. Pilots had to be operationally certified and have knowledge and experience in meteorology, communications, engineering, navigation, flight theory, and human relations. To meet the long-range requirements for international travel, aircraft manufacturers developed improved aircraft. The Gulfstream II, Dassault's Falcon 50, Canadair's Challenger, and Lockheed's Jet Star were internationally capable aircraft.

The aircraft manufacturers were willing to explore the general aviation market to a greater extent because of the demand for corporate/business aircraft. They knew that they had excellent products, provided excellent customer service, and gave personal attention. General aviation aircraft sales peaked in 1978 at 17,811, averaging 13,361 new aircraft sold per year during the 1970s.

The role of the helicopter also increased during the 1970s. When Bell developed a commercially certified helicopter in 1946, corporate managers were uninterested. Speed and range deficiencies had been overlooked in the 1970s for the flexibility to arrive and depart from limited areas. Surface travel was congested in many places; therefore, between 1970 and 1974, the number of helicopters used almost doubled; from 749 to 1,463. Executives increased their use by 69 percent and business flyers by 112 percent (Whempner, 1982). Improvements to the design, comfort, range, size, and speed have continued to encourage managers to use helicopters ever since. Additionally, the increase in heliports from less than 1,000 in 1970 to 4,500 today makes it easier to use this business tool.

1980s–1990s

During the early 1980s, three main factors affected the sale of general aviation aircraft. Sontag (Office of Aviation Policy and Plans, 1996, p. 103) reported that the "Arab oil embargo created a general recession and cutback on the availability of avgas." In addition, the federal air traffic controllers went out on strike and aircraft were limited on takeoffs and landings. Product liability was the third main factor causing a decline in general aviation aircraft sales. General aviation sales had peaked in 1978 at 17,811 and then started falling to 2,691 in 1983. Other economic factors affected the aircraft industry—the noise regulations and the Tax Reform Act of 1984. The latter was credited in a 45-day period during the first quarter of 1985 with precluding the sale of aircraft worth approximately $45 million (North, 1985). A fringe benefit tax rule fostered a wait-and-see attitude for corporate aircraft buyers. This tax rule will be reviewed in the next chapter.

Air traffic control constraints also caused some corporate managers to refrain from purchasing corporate aircraft. The major concern was the advantages commercial air carriers might have over general aviation users. In effect, these worries caused general aviation sales to drop dramatically to 899 in 1992 (Gilbert, 1993). GAMA (1996) reported the lowest level at 928 in 1994, but even then, this appears high due to counting military purchases of general aviation aircraft.

What had been realized by some managers who used general aviation aircraft during the 1940s to 1970s slowly began to be recognized by many business managers in the 1980s. Recessions, changes in commercial air transportation services, fuel crises, global competition in marketing, and other factors made many corporate managers and business people consider cutting costs, increasing productivity, and essentially making people manage their resources very tightly. Some corporate flight departments were the first activities to be affected; some closed down completely while others reduced. Some aviation managers traded up by obtaining a larger aircraft to replace two smaller ones. Downsizing, re-engineering, rightsizing, and other terms could be heard from top managers during the early 1990s. Overall, though, the concept that the aircraft was a business tool, to be managed like other company assets, became firmly realized by managers. Some top executives still use their company aircraft as privileged equipment, but the company aircraft today is known by all as a real *business tool*.

Since airline deregulation, there has been an exploding demand for air transportation, and the U.S. government reacted with "hassle factors," as stated by GAMA's president (Searles, 1990). He referred to the establishment of mechanisms designed to constrain capacity or air traffic control issues. An AOPA spokesman also reported that "general aviation will be the victim of limited aviation resources" (Searles, p.55).

The *National Airspace System Plan* (NASP) was developed in the 1980s (January 28, 1982) to be a blueprint of 97 programs to improve airport, airways, and air traffic control systems. Some improvements have been noted, but it appears that an expanded total aviation system has not been realized yet. Many people report that reliever airports are needed to compensate for the busy commercial airlines' use of the major U.S. airports.

Product liability effects have been severe in the loss of general aviation aircraft sales, and although the General Aviation Revitalization Act, known as GARA, was passed on August 17, 1994, aircraft yearly sales during the 1980s and 1990s have never reached the peak of 1978. The aircraft manufacturers have had to bear the burden of many lawsuits for their aircraft under a lifetime warranty. For example, in 1986, Cessna had over 500 product liability claims against it, even though it stopped aircraft production. Beechcraft was reported to have nearly $500,000 of expenses for processing each liability case, even though it was successful in 90 percent of the cases. Because of the high costs to do so, general aviation aircraft manufacturers had to increase their aircraft prices and reduce employment (Office of Aviation Policy and Plans, 1996). Now, at least a statute of limitations has been recognized.

GARA will be reviewed in the next chapter. Here, it is important just to recognize that general aviation aircraft sales, and similarly, corporate and business aircraft use significantly decreased because of the added costs involved in litigation and increases in aircraft prices. Aircraft had been greatly improved. *Technological advances* in engines, composite components, new materials, system programs for maintenance and flight, and many others resulted in the development of new aircraft with longer ranges, greater speed, more comfort, and ease of repair. Helicopter and fixed-wing designs have even been combined as seen from work on the tilt rotor. Many of the aircraft used today will be reviewed in Chapter 5. Performance and capacity specifications will also be covered.

Cessna developed its last piston-engine aircraft (Crusader) in the early 1980s, and Beechcraft reduced its production of piston aircraft. Piper declared bankruptcy after a series of product liability problems. The market at the time was in turboprop and turbojet aircraft. GAMA formed a Product Liability Committee in 1985 to help the aircraft manufacturers, with Frederick Sontag as its chairman (Office of Policy and Plans, 1996).

During the 1980s and early 1990s, corporate managers, for the most part, recognized the value of using a corporate/business aircraft. The NBAA has further promoted such use through its *No Plane-No Gain* program, a positive approach to determining company benefit (presented in Chapter 3 along with other agency programs).

Tax considerations were behind many of the decisions to buy general aviation aircraft; however, the Tax Reform Act of 1986, the empty seat rule (fringe benefit rule), and the loss of investment tax credit forced managers to consider operating costs and capital investment programs rather than tax savings. In 1997, the Airport and Airway Trust fund had over $15 billion in it, but still higher taxes on aviation users were levied. The investment tax credit of 10 percent when buying a new aircraft was replaced with a luxury tax. Whereas previously a company benefited from obtaining an aircraft, now unless the company uses the aircraft more than 80 percent of the flight time for business, a luxury tax of 10 percent must be paid—hardly an incentive to procure an aircraft and probably a loss of tax revenue for the government because of inhibiting sales.

Financing aircraft was an increasing option for corporate officials, and more interest was seen from business aircraft users rather than from personal aircraft users. Leasing, fractional ownership, and other means were undertaken by the aircraft operators.

During the 1980s and 1990s, corporate managers took a business-like approach to corporate/business aviation ownership. Managers exhibited real concerns for costs, safe operations, ease of repairs, training, international operations, joint use, and good capital investment; regulatory concerns, and a host of other issues. The concept of using an aircraft for business has been firmly accepted; what matters now are the availability, control, and benefits. Answering how the use of an aircraft will provide for a company's survival, profit, and growth has become the goal of potential operators. Charles Summa, president and chief executive officer of Piper Aircraft, Incorporated, identified several factors affecting the growth of the general aviation industry and, in turn, the use of business aircraft as the year 2000 approaches. These business factors were:

1. The loss of investment tax credit;
2. Elimination of accelerated depreciation;
3. Substantially higher insurance costs for owners, operators, and manufacturers above inflation;
4. Fuel costs escalating at a higher rate than inflation;
5. Decline in middle-class consumers and their ability to purchase products;
6. Cessation of the G.I. Bill;
7. Longlife of the product; and
8. Substantial over-production of aircraft in relation to consumer demand (Office of Aviation Policy and Plans, 1996).

There were several positive indicators for corporate and business aviation by the end of 1995. The number of pilots holding private certificates increased by 0.7 percent to 284,200, resulting in the reversal of a three-year decline. Additionally, the FAA handled more general aviation aircraft at its enroute centers for the fourth straight year which, in turn, indicated an increase in business flying; however, the total number of active pilots declined 1.7 percent to 654,000 in 1995 (Statistics and Forecast Branch, 1996).

As the 1990s draw to a close, business aviation has been characterized by a period of dynamic interactions. Technological innovations in the aviation industry led to efficient improvements to aircraft engines, airframes, and avionics, while business managers face changing regulations, capital costs, and global marketing proactively. Piston-engine aircraft are again being built, and there appears to be a slowly developing attitude that product liability impacts will be reduced significantly, thereby causing optimism for the aircraft manufacturers and users. Concerns, however, about the number of available pilots and aircraft mechanics are being discussed by those in the industry. General aviation and, in particular, corporate and business aviation, have developed into a viable and necessary industry. The advocates of this industry will be presented in the forthcoming chapters, along with many of its operating details.

Summary

Various definitions of corporate and business aviation are reviewed, but, essentially, corporate aviation is defined as a professional pilot flying corporate passengers, cargo, and/or mail on a company's aircraft. Business aviation is defined as an aircraft flown by a business person in furtherance of a firm's business, and piloting of the aircraft is incidental to the business. Many people designate the term business aviation to mean the use of an aircraft for business and do not distinguish any difference based on piloting. To help understand where corporate and business aviation falls within the U.S. aviation industry, the aviation categories are presented; corporate and business aviation falls under the general aviation category. How the FAA and GAMA report the general aviation groupings is shown in the tables, while an extensive review of the history and background of some of the developments of corporate and business aviation is made. Key people, aircraft manufacturers, associations, events, operations, and terms are presented, to include the NBAA, executive aviation, airline deregulation, world wars, and many others. Lastly, as the year 2000 approaches, specific factors affecting the general aviation industry and, in particular, business aviation, are identified. Today's business managers are considering the use of an aircraft as a business tool.

Activities

1. Review the federal aviation regulations and Federal Aviation Act to obtain various definitions of subject-related terms; for example, corporate aviation, commuter, common carrier, and others.
2. Research the literature and/or communicate with aviation historians to determine the development of corporate and business aviation.
3. Research the literature and communicate with regulatory agencies (e.g., DOT, FAA, NTSB) to determine current regulatory impacts on business aviation.
4. Interview a business aviation operator or corporate flight department member to determine his or her views on the use of general aviation aircraft.
5. Visit or write to a U.S. Government Printing Office to review or obtain a copy of the Federal Aviation Regulations (FARs), especially FAR Part 91. A listing of the U.S. Government Printing Offices is provided in Appendix B.
6. Use the Internet to obtain information from the NBAA, GAMA, FAA, or other aviation group activities about general aviation (a listing of Internet addresses is provided in Appendix C).

Chapter Questions

1. Define corporate and business aviation.
2. Explain where corporate and business aviation is placed within the U.S. aviation industry.
3. List five major contributing factors that promoted the positive development of corporate and business aviation.
4. Discuss five negative factors that inhibited the development of corporate and business aviation.
5. Explain the "royal barge" concept.
6. Choose a period of aircraft use (beginning from the 1920s) and discuss the development of aircraft used for business aviation.
7. Explain why an aircraft is considered a business tool.
8. Discuss the FARs that govern the use of corporate and business aircraft.
9. Identify the period in history which had the greatest impact on business aviation and discuss your selection.
10. Discuss the prevailing attitude and actions of business managers when considering the use of aircraft for business.

CHAPTER 2

Regulations and Associations

Overview

Any study of an aviation subject must include a regulatory review of pertinent rules; this is provided for corporate and business aviation in this chapter. The Federal Aviation Regulations (FARs) are numerous, but only specific rules applicable to business aviation will be defined and/or explained. Additionally, various important acts affecting this industry will be reviewed, as well as important state and local rules. IBAC, NBAA, and the other important business aircraft associations are also presented, along with a listing of key aviation regulatory offices and agency addresses.

Goals

✈ Identify the major regulations applicable to corporate and business aviation.
✈ Understand the development of aviation regulations.
✈ Know the major business aircraft associations.
✈ Become familiar with functions and activities of business aircraft associations.

Regulation Development

The *source* of aviation law in the United States is the U.S. Constitution. Although aviation wasn't an industry at the writing of this document, the founding fathers provided for future laws under Article I, Section 8, of the Constitution. Two subclauses of Article I, Section 8, the commerce clause (clause 3) and the necessary and proper clause (clause 18) provided the source for legislators to develop aviation rules. When the U.S. government provided federal monies in 1918 for airmail contracts, the only activities controlling U.S. aviation were private aero clubs. U.S. federal control resulted from the passage of the Air Commerce Act of 1926, which established the Aeronautics Branch under the Secretary of Commerce, later renamed the Bureau of Air Commerce in 1934. Safety in aviation was the predominant issue at the time (and still is). Pilots and planes then had to be licensed and registered by the federal government in compliance with civil aviation regulations (CARs).

Through airmail legislation, the airline industry formed and developed. The Airmail Act of 1925 and the Commerce Act of 1926 provided the *foundation* for aviation laws. When the Airmail Act of 1934 was passed to resolve many of the airmail problems at the time, further issues developed. Aviation was then in the control of the Post Office Department, the Interstate Commerce Commission, and the Department of Commerce.

With the passage of the Civil Aeronautics Act of 1938 (June 23, 1938) and an amendment in 1940, the Civil Aeronautics Board (CAB) and the Civil Aeronautics Authority (CAA) were established to manage the economic and non-economic activities of U.S. aviation, respectively. The next major aviation legislative act was the Federal Aviation Act of 1958 (August 23, 1958); it remains the dominant aviation law today. This important aviation law resulted from people's concern for air safety and the operation of new jet aircraft. The CAA was replaced by the Federal Aviation Agency, but the CAB maintained economic control. At this time, the Federal Aviation Regulations (FARs) replaced the Civil Aeronautics Regulations (CARs). See Appendix D for a listing of the FARs.

The Department of Transportation Act of 1966 was passed on October 15, 1966, and was listed as Public Law 89-670. At this time, the FAA became known as the Federal Aviation Administration, and its control came under the Secretary of Transportation (see Figure 2-1 for the FAA regional boundaries and Table 2-1 for the FAA Regional Offices). A National Transportation Safety Board (NTSB) was also established to investigate transportation accidents and determine their probable cause, but later this organization was placed under the president by the Transportation Safety Act of 1974.

These, then, are the basic laws that had and are having an immense effect on aviation, not only domestically but internationally as well. Many other federal rules also affect the aviation industry. For example, the Airport and Airway Development Act of 1970 affected airport management and initiated a revenue tax (known as the Airport and Airway Trust Fund). Notably, one of the major amendments to the Federal Aviation Act of 1958 was the Airline Deregulation Act of 1978 (signed October 24, 1978). One provision of this act was Title XVI or the Sunset Provisions. As a result of this title's provisions, the Civil Aeronautics Board was terminated on January 1, 1985, and its economic functions were transferred to various offices in the Department of Transportation, Federal Aviation Administration, U.S. Postal Service, and other agencies.

Aviation Regulations

Aviation, as a transportation mode, is more regulated than the other four transportation modes. Legislation involving noise (Noise Control Act of 1972), climate (National Climate Program Act of 1978), airports (Federal Airport Act of 1946), flight (National Aeronautics and Space Act of 1958), and many other issues have been passed. Rules pertaining to aircraft ownership, aircraft registration, financial responsibility, aircraft maintenance, taxation, accident reporting, aircraft leasing, labor relations, airmen, aircraft manufacturing, and many others exist.

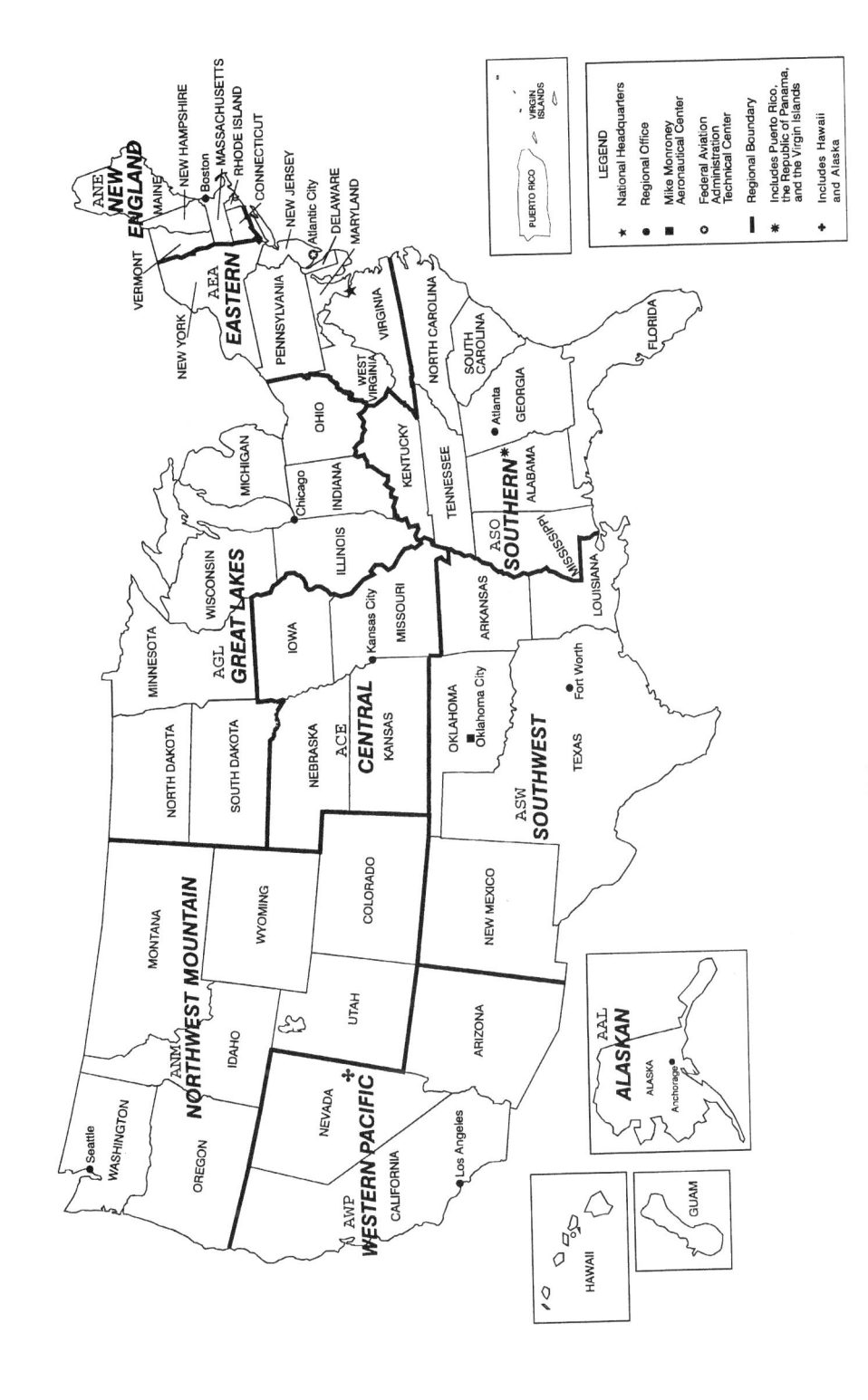

Figure 2-1. FAA Regional Boundaries. Source: FAA.

Table 2-1. Federal Aviation Administration

Name	Address
FAA National Headquarters	800 Independence Avenue, S.W. Washington, D.C. 20591
FAA Regional Offices	
FAA Alaskan	Anchorage Federal Office Building 222 West 7th Avenue, Box 14 Anchorage, AL 99513
FAA Central	Federal Building 601 East 12th Street Kansas City, MO 64106
FAA Eastern	Fitzgerald Federal Building Room 329 John F. Kennedy International Airport Jamaica, NY 11430
FAA Great Lakes	2300 East Devon Avenue Des Plaines, IL 60018
FAA New England	12 New England Executive Park Burlington, MA 01803
FAA Northwest	1601 Lind Avenue, S.W. Renton, WA 98055-4056
FAA Southern	3400 Norman Berry Drive East Point, GA 30344
FAA Southwest	4400 Blue Mound Road Fort Worth, TX 76106
FAA Western-Pacific	P.O. Box 92007 Worldway Postal Center Los Angeles, CA 90009

Owning/operating a business aviation aircraft requires a manager to become familiar with the federal, state, and local laws. A corporate aviation manager must at least review the applicable rules relating to the issues identified in Table 2-2. Add to these (for the international operators) international rules such as the International Civil Aviation Organization (ICAO) Standards and Recommended Practices (SARPS) and Procedures for Air Navigation (PANS). Meeting the rules becomes a formidable task for any aircraft operator.

In Appendix D, the non-economic federal aviation regulations are listed as FARs 1–199, while the economic FARs are identified as FARs 200–1199. Both groups of rules fall under

Table 2-2. Regulatory Issues for Corporate Flight Managers (Examples)

Federal	State/Local	International
Aircraft 　　Certification 　　Airworthiness 　　Maintenance 　　Registration 　　Titles/security 　　Finances	Aircraft Registration Insurance Labor rules	Standards & Recommended Practices (SARPS) Procedures for Air Navigation (PANS)
Airmen 　　Certification 　　Medical standards 　　Type ratings 　　Responsibilities	Environmental Protection Noise & Air Fuels	International agreements
Airspace Control	Airports	Eurocontrol
Air Traffic and General Operating Rules	Security	Security Terrorism
Airports	Warranties	Liabilities
Administrative	Operating Procedures	
Safety	Safety	Safety
Accidents	Accidents	Sovereign Rights
Liability	Liability	
Federal Taxes	State/local taxes	

Chapter II, Title 14 (Aeronautics and Space), of the Code of Federal Regulations. The major difference is that non-economic rules relate to safety, licensing, registration, and similar matters. The economic rules pertain to routes, tariffs, rates, reporting, organization, and other subjects having to do with the distribution and consumption of wealth, or the ways of doing business.

FAR 91, General Operating and Flight Rules, is the appropriate regulation for the corporate and business aircraft operator not using an aircraft to earn money for carrying passengers or moving cargo for compensation. Its major subparts include:

1. Subpart A-General;
2. Subpart B-Flight Rules;

3. Subpart C-Equipment, Instrument, and Certificate Requirements;
4. Subpart D-Special Flight Operations;
5. Subpart E-Maintenance, Preventative Maintenance, and Alteration;
6. Subpart F-Large and Turbine-powered Multi-engine Airplanes;
7. Subpart G-Additional Equipment and Operating Requirements for Large and Transport Category Aircraft;
8. Subpart H-Foreign Aircraft Operations and Operations of U.S. Registered Civil Aircraft Outside of the United States;
9. Subpart I-Operating Noise Limits; and
10. Subpart J-Waivers.

Extracts from FAR 91's rules are important to review to determine applicability (Office of Federal Register, 1994). These extracts include:

1. **Subpart A-General—91.1 Applicability.**
   a. This part prescribes rules governing the operation of aircraft (other than moored balloons, kites, unmanned rockets, and unmanned free balloons, which are governed by part 101 of the chapter, and ultralight vehicles operated in accordance with part 103 of the chapter) within the United States, including the waters within three nautical miles of the U.S. coast.

   While the Federal Aviation Administration prescribes many aviation rules and procedures, other federal agencies set requirements for corporate aviation managers that may not directly involve aviation, but at the same time, affect aviation operations. Some of these other federal regulators are identified in Table 2-3.

   b. Each person operating an aircraft in the airspace overlying the waters between 3 and 12 nautical miles from the coast of the United States shall comply with sections 91.1 through 91.21.

2. **Subpart B-Flight Rules—91.101 Applicability.**
   This part prescribes flight rules governing the operations of aircraft within the United States and within 12 nautical miles from the coast of the United States.

A review of these rules and the many other important regulations of FAR Part 91 can be accomplished by obtaining a copy of the FARs through a U.S. Government Printing Office (Appendix B) or through any one of several aviation associations or institutions; for example, Aviation Supplies & Academics, Incorporated (7005 132nd Place SE, Newcastle, WA 98059-3153, phone 1-800-ASA-2-Fly or 1-800-272-2359). This company provides the standard reference for the aviation industry and free midyear updates.

Jackson and Brennan (1995) provided an excellent text, titled *Federal Aviation Regulations Explained, Parts 1, 61, 91, 141, and NTSB 830*. In this informative text, the answer to the question *What does it mean?* is provided to pertinent regulations. It is available by contacting Jeppesen, 55 Inverness Drive East, Englewood, CO 80112-5498, phone 303-799-9090.

Table 2-3. Other Federal Regulators

Policy Maker

Civil Rights Commission–enforces airline minority hiring laws and minority business enterprise laws.
Department of Agriculture–enforces airline produce shipments and airport agricultural pest control laws.
Department of Commerce–encourages, serves, and promotes nation's international trade, economic growth, and technological advances. Provides social and economic statistics for businesses.
Department of Energy–responsible for the development of standby aviation fuel allotment and price-control rules.
Department of Justice–enforces the antitrust rules for all industries, including aviation.
Department of Labor–Occupational Safety and Health Administration–responsible for regulating the safety of the workplace. Fosters, promotes, and develops policies to improve working conditions and workers' rights.
Department of State and the President–responsible for approving bilateral agreements governing air service to other countries. Formulates and executes foreign policy.
Department of the Treasury–formulates and recommends economic, financial, tax, and fiscal policies. Enforces the corporate tax laws relating to depreciation, business-cost deductions, etc.
U.S. Customs Service-enforces the law relating to illegal aliens/fringe benefit rule.
Environmental Protection Agency–controls and abates pollution in air, water, radioactive, and other substances. Regulates aircraft noise, etc.
Equal Opportunity Commission–works to eliminate discrimination based on race, color, religion, gender, national origin, and age.
Federal Communications Commission–issues licenses for radio operators and transmission equipment, avionics repair, and others.
National Labor Relations Board–assists in achieving settlements of contract labor disputes.
National Oceanic and Atmospheric Administration–U.S. Department of Commerce–responsible for official air navigation charts, maps, and approach plates.
National Transportation Safety Board–determines probable cause of transportation accidents and recommends corrective actions.
Securities and Exchange Commission–regulates the reporting of financial interests in aviation companies through stock sales, reports, etc.
U.S. Postal Service–develops and enforces regulations concerning the handling of the United States mail.

Another important federal regulation that must be followed by some corporate managers is FAR Part 135, Air Taxi Operators and Commercial Operators. Its provisions apply when a corporate manager decides to use an aircraft to earn compensation while operating as a commercial operator. The aircraft must have a seating capacity of less than 20 passengers or a maximum payload capacity of less than 6,000 pounds. Other rules apply under Subpart A,

135.1, but this ruling appears to be the main consideration for corporate operators. Changes to the FARs occur frequently; for example, in March 1997, all FAR 135 scheduled air carriers carrying over 10 passengers must comply with the more stringent FAR 121 safety rules. The major difference here is that a company must be more strictly regulated when operating to carry traffic for compensation. Traffic, here, relates to carrying people, cargo, or mail.

Most, if not all, states have some form of aviation control that is administered through an aviation department, transportation department, or other regulatory body (see Appendix E). The norm that all rights not held by the federal government revert to the states provides the basis for these controls. Rules pertaining to airports, taxes, financing, labor, insurance, noise, operations, and many other factors must also be complied with, in addition to the federal rules. For example, a state may impose a fine and/or prison sentence on a pilot for alcohol or drug abuse, even though the FAA may have taken action.

State airport planning is greatly assisted by the federal government, but state controls or rules may have a major bearing on development and operations. For example, in some states, pilot licenses may have to be obtained, aircraft registration provisions may apply, state aviation fuel taxes have to be paid, and financial responsibilities may be imposed by state authorities. Local rules may involve city ordinances, noise and air pollution, taxes, and other pertinent requirements. Any owner/operator of an aircraft must contact all federal, state, and local authorities having jurisdiction and determine applicable business and aviation regulations that have a bearing on the aircraft's operation.

Airline Deregulation

While reference has previously been made to airline deregulation in 1978, further impacts of this important regulation must be reviewed. Many people associated with corporate and business aviation believe that this regulation has been one of the industry's top two major influences. The other major impact was World War II. How, then, did deregulation of the passenger aviation industry affect corporate and business aviation?

Firstly, executives, corporate officials, and business people had to extend their travel times in many instances when the commercial airlines developed their hub-and-spoke operations. The airlines optimized their use of major transportation centers in order to gain increased revenues, provide for aircraft servicing, improve scheduling efficiencies, eliminate costly service to small communities, change fare levels, and initiate other revenue and service-enhancing operations. An immediate effect of the hub-and-spoke system was the expansion of service to more city-pairs for the airlines (see Figure 2-2). With the same number of aircraft, an airline manager could offer more city-pair services. As can be seen in Figure 2-2, eight aircraft used to provide point-to-point service prior to the use of hubbing provide 28 city-pair services after implementation of this system.

The negative aspect for the business traveler who isn't interested in additional city-pair service is an extended journey time because an airline operates through a hub. A connection to another flight usually has to be made by the traveler. A two-hour journey might be lengthened to three-and-a-half hours, or even longer in some instances. While the leisure traveler has gained some advantages using discount fares because of the commercial airlines' pricing

$$X = \frac{N \cdot (N-1)}{2}$$

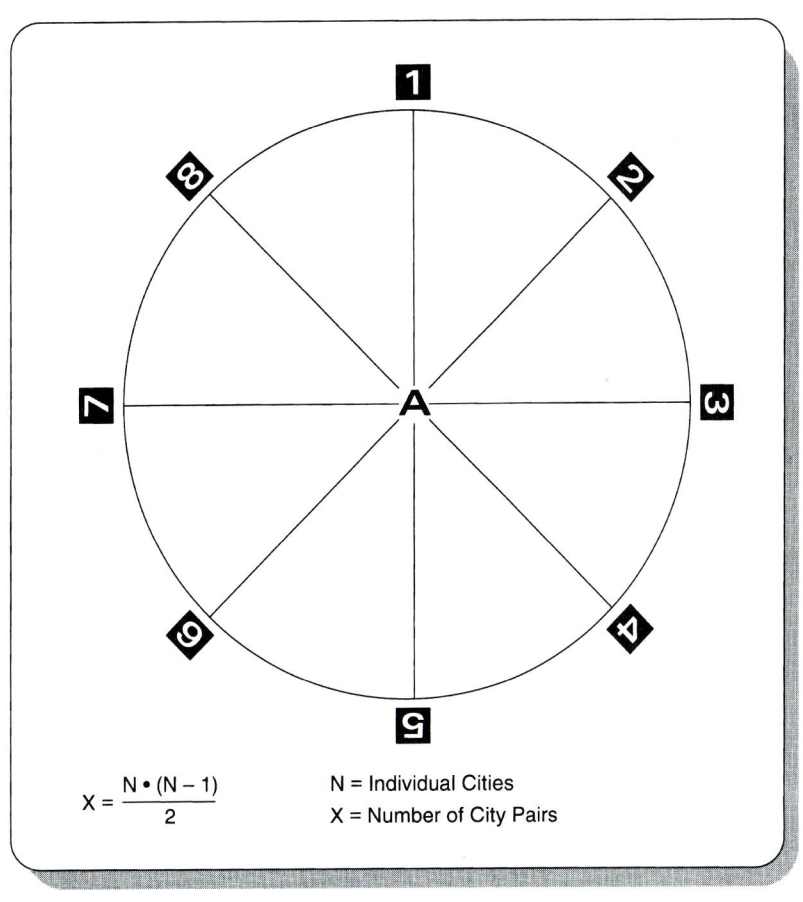

N = Individual Cities
X = Number of City Pairs

Figure 2-2. Hub-and-Spoke Model.

approach, business flyers have had to maintain and even pay higher fares for many services because they have had to use the commercial airlines.

Service to the small communities was drastically cut by the larger airlines rerouting their aircraft, and if the U.S. government had not provided an essential air service program under the deregulation act, some communities would have lost all air transportation services. Smaller commuter airlines and air taxis had to be used by the business traveler, and for access to communities that lost all flight services, another way had to be found to get to his or her final destination. This change was a catalyst for some managers to purchase aircraft for business use and forego using commercial airlines except on major routes where the airline service was an advantage. Business managers also increased their purchasing of business aircraft for international journeys, especially because of increased improvements in aircraft technology.

Other impacts of deregulation affected airport managers, aviation fuel buyers, travel agency managers, labor leaders, postal officials, and others. The main purpose of airline passenger deregulation was to let the commercial airlines compete, thereby benefiting the travelers with better service and rates. Many travelers did receive such benefits, but for the business traveler, fare decreases were not realized and travel times lengthened–some business fares actually increased due to airline managers gaining a larger share of specific, regional markets. These impacts, coupled with a recession in the early 1980s, meant that business travel was severely affected.

Fringe Benefit Rule

Although not titled the fringe benefit rule, under the Internal Revenue Service (IRS) Title 26, Code of Federal Regulations, personal use of a company's aircraft falls under Section 1.61-21 after January 1, 1989 (under Section 1.61-2T between 1985–1988). Fundamentally, any employee who receives a service (free ride) on a company aircraft must declare income on his/her individual tax form.

This tax ruling becomes somewhat complex when one has to consider whether the employee is labeled a control or non-control employee, the weight of the aircraft used, and the current Standard Industry Fare Level (SIFL) set by the Department of Transportation. An example of a control employee is a Board member or an employee who owns a 5-percent or greater equity, capital, or profit interest in the company (other categories exist). If such a person received a free flight and the aircraft weighed over 25,000 pounds, then the person must declare 400 percent of the value assigned in the SIFL. If the person is not a control employee and the flight is made on an aircraft weighing less than 6,000 pounds, then the person declares only a 15.6-percent value. The range can be quite extensive; however, there is an exemption that applies to the free-riders.

The seating capacity rule applies when 50 percent or more of the regular passenger seating capacity is used for business reasons; the remaining seats are valued at zero. The SIFL is determined twice a year (usually in January and July) and is computed for three categories:

   a. up to 500 miles,
   b. 501–1500 miles, and
   c. over 1,500 miles.

A terminal charge also applies. For example, the June 1997 SIFL rates were $0.1735 per mile for the first 500 miles, $0.1323 per mile for 501 to 1,500 miles, and $0.1272 per mile for over 1,500 miles. The terminal charge was $31.73. Review the NBAA web site (http://www.nbaa.org/nonmember/library/digest/sepdi97/ sepdi97.htm) for the current rates.

General Aviation Revitalization Act

A major legislative act that also significantly affected corporate/business aviation was the General Aviation Revitalization Act of 1994 (August 17). While GAMA and other agencies report that product liability affected aircraft sales beginning in 1979 (the year after peak aircraft sales), liability rules about negligence had begun much earlier. In 1963, the California Supreme Court set a precedent in its ruling (Greenman versus Yuba Power Products, Incorporated) that the injurer was liable for damages even if reasonable care was taken. Under strict liability rules, then, the fact that the injurer's behavior was reasonable and free of fault was not a defense (Truitt & Tarry, 1995). Courts across the nation established a rule of strict liability and shifted the burden to the manufacturers.

Around 1979 there was a major increase in insurance premiums, and in the early 1980s, a recession in the economy existed. Insurance premiums increased about tenfold, from $24 million in 1978 to $210 million in 1985. The general aviation aircraft manufacturers realized costs for product liability at a range of $70,000 to $100,000 per aircraft produced and shipped. Industry employment dropped from 40,000 employees in 1980 to 21,580 in 1991. A study conducted by Beech resulted in the finding that the average cost for each accident to a manufacturer was $530,000 (Truitt & Tarry, 1995). Since 1978, there has been no doubt that product liability had a very negative effect on general aviation aircraft sales.

The majority of aircraft flown falls within the general aviation category. Figure 2-3 depicts the U.S.-manufactured general aviation aircraft and billings as provided by GAMA, while Figure 2-4 is a breakdown of general aviation shipments by type of aircraft. For 1996, general aviation aircraft shipments were only slightly ahead of 1995 numbers by 4.7 percent—1,130 to 1,077 aircraft. There was a fall in the production of business jets by 2.1 percent (241 from 246), but an increase of 11.2 percent in turboprops (289 from 255) from 1995 to 1996. The overall total of new general aviation (GA) aircraft shipments for 1996 was reported at 1,130 units (GAMA, n.d.).

What can be gathered from these statistics is that general aviation sales have fallen drastically since 1978 because of the major reasons already presented (i.e., recession, oil embargo, air traffic control operations, costs of owning and operating aircraft, repeal of the Investment Tax Credit, product liability, and others). The effects of product liability were severely felt by the aircraft manufacturers, and GAMA took the initiative in 1985 to develop the Product Liability Committee after U.S. lawyers had successfully prevented any regulations being passed to help the aircraft manufacturers. During its early work, the committee focused on data sharing and defense in liability cases. In 1986, Congressman Dan Glickman and Senator Nancy Kassenbaum teamed with members of AOPA, NBAA, National Air Transportation Association (NATA), and other aviation associations to fight for limitations in product liability.

President Clinton signed the General Aviation Revitalization Act (S. 1458, PL 103-298) on August 17, 1994, and established an 18-year statute of repose for product liability lawsuits against the manufacturers of airframes, engines, and components for piston-, and turbine-powered aircraft with 20 seats or less. A new statute begins at the time a new engine or part is installed.

Immediately, the repose exempted more than 90,000 general aviation aircraft (Phillips, 1994). Another effect was that about one-third of the piston-engined aircraft fell under the exemption because of their average age of 27 years. The 18-year statute of limitation also affected many of the aircraft built during the peak year of 1978 (17,811 aircraft) and, of course, the many built prior.

The impact of product liability on the manufacturing of general aviation aircraft had been severe, especially in research and design of new aircraft. Price, not design or performance, has been a major concern for general aviation aircraft buyers. Optimism exists that a slow revival will be seen, but product liability has been one legislative action that caused unemployment, legal battles, and a downside for general aviation for many years.

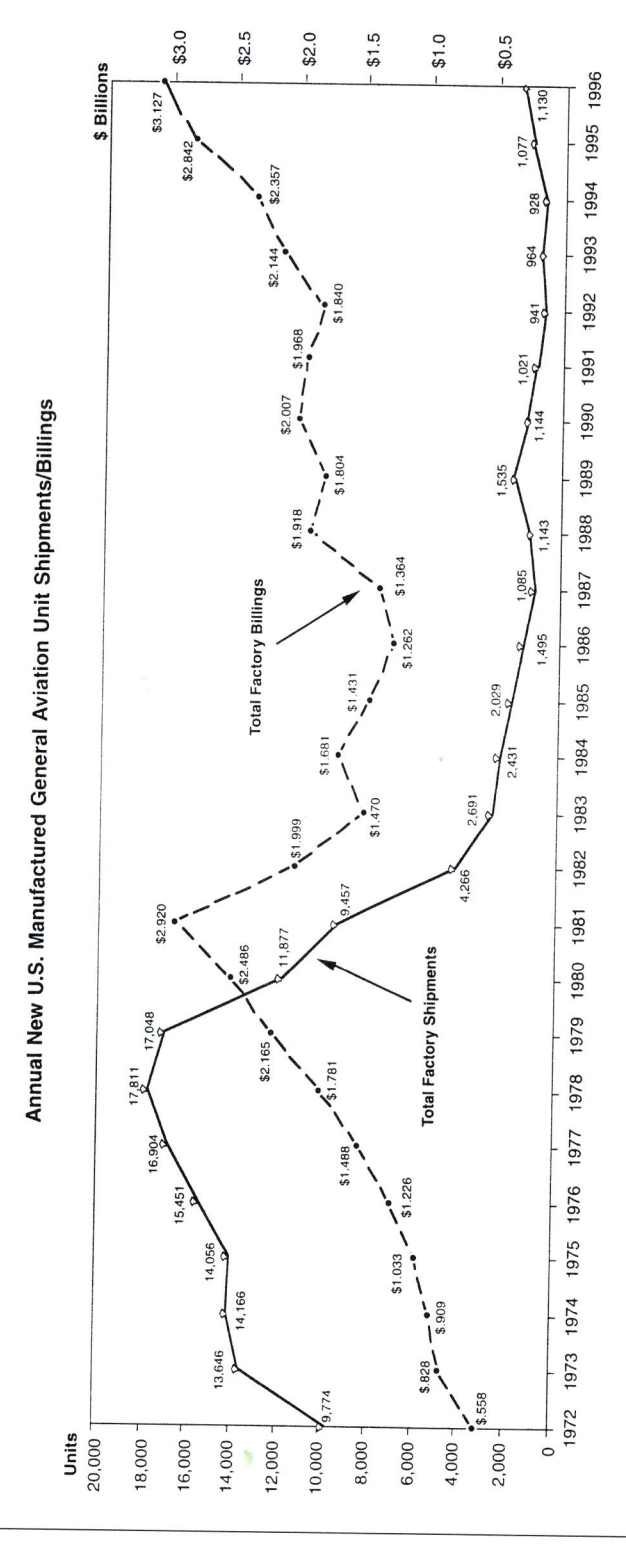

Figure 2-3. Annual New U.S. Manufactured General Aviation Unit Shipments/Billings. Courtesy of GAMA.

Annual New U.S. Manufactured General Aviation Aircraft Shipments By Type of Aircraft

Year	Total	Single-Engine	Multi-Engine	Total Piston	Turboprop	Jet	Total Turbine
1962	6,697	5,690	1,007	6,697	0	0	0
1963	7,569	6,248	1,321	7,569	0	0	0
1964	9,336	7,718	1,606	9,324	9	3	12
1965	11,852	9,873	1,780	11,653	87	112	199
1966	15,768	13,250	2,192	15,442	165	161	326
1967	13,577	11,557	1,773	13,330	149	98	247
1968	13,698	11,398	1,959	13,357	248	93	341
1969	12,457	10,054	2,078	12,132	214	111	325
1970	7,292	5,942	1,159	7,101	135	56	191
1971	7,466	6,287	1,043	7,330	89	47	136
1972	9,774	7,913	1,548	9,446	179	134	313
1973	13,646	10,788	2,413	13,193	247	198	445
1974	14,166	11,579	2,135	13,697	250	202	452
1975	14,056	11,441	2,116	13,555	305	194	499
1976	15,451	12,785	2,120	14,905	359	187	546
1977	16,904	14,054	2,195	16,249	428	227	655
1978	17,811	14,398	2,634	17,032	548	231	779
1979	17,048	13,286	2,843	16,129	639	282	921
1980	11,877	8,640	2,116	10,756	778	326	1,104
1981	9,457	6,608	1,542	8,150	918	389	1,307
1982	4,266	2,871	678	3,549	458	259	717
1983	2,691	1,811	417	2,228	321	142	463
1984	2,431	1,620	371	1,991	271	169	440
1985	2,029	1,370	193	1,563	321	145	466
1986	1,495	985	138	1,123	250	122	372
1987	1,085	613	87	700	263	122	385
1988	1,143	628	67	695	291	157	448
1989	1,535	1,023	87	1,110	268	157	425
1990	1,144	608	87	695	281	168	449
1991	1,021	564	49	613	222	186	408
1992	941	552	41	593	177	171	348
1993	964	516	39	555	211	198	409
1994	928	444	55	499	207	222	429
1995	1,077	515	61	576	255	246	501
1996	1,130	530	70	600	289	241	530

Source: GAMA

Figure 2-4. Annual New U.S. Manufactured General Aviation Aircraft Shipments by Type of Aircraft. Courtesy of GAMA.

There have been many regulations affecting the aviation industry, in particular general aviation. Because of the negative impact of many rules, aircraft owners/operators had to endure added costs and more strict operations; therefore, they have had to understand and evaluate the value/benefit of using aircraft for business rather than just for personal use. In the next chapter, this topic will be addressed, but a review of the major associations that help promote the use of aircraft for business is provided.

IBAC and Regional Associations

International Business Aviation Council

The International Business Aviation Council (IBAC) "represents, promotes, and protects the interests of business aviation in international forums" (International Business Aviation

Council [IBAC], 1988, p. 1). IBAC was incorporated on June 15, 1981, by five founding members. As many organizations begin, IBAC was started to serve the common good of its worldwide members and to serve as a voice and representative for them. IBAC promotes international business aviation through its lobbying efforts with the International Civil Aviation Organization (ICAO) and other cooperative efforts with its members. Nine regional groups (Table 2-4) form the alliance. IBAC's headquarters was in Montreal, Canada—the same location of ICAO—but IBAC's current director general (as of 1997) is Dennis Wright, and his office is now located in Frederick, Maryland.

From *What is IBAC?* (IBAC, 1988), the main function and activities included:

1. Representing international business aviation operators in international forums that influence flight operations, foreign or domestic;
2. Developing policy/position papers (PPP) for the business aviation community;
3. Developing air crew member identification cards for use in international environments (ICAO Annex 9 on Facilitation);
4. Collecting data on business aviation and the fleet;
5. Liaisoning with international aviation organizations such as IATA, IAOPA, IFALPA, AACC, AECMA, IFATCA, IACA, and others (see Appendix A for titles); and
6. Publishing the International Update, the latest information on time-critical operational flight information.

Overall, IBAC helps over 11,000 operators worldwide involved in business aviation. These firms operate over 16,000 turbine-powered aircraft and have an estimated annual gross turnover in sales in trillions of dollars (Woolsey, 1996). During 1996, 11,057 business aircraft operators were reported by the NBAA (1997) to use 17,791 turbine-powered business aircraft worldwide—7,507 of these operators using 11,285 aircraft in the United States. Costs of IBAC are shared among its members, and the total estimated annual gross turnover of the members is $4 trillion.

National Business Aviation Association

The National Business Aviation Association (NBAA) is the largest of the regional IBAC members with about 4,500 U.S. companies operating nearly 6,000 turbine-powered aircraft to over 5,500 U.S. airports, compared to about 550 airports served by the commercial airlines (NBAA, 1997). Its headquarters is in Washington, DC, and its president is John W. Olcott. Rossier (1994) reported that Bud Lathrop, assistant vice president at Bristol-Meyers, and Sidney Nesbitt, president of Atlantic Aviation Corporation, organized a group of 13 men at the Biltmore Hotel in New York on May 17, 1946, to form the Corporation Aircraft Owners Association (19 member companies), later changed to the NBAA in 1953. Dedicated to increasing safety, efficiency, and acceptance of business aviation, the NBAA promotes the concept of the aircraft as a business tool.

Table 2-4. Members of IBAC

Association	Address	Phone
IBAC	Administration Office 7508 Rockwood Road Frederick, MD 21702-3648 Dennis Wright, Director General	1-301-473-4755 Tel 1-301-473-4756 Fax
ABAA	Australian Business Aircraft Association c/o Southern Commander Pty. Ltd Hangar 6, Wirraway Street Essendon Airport, VIC 3041 Australia John South, Director	61-3-9374-2044 Tel 61-3-9379-8460 Fax
ABAG	Associaco Brasileira De Aviacao Geral Rua Simoes Magro, 155 Sao Paulo 04342-100 Brazil Sr. Fabio Rebello, Director Executive	55-11-5583-3032 Tel 55-11-5583-2405 Fax
BAUA*	Business Aircraft Users Association, Ltd. Crossmount House, Dunalastair Kinloch Rannoch Perthshire PH16 5QF Scotland Derek Legget, CEO	+44 1882 632252 Tel +44 1882 632454 Fax
CAA/SA	Business Aviation Division of Commercial Aviation Association of South Africa P.O. Box 807 1620 Kempton Park REPUBLIC OF SOUTH AFRICA Pierre F. De Bruyn, Executive Director	27-11-394-5316 Tel 27-11-975-5147 Fax
CBAA*	Canadian Business Aircraft Association Suite 1317, 50 O'Connor Street Ottawa, ON KIP 6L2 Canada John D. Lyon, President and CEO	613-236-5611 Tel 613-236-2361 Fax cbaa@istar.ca
EBAA*	European Business Aircraft Association Brussels National Airport Building 27(JBA) B-1930 Zaventem Belgium Fernand Francois, CEO	32-2-721-4272 Tel 32-2-721-2158 Fax 100700.3521@compuserve.c

Table 2-4. (Cont.)

Association	Address	Phone
GBAA *	German Business Aircraft Association c/o BDI, Abteilung Verkehrs-und Kommmunikationspolitik Gustav-Heinemann-Ufer 84-88 50968 Koeln 51 GERMANY Andreas Kraus, Chief Executive Officer	49-221-370-8424 Tel 49-221-370-8575 Fax
IBAA	Italian Business Aircraft Association Via Aldo Moro 4 20080 Ozzero, Milan ITALY Vittorio Sacchi, Corporate Secretary	39-2-9407358 Tel 39-2-9407358 Fax
NBAA*	National Business Aircraft Association 1200 Eighteenth Street Washington, DC 20036-2598 USA John Olcott, President	202-783-9000 Tel 202-331-8364 Fax jolcott@nbaa.org

Source: Multiple. Note: Data compiled November 1997. Names and addresses may change. * indicates founding member.

NBAA represents the interests of approximately 4,500 member companies that own, operate, or support approximately 5,500 general aviation aircraft used as a tool in the conduct of their business. NBAA member companies employ millions of people worldwide and earn annual revenues in excess of $3 trillion—a figure that is equal to more than half of the U.S. gross national product (NBAA, 1997).

Who are NBAA's Members?

With the permission of John W. Olcott (personal communication, January 23, 1996), information pertaining to NBAA's membership and other interesting statistics are provided. Membership consists of corporate, business, associate, and affiliate members. *Corporate members* are "defined as any commercial or industrial enterprise that is engaged in business, commerce, trade or industry, and which owns or operates U.S.-registered aircraft, primarily not for hire, as an aid to the conduct of its business" (NBAA, 1997, p. 7). A corporate member must also:

1. Own or operate a multi-engine aircraft;
2. Use an operations manual and maintenance program;

3. Use two professional pilots when passengers are flown, one pilot must be airline transport rated and the other at least have a commercial license and be instrument rated;
4. Certify recurrent training and annual proficiency checks for pilot and air crew member members; and
5. Have less than 50 percent of its total corporate sales, including all affiliates and subsidiaries, from products or services sold to business aviation clients.

Business members have the same definition as corporate members, but differ in that they must:

1. Not qualify for corporate membership,
2. Use pilot(s) who have a commercial license and valid instrument rating, and
3. Certify annual proficiency checks for air crew member members.

An *associate member* is "any commercial or industrial enterprise that derives 50 percent or more of its dollar volume from the field of business aviation or owns or operates aircraft that are not flown by pilots meeting the criteria set forth" (NBAA, 1997, p. 7) for corporate or business members. Companies owning or operating aircraft not registered in the United States may be *affiliate members*.

Corporate members include the large firms with flight departments while business members include firms with more regional air travel requirements. Associate members include airframe, engine and avionics manufacturers, and other companies involved with aviation products or services. The number of NBAA member companies from 1979 to 1996 is presented in Figure 2-5 while membership by state is shown in Figure 2-6.

Diversity of NBAA Membership

NBAA reported that the number of member companies has grown 128 percent since the late 1970s and 45 percent since 1990. One-third of the members are manufacturing firms, one-third are service-sector businesses, and the remaining portion includes other groups such as mining and construction, transportation, utilities, communications, and wholesale/retail trade firms. Service-sector firms include banks, insurance companies, and real estate developers. The NBAA member firms operate from one aircraft to large fleets. The average aircraft fleet size per member company is less than two. For two-aircraft operations, the average number of employees is 6.8. More than 50 percent of the firms use only one aircraft and employ an average of 3.1 workers. NBAA reported 1,004 business members as of December 27, 1996 (NBAA, 1997).

Corporate members have a larger aircraft fleet operation and have flight departments involving management, operations, and maintenance. As of December 27, 1996, 1,415 corporate members were reported, and together with the business members these groups comprise 53.8 percent of the membership. Associate members accounted for nearly 45 percent of the membership with 2,018 companies involved—149 companies were reported located outside of the United States. Fifty-eight affiliate members were reported, and combined with all members, NBAA member firms can be found in all 50 states (NBAA, 1997). Notice the size of membership by state in Figure 2-6.

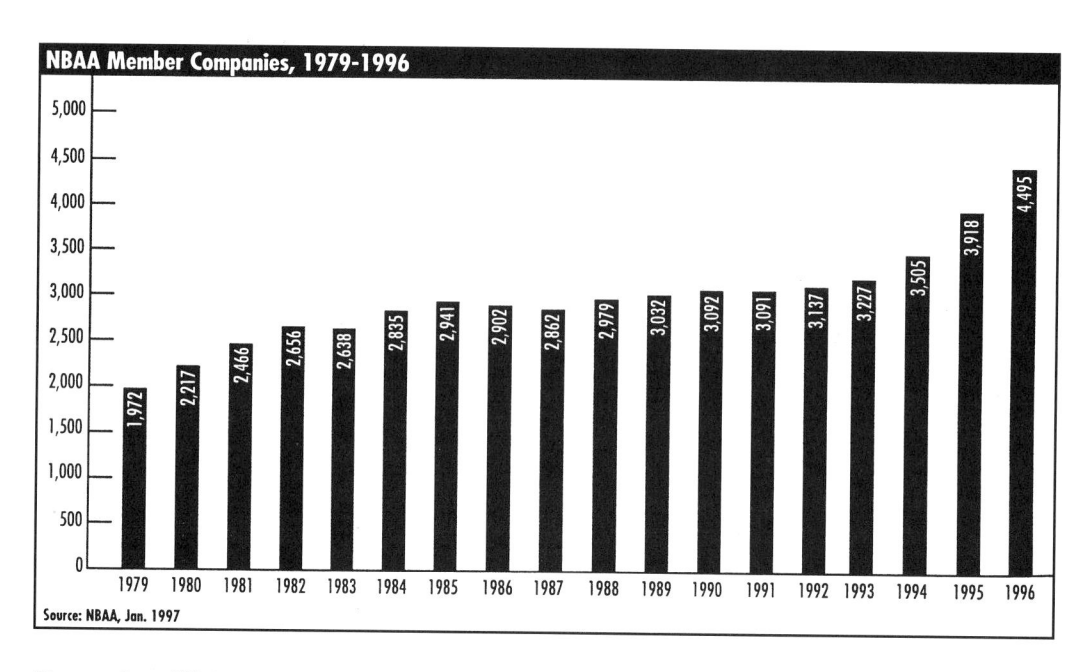

Figure 2-5. NBAA Member Companies, 1979–1996. Courtesy of NBAA.

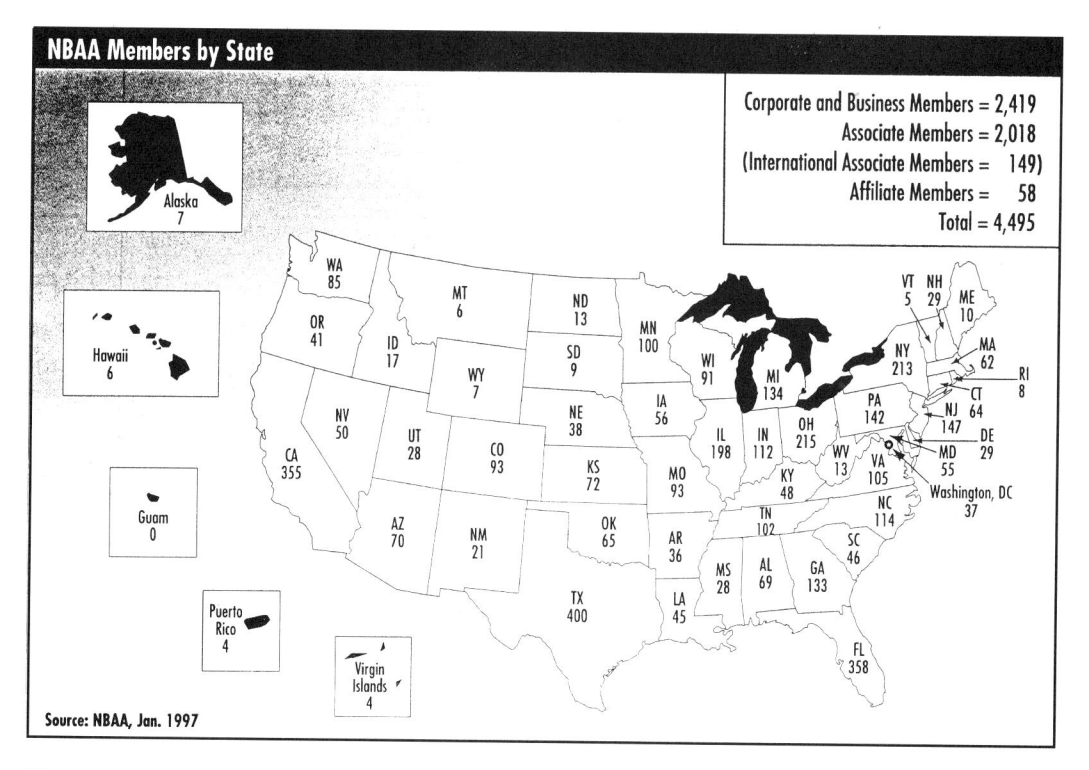

Figure 2-6. NBAA Members by State. Courtesy of NBAA.

38   *Chapter 2*

NBAA Activities

NBAA representatives are active in federal, state, and local governments, representing business aviation interests within the United States and in international business aviation matters. They are involved with Congress, the executive branch, regulatory agencies (such as the FAA, IRS, and others), and with IBAC. Through its operations department and standing committees, NBAA contributes to aviation forums that discuss air traffic control procedures, weather, air crew member training, airspace access, performance standards, and other key issues affecting business aviation activities.

NBAA standing committees include:

1. Airports/Heliports,
2. Airspace/Air Traffic and its Aviation Weather Subcommittee,
3. Associate Member Advisory Council,
4. Corporate Aviation Management and its Subcommittee on FAR 135 Operations,
5. Government Affairs,
6. International Operations,
7. Operations,
8. Schedulers and Dispatchers,
9. Tax, and
10. Technical and its subcommittees on specific aircraft and engines (NBAA, 1997).

The highest level of flight safety is NBAA's top priority. It promotes safety through various programs, awards, and publications (such programs are presented in Chapter 8). The NBAA annual meeting and convention is an example of a forum to discuss the variety of issues affecting corporate and business aviation. At such venues, an increasing number of attendees participate in management discussions, view the latest equipment, talk to experts, exchange information on maintenance, dispatching, scheduling, and other interest areas, and generally develop their skills and knowledge to further business aviation activities. In 1996, for example, an earlier record number of attendees (24,884 in 1995) was nearly equaled with 24,565 attendees who exhibited products and services, including numerous aircraft.

To provide assistance in helping a manager begin a business aviation department, the NBAA develops many publications. Some of these are presented in Table 2-5.

Canadian Business Aircraft Association

The Canadian Business Aircraft Association (CBAA) was established in 1961 and is the main representative and voice of business aviation in Canada. The CBAA defines business aviation as the "operation of owned or leased aircraft to transport the personnel or cargo of an enterprise to further its business pursuits, with professional pilots receiving a direct salary or compensation for piloting" (Canadian Business Aircraft Association [CBAA], n.d., p. 3). CBAA is a founding member of IBAC and is affiliated with other aviation associations such

Table 2-5. NBAA Publications

Name	Description
Airport Noise Summary	Compilation of noise regulations at approximately 400 U.S. airports.
Aviation Financial Analysis Package	Five descriptive manuals justifying aircraft acquisition; lease, buy, or charter; aircraft cost allocation and recharge rates; business aircraft fleet planning; and classification and management usage of costs in business aviation.
Financial Benefits and Intangible Advantages, Business Aircraft Operations	Reasons for using company aircraft.
NBAA Business Aviation Management Journal	Annual publication containing articles on flight department management, including finance, budgeting, planning, and personnel.
NBAA Business Aviation Safety Journal	Annual publication relating to safety of flight.
NBAA Federal Excise Tax Handbook	Elements to consider in application of appropriate taxes. Reference on various state taxes.
NBAA Directory of Member Companies, Aircraft and Personnel	Membership information.
NBAA Management Guide	Guide for establishing a new aviation department that describes administration, flight operations, international operations, and maintenance.
NBAA State Aviation Tax Report	Reference on various state taxes on aviation fuel, personal property taxes, sales and excise taxes, and aircraft registration fees.
Oceanic Operations	Background information for aircraft operators transiting oceanic airspace.
Operating Costs	Survey results of actual business aircraft operating costs.
Range Formats	Standardized mission profiles providing a basis for comparing the range of jet, turboprop, and helicopter aircraft.
U.S. Customs Guide for International Business Aviation	Details customs requirements and airports.

Source: National Business Aircraft Association. (1995). *How to Start a Corporate Flight Department.* Washington, DC: Author. Used with permission of John Olcott, NBAA president.

as the Air Transport Association of Canada and Aviation Councils. It actively works on various important issues, including

    a. aircraft emissions committees,
    b. airworthiness committees,
    c. AME training and licensing,
    d. aviation inquiries,
    e. Canadian Airspace Review,
    f. Canadian Airspace System Plan,
    g. Cost Recovery program,
    h. Environmental review processes,
    i. ICAO representation,
    j. international liaison,
    k. open skies,
    l. rationalization of air navigation services,
    m. Royal Commission on Transportation, and
    n. substance abuse programs.

President and Chief Executive Officer is John-David Lyon (see Figure 2-7 for CBAA's organizational chart) while its membership includes business and commercial, affiliate, and associate members. The first group includes companies that own and operate aircraft for transportation purposes to advance their business objectives, companies in the second group own and operate foreign-registered aircraft, and the third group consists of manufacturers, product-support, and service companies. Around 180 members operate more than 70 percent of Canadian business aircraft.

Membership includes eleven categories:

1. petroleum and energy,
2. manufacturing,
3. resources,
4. aircraft and parts producers,
5. business service,
6. FBO/charter,
7. retail,
8. utilities,
9. construction,
10. government, and
11. other.

The petroleum and energy group is the largest with 25 percent of operating members, and manufacturing next with 20 percent. Alberta's membership (see Figure 2-8) comprises one-third of the members—high due to its oil industry. The U.S. distribution of members is shown in Figure 2-9, with Texas having the highest number of members (CBAA, n.d.).

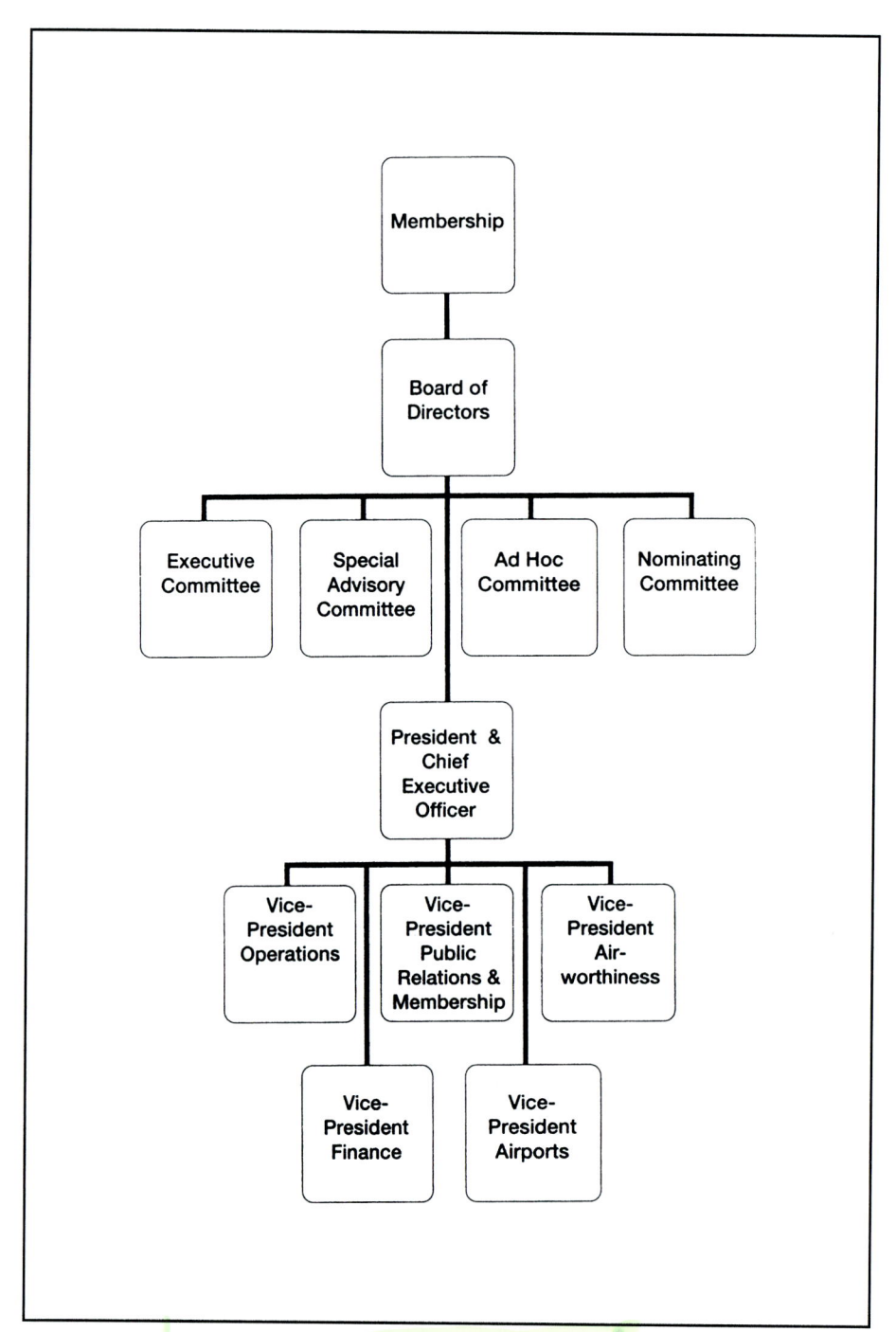

Figure 2-7. CBAA Organization Chart. Courtesy of CBAA.

	\multicolumn{6}{c	}{Membership Types}	
	Business	Associate	Affiliate Foreign	Commercial	Regional Chapters	Affiliate Organizations	Totals
Alberta	18	17		1	2	1	39
British Columbia	9	17			1	1	28
Manitoba		3					3
New Brunswick	5				1		6
Newfoundland	1	1					2
Nova Scotia		1					1
Ontario	18	17	1*		1	2	39
Quebec	8	14		3	1	1	27
Saskatchewan						1	1
Totals	59	70	1	4	6	6	146

* Barbados

Figure 2-8. CBAA Canadian Membership by Province, Summer 1991. Courtesy of CBAA.

	\multicolumn{6}{c	}{Membership Types}	
	Business	Associate	Affiliate Foreign	Commercial	Regional Chapters	Affiliate Organizations	Totals
Arizona		1					1
California			1				1
Colorado		2					2
Connecticut			1				1
Florida		1	1				2
Georgia			1				1
Illinois		1	3				4
Indiana			2				2
Kansas		2					2
Missouri			1				1
New Jersey		2					2
New York		1	1				2
North Carolina			1				1
North Dakota			1				1
Pennsylvania		1	1				2
Texas		3	2				5
Virginia		1	1				2
Totals		15	17				32

Figure 2-9. CBAA U.S. Membership by State, Summer 1991. Courtesy of CBAA.

EBAA POSITION POLICY PAPERS

PPP No 1 - Business Operations in Europe

PPP No 2 - Aircraft Certification

PPP No 3 - Air Crew Licensing

PPP No 4 - Business Aircraft Airport Requirements

PPP No 5 - Energy and Business Aviation

PPP No 6 - Business Aircraft and Environment

PPP No 7 - Aircraft Services (ATS) User Charges

PPP No 8 - Cockpit Voice Recorders (CVR) and Flight Data Recorders (FDR)

PPP No 9 - Communications, Navigations and Surveillance Systems (CNS)

PPP No 10 - Instrument Approach Landing Systems (IALS)

PPP No 11 - CAT II/CAT III Approach and Landing Criteria

PPP No 12 - Access for Business Aviation to Congested Airports and Airspace

PPP No 13 - Aircraft Noise

Figure 2-10. EBAA Position Policy Papers. Courtesy of EBAA.

European Business Aviation Association

The European Business Aviation Association (EBAA) represents a large number of private and commercial business aircraft operators, and its stated primary objective is to represent these groups either by national associations or individual members with regard to:

1. National and international regulations,
2. Public recognition of business aviation, and
3. Fair access to airspace, airports, ground facilities, and services (EBAA, n.d.).

EBAA began in 1975 when the management of the Phillips Group acted to form the association. Approximately 85 members were active in 1994, and they have attended seminars and meetings to exchange their views on issues affecting their business. Figure 2-10 depicts the major position papers EBAA works on to proactively represent business aircraft users. The president of the EBAA, Rinaldo Piaggio, stated that over 500 aircraft are used by EBAA members and that major concerns for EBAA are the congested European airspace and airports.

German Business Aircraft Association

The German Business Aircraft Association (GBAA) was founded on June 5, 1985, in Wolfsburg, Germany. Its main task is "to promote political, economical, and legal interests of business aviation in Germany, Europe, and worldwide" (personal communication, K. Lammich, March 5, 1996). GBAA works along with the Committee for Business Aviation (Arbeitskreis Geschaftsfliegerei), which was established on June 22, 1970, in Cologne, Germany. Fourteen business members comprised GBAA's organization in 1996, to include Robert Bosch GmbH, Daimler-Benz AG, SIEMAG Verwaltungs-GmbH & Co., Dornier Luftfahrt GmbH, and others.

Business Aircraft Users Association, Ltd.

Derek Leggett, chief executive of the Business Aircraft Users Association (BAUA), reported that BAUA was formed in 1961 to ensure that the requirements of business aircraft users were clearly presented and understood by government departments and civil authorities, at any level, that influenced the efficiency, safety, and economical use of business aircraft. BAUA is also a founding member of IBAC and has a permanent working member on the United Kingdom (UK) Civil Aviation Authority's Operations Advisory Committee and on its Finance Advisory Committee.

The organizational structure of BAUA is in two bodies, the BAUA Council and the BAUA Operations Committee; a Secretariat/Administration also exists. Every other month, the Council meets to consider policy, politics, and finance, while in the intervening months the Operations Committee meets at different airports around the UK to discuss technical and operational issues.

In February 1996, BAUA had 32 members whose main issues at the time were access to airports and controlled airspace, RVSM (reduced vertical separation minimum), and ETOPS (extended range operations). Success at resisting legislative attempts to restrict business aircraft access has been realized, but Leggett reported that airlines are again attempting to establish a priority system for access to controlled airspace (personal communication, D. Leggett, February 9, 1996).

Other Business Aircraft Associations

The remaining IBAC business aviation associations include those in Brazil, Australia, and Italy. Additionally, the Business Aviation Division of Commercial Aviation Association of South Africa (CAA/SA) has much the same function as the NBAA in the U.S. and the BAUA in the UK. Its previous executive director, C.Z.A. Beek, stated that in South Africa over "900 aircraft are used in business aviation roles, including at least 100 helicopters" (personal communication, C.Z.A. Beek, February 1991). Because of the size of Africa and the lack of excellent transport infrastructures, the business aircraft has an important role, and there is a continuing need for aviation training and support.

Besides the business aircraft associations already identified, there are many other professional groups that contribute to the development and promotion of safe and efficient aviation operations. Everyone involved in studying or working in aviation should participate in a professional association. The main purposes of such membership are to increase one's subject knowledge and to gain understanding of the issues that affect the industry. Attending meetings, reading professional journals, discussing organizational and operational matters, and making contacts are but a few of the activities realized through association membership. A listing of various professional associations is provided in Appendix F.

Summary

The source and foundation of U.S. aviation laws are identified, namely the U.S. Constitution (Article I, Section 8) and the Airmail Act of 1925/Commerce Act of 1926, respectively. Through a series of legislative acts, aviation has become primarily regulated by the Department of Transportation and the Federal Aviation Administration, but many other federal regulators are identified. In addition to these federal controls, state and local regulatory agencies and associations are presented; for example, state aviation departments, city authorities, and the NBAA. See Appendix F for a listing of other aviation associations and governmental contacts. Aviation is considered more heavily regulated than any other transportation mode. Specific to corporate and business aviation is FAR Part 91, General Operating and Flight Rules; however, some managers also follow FAR Part 135 when using their aircraft to earn monies from air transportation services.

Details about the Airline Deregulation Act of 1978, the fringe benefit tax rule, and the General Aviation Revitalization Act are provided. IBAC, NBAA, and the other regional business aircraft associations are identified, with NBAA, CBAA, EBUA, GBAA, and BAUA presented in detail. Various chapter tables and figures provide specific statistics and communication information.

Activities

1. Review any information source that provides historical data about the development of aviation law in the United States or in other countries. Prepare a briefing or outline on this development.
2. Review a major aviation transportation act and determine how it affected the business aircraft operator.
3. Visit a National Aeronautics and Space Administration center or other national/local aerospace location to gain a fuller appreciation of the development and value of air transportation.
4. Research or write to a U.S. Government Printing Office (Appendix B) to obtain a copy of rules pertaining to aviation.

5. Write or call a business aircraft association to learn more about its role and functions (see Table 2-4).
6. Search the Internet for information about corporate and business aviation issues (see Appendix C).

Chapter Questions

1. Review one of the major acts that affected U.S. aviation and discuss how it affects the U.S. airline industry. Consider the following: Airmail Act of 1934, Civil Aeronautics Act of 1938, Federal Aviation Act of 1958, and the Department of Transportation Act of 1966.
2. Explain the effects of the Airline Deregulation Act of 1978 on the airlines versus the business aircraft owners and operators. Identify pros and cons for each.
3. Review Article I, Section 8, of the U.S. Constitution and discuss how it can be used to develop aviation laws.
4. Explain why the Air Commerce Act of 1926 was so important to U.S. aviation.
5. Describe the results from the Civil Aeronautics Act of 1938 (as amended in 1940) and indicate their impact on corporate and business aviation.
6. Describe the results of the Federal Aviation Act of 1958 and indicate their impact on corporate and business aviation.
7. Review the Airline Deregulation Act of 1978 and argue for or against its enactment from the standpoint of a commercial airline manager. From the standpoint of a business aircraft operator.
8. Select IBAC or one of its regional aircraft associations and explain its role and functions.
9. Discuss the main issues being worked on by one of the business aircraft associations and explain their significance.
10. Discuss why a corporate/business aircraft operator would want to belong to an aircraft association. Explain the advantages and any disadvantages.

Value/Benefit of Using Business Aircraft

CHAPTER 3

Overview

The value and benefits of using a business aircraft are realized by individuals, companies, government, and society; an overview of the social, economic, and political effects of business aviation is presented. Various tangible and intangible benefits of such use are identified, along with many reasons why business aircraft use involves advantages and disadvantages for managers. A brief look at some of the major reports, such as No Plane-No Gain report, Fortune 500 Study, Travel$ense® program, and others, provides information about managerial considerations for business aircraft.

Goals

- ✈ Understand the value and benefits of using business aircraft.
- ✈ Discuss pertinent reports concerning the benefits of corporate and business aviation.
- ✈ Evaluate the major factors corporate managers consider when making business aircraft decisions.
- ✈ Explain the advantages of corporate aviation to firms.
- ✈ Distinguish between tangible and intangible benefits.
- ✈ Compare corporate aviation services to commercial aviation services.
- ✈ Compose a cost/benefit analysis of corporate aviation versus commercial aviation.
- ✈ Describe social, economic, and political effects of business aircraft use.

Introduction

Business aircraft make business sense, according to John Olcott, president of the NBAA, and many others; however, why do many managers not operate corporate or business aviation activities? An old adage is that it is *all in the eyes of the beholder!* Here, then, is one reason why the current approach of aviation users is to influence managers to use general aviation aircraft for business. In other words, these promoters believe that using an aircraft will improve company profits, increase employee productivity, improve morale, give a better corporate image, and provide other benefits. The value/benefits for using such aircraft are significant in business, according to IBAC, NBAA and the other regional business aircraft associations, and GAMA. Not every business manager is encouraged to go out and buy an aircraft, but every manager is encouraged to determine whether the use of an aircraft would improve employee productivity and contribute to managing a firm's most important assets—people and time. The difference is in the eyes of the beholder! During the turn of the millennium, business managers are actively involved in reducing costs to beat their competitors through a long-term campaign of working smarter and harder. Their focus is on key company resources and improved productivity.

A review of the number and use of general aviation aircraft showed that, for the most part, these aircraft have been used not to gain compensation for flying, but to provide scheduled or on-demand private flying for company personnel. Salespeople, doctors, maintenance repair workers, and others routinely fly general aviation aircraft hundreds of miles around their home bases in the conduct of their business. Large companies use aircraft on scheduled flights to move their employees and executives to frequently visited sites/operations. A number of flights are flown on a call-up, or on-demand, basis. When there is a need to be on location, no matter where, general aviation aircraft can be used to provide timely and, in some instances, the only available service. For example, a company manager who must provide a critically needed part for an oil rig platform operation values the use of a company helicopter immensely. Even airline-type aircraft (e.g., B-727s) are used to provide corporate flights for increased or regularly scheduled movements of large numbers of company personnel, cargo, and mail requirements.

Value/Benefits of Business Aviation

The NBAA (1997, p. 2) reported that "of all the benefits of business aircraft, *flexibility* is probably the most important." This indicates that company managers can control all aspects of their travel plans; make itinerary changes; fly to many airports not served or inadequately served by the commercial airlines; set their own flight times; eliminate worries about missed connections, lost baggage, and overbooking; and use their aircraft as a business tool in meeting a mission requirement. The NBAA listed the benefits of using business aircraft in 10 categories (Table 3-1).

These 10 benefit categories have significance for business managers considering using general aviation or even airline-type aircraft for business. Other interesting reasons for using business aircraft were identified during a personal survey of 79 corporate aviation managers, pilots, dispatchers, and others who were involved in corporate and business aviation studies during the period 1995–1997 (see Table 3-2).

Several items reported by these respondents also relate to the NBAA's listing; for example, privacy, comfort, confidentiality, flexibility, and others, but some items include

Table 3-1. Benefits of Using Business Aircraft

Benefit	Description
Time Savings	Reduce flight time by providing point-to-point service; utilizes smaller airports closer to final destinations; allows greater productivity while flying.
Flexibility	Freedom to set own travel schedules, change course enroute, and fit aircraft use to business plans, not vice versa.
Reliability	Business aircraft are maintained to the highest standards and readiness.
Safety	Outstanding safety record.
Improved Marketing Efficiency	Allows sales force to extend its areas and to bring customers to the point of sales.
Facilities Control	Allows for visits to outlying locations.
Personnel and Industrial Development	Mobility provides for accelerated training, orientation, and teamwork.
Privacy and Comfort	Allows for confidential communications on board aircraft with cabin configurations as required.
Efficiency	Maximizes use of people and time.
Security	Allows for the control of all aspects of air travel, for the users and equipment.

Adapted from NBAA. (1997a). *Business Aviation Fact Book 1997.* Washington, DC: Author. Used with permission of John Olcott.

interesting intangible as well as tangible benefits. While the respondents involved in aviation studies reported some interesting advantages, like an executive being able to carry a dog on his lap during a trip, most of the advantages can be easily factored into one of the 10 benefits identified by the NBAA. Not having to sit next to people who might be carrying contagious germs, be ill, or even become sick during a flight may be a reason not to fly on a commercial airline for some travelers. The ability of a business manager to redirect a flight and to use a greater number of U.S. airports is a distinct advantage (flexibility). If weather problems prevent landings at a major airport, a business aircraft operator could go to an available airport; for example, if Newark airport was under bad weather, the business flyer could divert to Teterboro airport.

Table 3-2. Listing of Benefits of Using Business Aircraft (personal survey, 1995–1997)

Flexibility	Comfort
Improved image	Airborne office
Advance business opportunities	More space on aircraft
Carry dog on seat	Can smoke on board
Privacy	Luxury
Recruitment tool	Moveable office
Security	Better safety record
Able to change flight plans	Carry more baggage
Tax benefits	Enables face-to-face meetings
Income from 135 ops	More speed in flight
Configuration ability	Know the flight crew
Eat what you want	Confidence of availability
Know crew is well-trained	Respond to customer's needs
Bring business to you	Inflight changes possible
Confidentiality	Able to fly in bad weather
Avoid ill/sick passengers	Care for personal illness
Customer feels important	Quick response to meetings
No airline strike problem	Use more airports
Convenience for family	Community support
Redirect in emergencies	Eliminate ATC problems
Productivity enroute	Airline crash would hurt firm
Leave heavy luggage on aircraft	Better international travels
Better health under stress	Sense of loyalty
Competitive edge	Contingency response
Reduced passenger travel costs	Saves billions by being there

Note: These are verbatim items from reports received from October 1995 to March 1997.

Still another listing of benefits provided by the NBAA and GAMA is shown in Table 3-3. David Almy, NBAA, paraphrased the informal survey responses of passengers being flown on trips considered typical for business travelers. While the wording appears slightly different, the main benefits again focus on flexibility, costs, time savings, and convenience. It can be seen from the listings provided that using a private aircraft means that the operator has control and can decide how and when to use the aircraft owned.

Managers can compare the cost of an airline flight and a corporate flight quite easily—helped by new programs like NBAA's Travel$ense®. They can also compare the time involved in making a business trip via commercial and corporate aircraft (example is provided later in the chapter).

Costs and time, then, are classed as tangible variables. A tangible benefit is one that can be measured or specifically determined. Costs of food, lodging, cargo shipments, insurance,

Table 3-3. Benefits of Business Aircraft According to the Passengers

1. Direct flights cut travel time from 10 to 60 percent or more, increasing productivity.
2. Complete scheduling flexibility/even changing course enroute/greatly improves work.
3. Privacy/quiet encourages/allows work enroute.
4. Typically costs less than the airlines if half the seats are occupied.
5. Peace of mind/safety advantage of knowing pilots/passengers/aircraft.
6. Gives an advantage over competition.
7. Eliminates hassles/delays of main airport terminal.
8. Operation out of sight of competition.
9. Predictable travel budgeting versus volatile/irrational airline fares.
10. Provides access to markets/customers we couldn't otherwise reach.

Source: National Business Aircraft Association & General Aviation Manufacturers Association. (n.d.) *Face to Face* [Brochure]. Washington, DC: Author. Note: Paraphrased by David Almy. Listing of top ten benefits of passengers by informal survey.

labor, aircraft operations, and other items can be quantified objectively. An intangible benefit, on the other hand, cannot be easily measured. These benefits are subjectively assessed; for example, what is the benefit of the convenience of using a firm's aircraft? How are factors such as employees' comfort, privacy, fatigue, morale; the firm's prestige, image, and pride; and market accessibility costed?

A study revised by PRC Aviation (1995, p. 2-1) for GAMA and NBAA resulted in the categorization of eight tangible benefits for company managers to consider. These quantifiable benefits included:

1. Personnel time savings,
2. Time saved by scheduling trips according to business needs rather than airline schedules,
3. Feasibility of last minute trips without constraint of reservation availability,
4. Time saved because of operational reliability,
5. Avoided costs of alternate transportation,
6. Enhancement of mental performance and physical endurance,
7. Increased work productivity while enroute, and
8. Transportation of company mail and cargo.

An example of increased reliability and flexibility is given by PRC Aviation through its comparison of airports serving commercial aviation and those serving general aviation. In Figure 3-1, it can be seen that there are increased opportunities for general aviation aircraft users to conduct business in the Atlanta, Georgia, area. Thirteen airports versus only one for the commercial airlines are available. In Figure 3-2, PRC Aviation compared the business aircraft and commercial airline service between Hartford, Connecticut, and West Palm Beach, Florida. A time savings of 6 hours and 43 minutes was realized through the use of a business

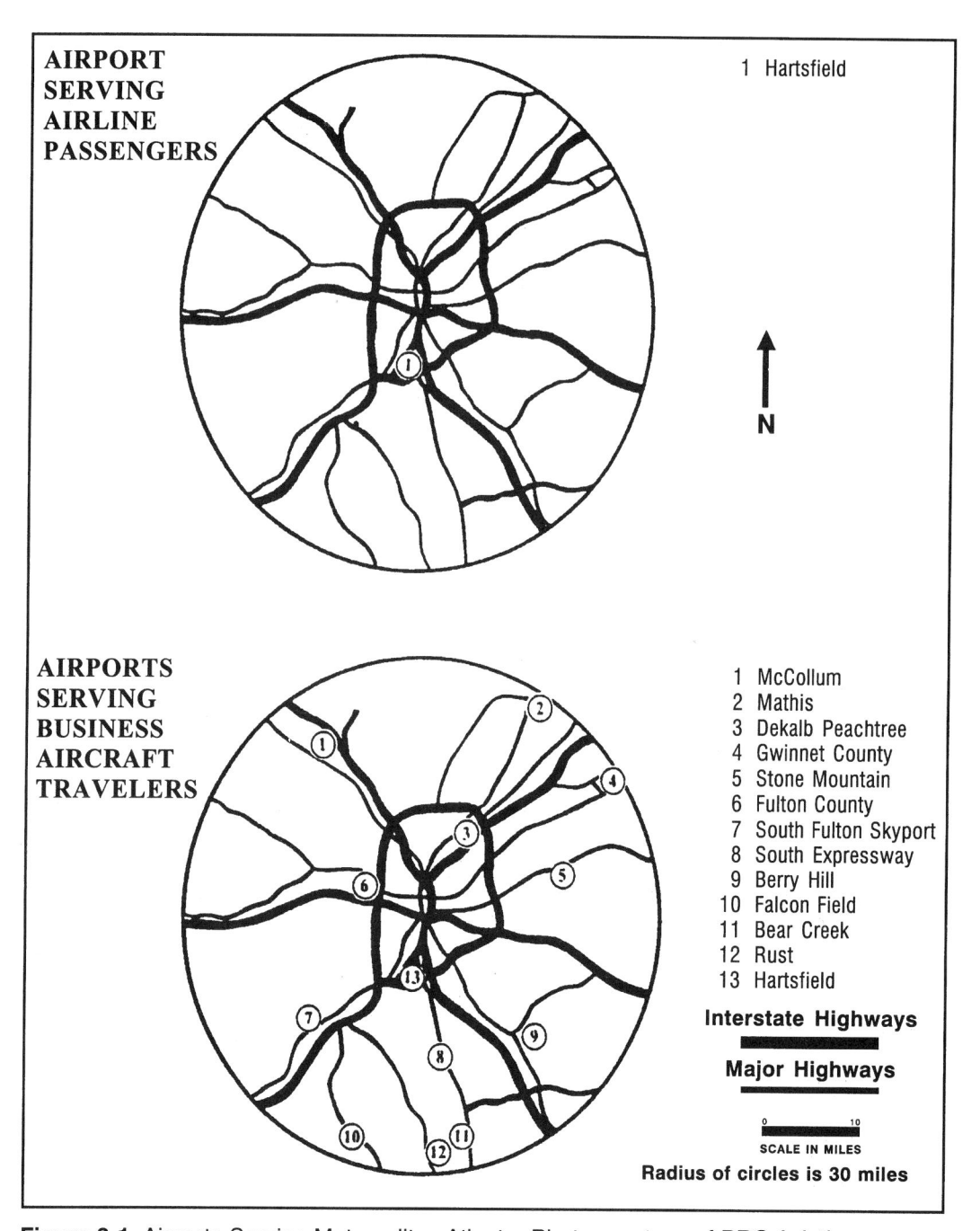

Figure 3-1. Airports Serving Metropolitan Atlanta. Photo courtesy of PRC Aviation.

54 *Chapter 3*

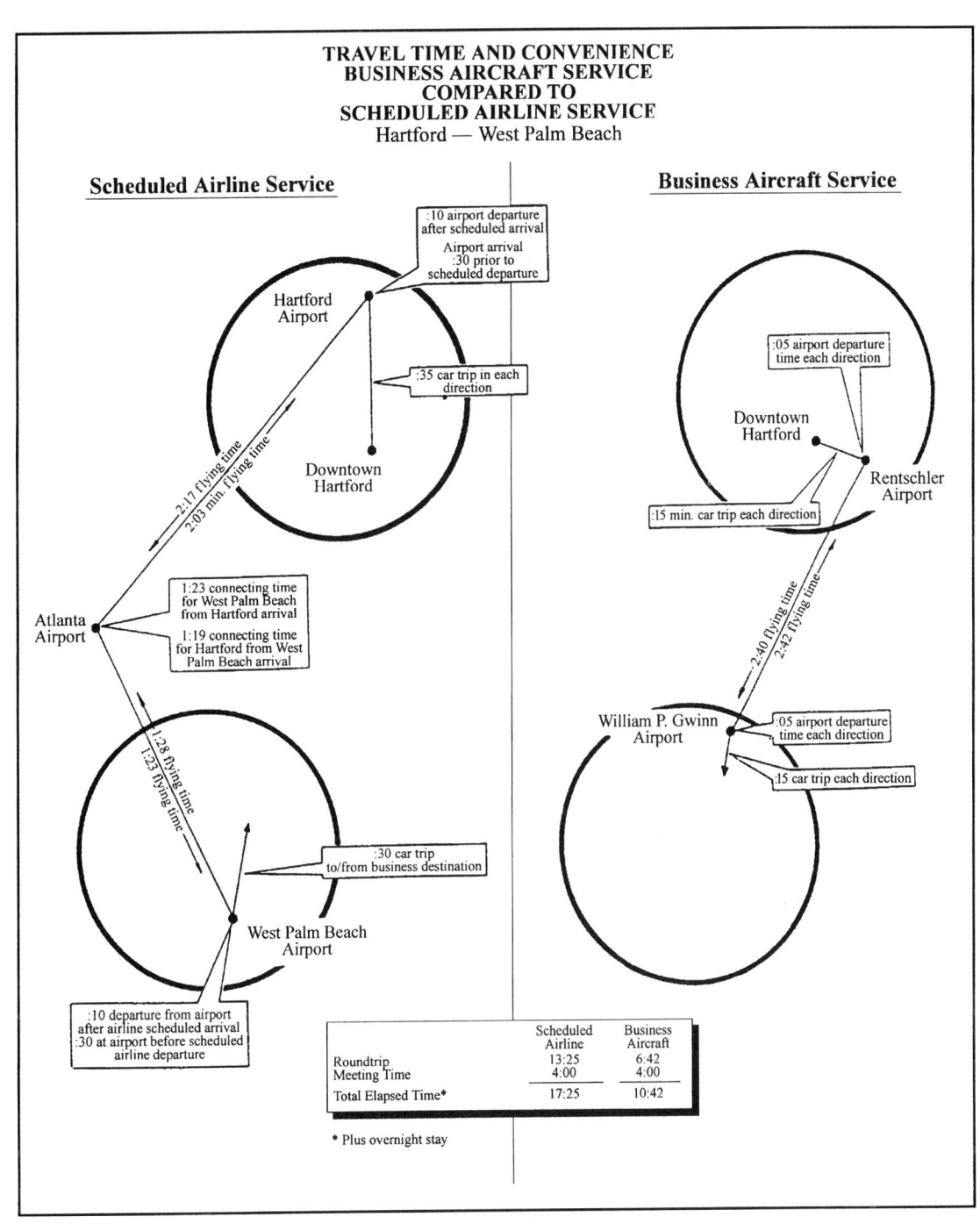

Figure 3-2. Travel Time and Convenience Business Aircraft Service Compared to Scheduled Airline Service. Courtesy of PRC Aviation.

3: Value/Benefit of Using Business Aircraft 55

Table 3-4. Ground Transportation Elements

1. Residence or office to airport
2. Airport arrival to check-in counter
3. Check-in counter to security barrier
4. Clearing through security barrier
5. Security barrier to airplane gate
6. Airplane boarding to airplane departure
7. Arrival to departure at connecting airport
8. Airplane arrival to baggage retrieval area
9. Awaiting baggage
10. Baggage in hand to airport departure
11. Airport departure to residence or office

Source: PRC Aviation. (1995). *Business aircraft operations: Financial benefits and intangible advantages.* Tucson, Arizona: PRC Aviation. Adapted from pages 3–8. Reproduced with permission.

Table 3-5. Ground Times Comparison—Airline versus Corporate

Element	Airline	Corporate
Drive time to airport	45 min	25 min
Departure processing	35 min	5 min
Arrival processing	25 min	10 min
Drive time from airport	35 min	20 min
Totals	**2 hr 20 min**	**1 hr 00 min**

Note: Drive time differs when using major commercial airport versus closer regional airport.

aircraft. In Table 3-4, PRC Aviation identified 11 elements of ground transportation that should be considered when evaluating executive travel, while Figure 3-3 includes the ground-time requirements for these elements. To compare the ground times using a corporate aircraft versus a commercial aircraft, the trip data in Table 3-5 are shown for a general arrival and departure plan. Different times may be planned based on more specific categories as presented by PRC Aviation. However, for the purpose of providing a general overview of the difference in times that a manager may realize, the categories in NBAA's Travel$ense® program are presented, along with its default times. The time driving to an airport may differ because a business flight may operate out of a smaller, more readily accessible general aviation airport. Parking at an airport, traveling to the terminal, checking in, processing through airport security, and other arrival functions will take more time at a commercial airport than processing through a fixed-base operator and boarding a corporate aircraft. Additional sav-

GROUND TIME REQUIREMENTS
BUSINESS AIRCRAFT TRAVEL
COMPARED TO AIRLINE TRAVEL

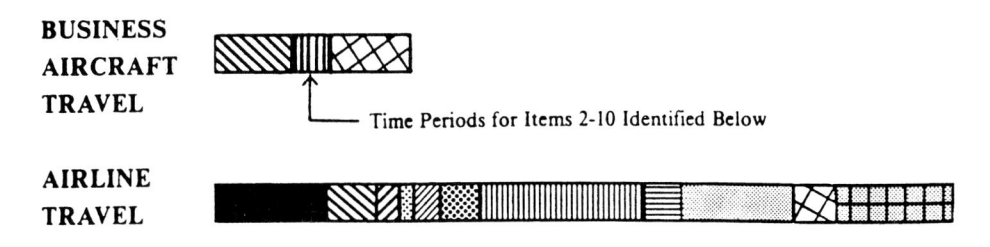

BUSINESS AIRCRAFT TRAVEL

— Time Periods for Items 2-10 Identified Below

AIRLINE TRAVEL

Ground Time Item	Code	Airline Travel	Business Aircraft Travel
1. Residence or office to airport		:30	:20
2. Airport arrival to check-in counter		:15	:05
3. Check-in counter to security barrier		:04	:00
4. Clearing through security barrier		:02	:00
5. Security barrier to airplane gate		:04	:02
6. Airplane boarding to airplane departure		:10	:01
7. Arrival to departure at connecting airport		:50	:00
8. Airplane arrival to baggage retrieval area		:10	:02
9. Awaiting baggage		:30	:05
10. Baggage in-hand to airport departure		:10	:02
11. Airport departure to residence or office		:30	:20
Total ground time requirements		3:15	:57

Figure 3-3. Ground Time Requirements. Courtesy of PRC Aviation.

ings in time result upon arrival at a destination airport. Planning times depend on specific itineraries and airports used; however, savings in ground transportation times may be significant when using a corporate aircraft, especially when compared to flying on busy commercial airlines. In some instances, up to two hours have been spent in waiting after checking in with a commercial carrier before boarding. The flight hasn't even started yet.

Still other factors should be considered in valuing the benefit of using a business aircraft. The Canadian Aircraft Business Association (n.d.) stated that business aircraft advantages included:

   a. using the most direct routes while avoiding inconveniences of changing commercial airline operations,
   b. privacy and security,
   c. reducing time away from home with greater flexibility and convenience,
   d. attracting key executives and professionals, and facilitating improved customer relations through more frequent contact.

Seven mental and physical benefits derived from business aircraft travel by executive and professional travelers were reported in the PRC Aviation (1995) study. These were:

   1. Utilization of intellect,
   2. Preservation of stamina,
   3. Exercise of initiative,
   4. Maintenance of patience,
   5. Reliability of perceptions,
   6. Alertness to opportunities, and
   7. Measure of total effectiveness.

Earlier estimates by the NBAA of aircraft enroute productivity time for travelers resulted in a value of 75 percent when using a corporate aircraft and 15 percent when flying on a commercial airline. Under NBAA's Travel$ense® program, enroute productivity time varies according to the type of aircraft as well as the travel method. For example, a business traveler is assumed to be productive 20 percent of the time when flying on a turboprop airliner and 40 percent of the time on a jet. The percentage difference can be attributable to aircraft noise and personal space.

The NBAA noted that the difference for senior managers in enroute productivity time possible on business aircraft versus commercial airliners approximates 40 percent. Factors such as privacy and quietness, lack of interruptions, more seating and table space, and office equipment availability (phones, modems, fax, printers, etc.) contribute to greater productivity on business aircraft (NBAA, 1997b).

Additionally, differences in level of management (top, middle, or operating), type of work (e.g., technician, salesperson, director, etc.), purpose of trip (emergency repairs, sales presentation, casual meeting), and other factors need to be considered in determining a person's productive time during a flight. Previously, the NBAA established a productive percentage of 75 percent while flying on a corporate aircraft; however, traveler productivity varies by type of aircraft today. Individual corporate assessments are made as to the amount of productive time employees have on a given trip, but it is generally accepted that a person flying on a corporate aircraft is more time productive than flying on a commercial aircraft. The bottom line is that using a corporate aircraft enables a business traveler to increase his or her productive time while enroute by as much as 50 percent over time in a commercial aircraft—unless, of course, the business person is piloting.

Estimating the value of the benefits gained by using a business aircraft is specific to each company manager; the NBAA has developed the Travel$ense® program, which helps one determine an overall value (including increased productive time during other travel activities). Following a review of a time/cost comparison between airline and corporate air travel, a list of disadvantages for using a business aircraft will be presented.

For such a time/cost comparison, a total door-to-door elapsed time is appropriate because an executive or an employee's total productivity should be considered. In Table 3-6, a time/cost comparison is made for a business trip that involves a one-stop airline flight. The commercial airline fare and travel time can be easily determined through the use of travel programs or by calling an airline. Corporate flight times are also easily computed; however, set travel fares may not be established for corporate flights. Comparison, then, involves a determination of the total time for each method of travel and the use of productivity assessments.

What must also be considered when evaluating the costs of air travel are many applicable tangible and intangible benefits. It can usually be shown that a commercial air flight costs less than the operating and ground element costs of a corporate aircraft; however, when more business travelers use the corporate aircraft at the same time, economies of scale are realized. In Table 3-6, the breakeven number of travelers is two; therefore, when five or six managers are flown on the AASI 500 Jetcruzer, even greater corporate benefits/savings may be achieved. Additionally, a moderate productivity factor of 3.75 was used in this example; higher factors could be used for more senior executives. Still another factor is that the travelers could carry extra cargo (e.g., 200 pounds of equipment) on the corporate flight; this cargo would have to be paid for on a commercial airline.

There are many factors to consider when valuing a corporate flight versus an airline flight, and the use of the NBAA's Travel$ense® program helps managers in this assessment. Travel$ense® also enables a comparison of charter flight operations along with the airlines and corporate operations. Finally, the value of the intangibles cannot, and should not, be overlooked. What would the value be if the CEO arrived ready for negotiations in a major sales contract versus somewhat tired from an additional two hour plus journey? Could the time spent preparing for a briefing or finalizing important matters while in the firm's aircraft versus in more cramped and crowded passenger service be significant? Decision makers need to consider these very important intangible factors besides the tangible cost and time variables. Still, corporate flying isn't for every manager or firm, and a review of some disadvantages may be in order.

Table 3-6. Cost/Time Comparison (commercial versus corporate air travel)

Origin/Destination	Airline Time/Cost	Corporate Time/Cost
St. Louis, MO, to Roanoke, VA		
Drive time to airport	30 min /$4.80	30 min/$4.80
Parking	4 days/ $60	NC
Check-in	60 min	15 min
Flight to Charlotte	105 min/$608 total fare	
Layover	60 min	
Flight to Roanoke	45 min	110 min
Baggage/rental car	45 min/$30 per day	30 min/$30 per day
Airport to hotel	<u>30 min</u>	<u>30 min</u>
Total travel time	**6 hr 15 min**	**3 hr 35 min**
Subtotal cost	$792.80	$124.80

Productivity value computation:

    Time productive on airline (20%)   on corporate (75%)
        150 min × .2 = 30 min   110 min × .75 = 82 min

    Productivity factor 3.75 × hourly salary
    $52,000 (annual salary)/ 2400 hours (annual work hrs) = $21.67
    3.75 × $21.67 = $81.26 @ hr
        30 min = ($40.63)   4 hr 2 min = ($327.75)
                                                              (includes 160 min saved)

Corporate aircraft cost per mile
(AASI 500 Jetcruzer)
        $1.80 @ 619 sm one-way $2,228.40

Net total cost:	**1 employee**	**2 employees**	**3 employees**	**4 employees**
Airline	$752.17	$1,439.54	$2,126.91	$2,814.28
Corporate	$2,025.45	$1,942.50	$1,609.95	$1,402.20
Difference	($1,273.28)	($502.96)	$516.96	$1,412.08

Notes: Employees are mid-level managers; corporate aircraft use realizes economy of scale; net total computations include productivity values and costs per employee. Costs are current at the time. sm = statute miles

Table 3-7. Disadvantages of Using Business Aircraft

High Cost of acquisition	Product liability concerns
High costs of possession	Additional labor costs
Added insurance costs	Environmental concerns
Security problems	Safety concerns
Training costs	Domestic and foreign rules
Changing company mission	Internal scheduling conflicts
Operating inefficiencies	Possible image concerns

Source: Kovach, K. J. (1993). *Corporate and business aviation independent study guide.* Adapted from page 31. Daytona Beach, Florida: Embry-Riddle Aeronautical University.

Disadvantages of Using a Business Aircraft

A listing of some of the disadvantages of using a business aircraft was developed for Embry-Riddle Aeronautical University students. This listing is presented in Table 3-7.

Going back to the previously reported survey of 79 aviation managers, pilots, and other aviation-related professionals who studied corporate and business aviation, another listing of various disadvantages identified for using a business aircraft is provided in Table 3-8. Several of the items identified by the respondents indicated some concerns of the pilots, but other respondents were concerned about how company managers used or viewed the use of their aircraft. Depending on a firm's aircraft to go anywhere, anytime places severe pressures on aviation flight department managers.

Corporate executives who believe that just because they own an aircraft, they can use it without regard to air transportation rules, air traffic control agencies, weather problems, aircraft performance specifications, and safety concerns cause many managers to refuse to consider corporate aviation operations. Corporate pilots have told many stories about when their executives wanted to fly "his or her" aircraft, of the onboard aircraft celebrations after the "big sale," and of the conflicts between executives about who gets to go where. These issues may be reduced through the use of a corporate operations manual, but can probably never be totally resolved. There are disadvantages of owning and operating a corporate aircraft, and they cover more than just the high costs involved. The key is to determine and evaluate the value of the tangible and the intangible benefits for a specific operation and to have the full support of the corporate heads for the use of a business aircraft. Lastly, one disadvantage of the high costs of operating a corporate aircraft may be overcome by using the aircraft under FAR Part 135 operations, or earning revenue for the use of the company plane on a hire basis.

Table 3-8. Listing of Disadvantages of Using Business Aircraft (personal survey, 1995–1997)

Different destinations for executives
Royal barge identification
Renting surface transportation at out-of-the-way airports
Aircraft maintenance problems
Too many departments, not enough aircraft
Limited number of seats
Personal prejudices of executives
Adverse publicity
Limited support and equipment
Regulatory concerns
Exposure to illegal elements
Underutilization
Fuel costs
Dependency on aircraft by company managers
Poor FBO services
Enroute breakdowns and lack of support
Snobby travelers
ATC congestion
ATC priorities
Crash worriness of people
Maintaining a flight department
Impact of weather on flight operations
Logistical support at destinations
Amount of energy required for planning
Remote operations
Complex long-term planning
Loss of frequent-flyer miles
Liability
Terrorism and sabotage
Low morale if executives use only
Complexity of flight department
Segregates CEO from masses
Training problem
White-knuckle flying
Major disaster if accident

Note: These are verbatim reports of survey respondents received from October 1995 to March 1997.

Table 3-9. Social, Economic, and Political Benefits of Using Business Aircraft

Social

Comfort—Help provide community support through hospitality flights, awareness that business contributes to the community good, and instills pride in people.
Cultural—Provides opportunities for community people to experience different people and products.
Education—Makes people aware of the value of aircraft activities and creates interests in career in business and aviation.
Travel—Provides opportunities for people to go to and from the community to see and experience interests.

Economic

Essential Air Service—Enables people to go to and from areas which have no air transportation services.
Business—Develops and maintains business activities.
Employment—Brings jobs in the form of aviation-related services, i.e., airports, catering, fuel services, cleaning, maintenance, etc.
Industry Development—Helps develop new industries and enable others to grow.
Sales—Brings buyers of products to communities and promotes economic growth of industries. Improves commerce in and out of areas, including domestic and international markets.
Taxes—Provides income to federal, state, and local governments through taxes.
Subsidies—Brings government monies to communities to develop business and governmental programs, especially in airport and related services.

Political

Mobility - Provides quick response when needed for government and national requirements.
Technology - Continues to result in national development through technological innovations and uses.
Movement - Provides fast, efficient air transportation of political leaders to national and international areas.
Defense - Enables use of aircraft, training of pilots and other aircrew, and development of the airways system.

Overview of Benefits

The advantages and disadvantages of using a general aviation aircraft, or even an airline-type aircraft, have been presented. Although company officials obtain or use aircraft for their business, there are still other considerations for the use of aircraft. In Table 3-9, some key social, economic, and political benefits for using aircraft are presented. Not only do corporate employees and business managers benefit personally through air transportation services, society as a whole benefits because of people's increased awareness of people on the move, of the generosity of corporate managers in providing support to local communities, and of the many useful and good services, such as angel or mercy flights, helping people during emergencies. Some business managers want their aircraft operations to be known in order to show community leaders that their firms help their communities through employment, commerce, and other activities. Communities that demonstrate their pride in having major firms around do so because the people know and understand that their existence means life to the communities themselves. Individuals within societies realize that aircraft flown by corporate pilots and by individuals bring an awareness to others of the life of the society. Some communities have developed to provide reliever airports for general aviation aircraft. In all 50 states, there is some taxation on aviation. These may include fuels tax, state sales and use tax, aircraft registration fees, personal property taxes, or other service-related tax.

Programs and Promotions

Various studies have been made concerning the value of using a business aircraft; for example, the Fortune 500 Study by Aviation Data Services of Wichita, Kansas, and the Business Aviation Performance Study by Arthur Anderson. The NBAA's Travel$ense® program provides managers with details to evaluate the effectiveness and efficiency of air transportation services relative to an individual company. While some details of these programs/studies are provided, contact the appropriate authority to obtain current data and services.

Fortune 500 Study

The top 500 industrial and service companies have been reviewed by Aviation Data Services and Fortune 500. Previous studies have been published by *Business & Commercial Aviation*. Basically, the top companies were identified as aircraft operators and non-operators and were listed within ten categories in order to show financial performance. These categories included the

   a. number of employees,
   b. assets,
   c. stockholders' equity,
   d. assets per employee,
   e. net income per employee,
   f. sales,
   g. net income,
   h. sales per employee,
   i. net income as a percent of sales, and
   j. net income as a percent of stockholders' equity.

Table 3-10. Aircraft Operators versus Non-operators (Fortune 500, 1984-1990)

Year	Aircraft Operators	Aircraft Non-Operators
1984	347	153
1985	328	172
1986	336	164
1987	333	167
1988	339	168
1989	336	164
1990	321[a]	179[b]
7 year average =	334[c]	167[c]

a- Within 29 Standard Industrial Classification (SIC) categories, only miscellaneous investments and real estate did not operate an aircraft.
b- All motor vehicles and parts and jewelry/silverware SICs within the Fortune 500 operated aircraft.
c- Average figures are rounded.
Source: Multiple Fortune 500 reports.

From Table 3-10 it can be seen that the aircraft operators generally outnumbered the non-operators two to one. Reference is still made to this fact by NBAA, GAMA, and other promoters of business aircraft; however, these type study has been replaced by the No Plane-No Gain proactive approach of the NBAA.

The NBAA reported that in 1995 nearly twice as many aircraft-operating companies existed among the Fortune 500. The sales of the companies that operated aircraft were almost $3.8 trillion, while the non-operators totaled less than $0.8 trillion. Employment by the operators included 16.8 million workers (less than 3.5 million for non-operators), assets were $7.6 trillion (non-operators $0.3 trillion), operators' net income reached $198 billion in 1995 (about $45 billion for non-operators), and stockholders' equity exceeded $1.4 trillion for the operators versus less than $300 billion for the non-operators (NBAA, 1997a). As in the National Business Aircraft Association's film, *Wings at Work,* the question may be asked, Is it cause or effect? The bottom line is that business aircraft operators consistently outperform the non-operators in the Fortune 500 in key economic performance measures. Although the operators indicate greater economic performance, there are still non-operators in the top 500 companies. Perhaps a survey should be conducted to determine why business aircraft are not used in those Fortune 500 firms that have reported no aircraft operations.

Plane-No Gain

Arthur Anderson presented the No Plane-No Gain concept after a 1993 report of research involving over 700 companies. For this program, it was shown that companies that purchased business aircraft from 1986 to 1990 had higher sales performance than non-purchasers. Results supported the findings that "public companies that operate aircraft have better sales growth, earnings per share, long-term return to investors, and productivity (sales per employee) than companies that do not use business aircraft" (NBAA, 1997a, p. 14). Four major findings were presented by the NBAA in its report:

1. Companies experienced 7 percent higher average cumulative sales growth after buying business aircraft during the period 1986–1990 than those that didn't.
2. Earnings per share for Value Line companies were 30 percent higher for those that used business aircraft.
3. Ninety-two percent of the Fortune 500 firms that used business aircraft between the period 1982 to 1992 had the greatest return to their investors.
4. *Business Week*'s "Productivity Pacesetters" for 1993 indicated that 80 percent were business aircraft operators.

These facts present a sound basis for any manager to consider whether the use of a business aircraft may contribute to the economic well-being of a firm. No manager should ignore the point that business aircraft have been shown to be reliable and productive resources, and that the operator does not need to be a large firm. Consider the instance where a business manager flies him- or herself to various business locations to conduct company activities. Perhaps the person drove a car previously and made a few stops at customer locations. With the use of a small aircraft and a private pilot's license, the same manager could make several more stops and even return home by the end of one day. Would the increased business contact result in more sales, more orders, the development of long-term company relationships, or other business developments? One cannot ignore the issue, but should use whatever resources are available to evaluate the value/benefit of a business aircraft. Another recent NBAA program has been Travel$ense®.

Travel$ense® Program

Travel$ense® has been developed by the NBAA as a management tool to analyze business travel. While specific details will be presented in the next chapter, an overview of the program follows. Productivity tracking software enables managers to conduct a travel analysis for their operations. According to the NBAA, the program is intended to:

1. Be operated on a day-to-day basis by a flight department or to generate management-defined trip reports by a travel center;
2. Help managers analyze the true cost of business travel on a trip-by-trip, city-by-city, or passenger-by-passenger basis;

3. Quantify the true cost of business travel from the perspective of the financial officer in dollars and cents via three methods—two user-defined and the third via the airlines;
4. Generate data on trip information;
5. Help define business travel management strategy; and
6. Illustrate the value business aircraft and a flight department can provide to a company on a dollars and cents basis (1997b).

The program had been designed to provide a five-part travel analysis. These parts include

1. Air services,
2. Travel times,
3. Trip expenses,
4. Cost analysis, and
5. Benefit analysis.

It was designed to run under Windows 3.X or Windows 95, use a modem capable of at least 14.4 kbps or V.32 protocol, operate through a telephone line, and use a CompuServe account. The program has no limits with regard to trip numbers and can generate various reports for managers; for example, trip, detailed trip, summary, and defaults reports.

A *trip report* is an analysis of a specific trip and is designed for management's review. The detailed *trip report* provides data for the chief pilot or other evaluators. A *summary report* lists totals for all trips within a given date range by trip status. Lastly, the *defaults report* is a printout of the estimates and assumptions made in analyzing trips.

Travel$ense® helps managers to evaluate existing programs or facilitate new travel strategies. Importantly, Travel$ense® has been labeled "dispassionate" because it can help justify developing a flight department or help eliminate one. Data are provided by the user to analyze business travel; it often is reported that business aircraft are time-efficient and cost-effective. For those who have never heard of this program, source the NBAA through the Internet at http://www/nbaa/T$.htm.

Summary

Using an aircraft for business has been shown to be of significant value/benefit; however, many of the top companies remain successful without using an aircraft as a business tool. At the start of a new millennium, business managers have to not only work smarter and harder, but they must increase their focus on key company resources—people and time. The programs and tables presented provided definitions of the tangible and intangible benefits of using an aircraft as a business tool, while many of the advantages and disadvantages were also covered. The major benefit of such use has been stated by the NBAA to be flexibility; this benefit has enabled substantial time and cost savings for aircraft operators. Programs like Travel$ense® help managers analyze their travel requirements by comparing airline versus business aircraft travels. The Fortune 500 Study and the No Plane-No Gain reports indicate that the top companies that operate business aircraft have more sales than the non-operators.

Various social, economic, and political benefits have been identified. Although no statement has been made that using a business aircraft is best for every manager, the emphasis is on the need for managers to evaluate company travel on an individual basis and make an appropriate determination.

Activities

1. Review a Fortune 500 Study found in *Business & Commercial Aviation* and make a comparison of the aircraft operators to the non-operators. Can you determine if using a business aircraft makes a difference in company performance?
2. Review the chapter tables on advantages and disadvantages of using a business aircraft. Add to these listings any other factor appropriate.
3. Select an origin and destination for a hypothetical trip or one actually used in your company. Accomplish a time and cost analysis.
4. Use the Internet to retrieve the Travel$ense® program from NBAA (http://www.nbaa/T$.htm). Review this program to determine whether its use would provide your operation benefits.
5. Contact the NBAA or GAMA to review the Arthur Anderson report concerning the No Plane-No Gain concept.
6. Prepare your own list of the value/benefits of using a business aircraft for your firm. How many tangible versus intangible factors can you identify.
7. Consider how the use of a business aircraft benefits society, a company, and the government. Can you add additional items to the listings in the chapter?

Chapter Questions

1. Define what is meant by a tangible and an intangible benefit of using a business aircraft.
2. Discuss the most recognized factor for using a business aircraft.
3. List three advantages and three disadvantages for using a business aircraft and explain their significance.
4. Explain how time can be analyzed according to the travel requirements of a business manager.
5. Explain the Travel$ense® program and how it may help a business manager.
6. Discuss the Fortune 500 companies and their use of business aircraft. Does aircraft use help these firms?
7. Accomplish a cost/time comparison for business travel for a real or an imaginary company.
8. Should every business manager use business aircraft? Why or why not?
9. Identify social, economic, and political benefits of using business aircraft.
10. Discuss the major factors business managers need to consider when evaluating the use of aircraft for business travel.

CHAPTER 4

Corporate/Business Aviation Decision

Overview

Knowing the difference between corporate and business aviation, understanding their value and tangible and intangible benefits, and realizing that there are many types of aircraft available to use in supporting a firm's mission are important factors in evaluating whether managers become involved with business aviation. In this chapter, you will review the three Cs of the business and corporate aviation decision-making process. Choosing a transportation mode is not always a straightforward decision; therefore, a review of some of the major determinants in this process will be made. Additionally, you will learn how to accomplish a transportation analysis, evaluate an executive/employee's worth, and develop and present a travel program. A partial review of the NBAA's Travel$ense® program exemplifies one method of decision-making. Lastly, business strategies have to be considered when deciding transportation selection.

Goals

- ✈ Know the transportation choices available for corporate use.
- ✈ Identify determinants in choosing a mode of transportation.
- ✈ Understand the three Cs of the business aviation decision-making process.
- ✈ Complete a travel analysis.
- ✈ Discuss methods of evaluating an executive's worth.
- ✈ Prepare a travel spreadsheet, a travel pattern map, and a travel pattern graph.
- ✈ Discuss various business strategies affecting transportation selection.

Introduction

Why travel? Why not use teleconferencing or some other technological means to accomplish business? As presented in the first chapter, transportation is a necessary element of *communication*. Its purpose is to provide time and place utility, or to have the right person at the right place at the right time (this also applies to cargo and mail). As trade and commerce developed, the marketplace for many business managers has become worldwide. This *globalization of markets* for products and services has, in turn, created the need for fast, efficient, and effective transportation services. Business managers, technicians, salespeople, chief executives, and other company personnel need to see their customers. They need to see their production plants, obtain primary data, make personal contacts, repair vital equipment, or accomplish other important activities necessary to not only have a business survive, but profit and grow. To achieve success, managers need to focus on three important and key resources:

1. People,
2. Capital, and
3. Customers.

The decision to obtain a business aircraft is easy when the CEO wants an aircraft, and even easier if he or she says which one to obtain. This situation might occur when a manager either loves to fly, has the personal desire (and resources) to obtain an aircraft, or has already made a personal evaluation of the value/benefits an aircraft will bring. For example, a movie star like John Travolta began his love for flying at about age five. He was quoted in *Professional Pilot* as saying "Aircraft give me the ultimate privacy and allow for ultimate scheduling capability" (McLaren, 1996, p. 66). He based his aircraft selection on his mission, operational factors, and dispatch reliability and had the desire and resources to obtain and use a business aircraft. There was no need to convince or influence anyone else; he valued the benefits the aircraft operation provided him. Conversely, for those managers who refuse to fly or use a business aircraft, the decision is also made without further analysis.

The following information, however, will be most useful to those managers who need to make a decision only after careful analysis and who manage effectively and efficiently their people and capital while satisfying their customers. When the decision to use a business aircraft is under consideration, business managers have to weigh the advantages and disadvantages and then decide to use business aircraft versus commercial or charter airline services, or even another transportation service.

Three Cs

Consideration, cost, and comparative analysis are the three Cs that relate to making a decision to obtain a specific aircraft for business. *Consideration* of the tangible and intangible benefits and of the advantages and disadvantages of operating a business aircraft must be taken when trying to decide whether a business aircraft is best. Desirability may be important to a manager, but realistic appraisal of the use of an aircraft is essential.

Beginning with the firm's mission statement, a manager must evaluate the effectiveness, efforts, and resources of meeting the company's mission. Effectiveness differs from efficiency in that a manager may be meeting the mission, but may not be optimizing the costs to do so. *Optimizing* relates to accomplishing the mission at the most favorable costs, not

70   Chapter 4

necessarily minimal costs. In some instances, a business manager must decide to spend more on services than necessary in order to maintain a larger share of a business. Operating in a competitive environment means that managers do not have the monopolistic advantages of minimizing costs. Consideration of the firm's goals and objectives and reviewing costs/benefits relative to the overall company operation are activities managers need to continually accomplish.

The use of a computer in today's global business environment is an absolute necessity for most business managers. A manager can communicate using electronic mail just about anywhere in the business world, review business actions, review important information becoming available from a variety of sources, and prepare required documents through the use of a laptop or other computer equipment. Today's business manager *considers the computer a necessity*. This same consideration should be given in determining if using an aircraft is essential in business.

After considering the various advantages and benefits that using an aircraft brings (e.g., cost efficiency, confidentiality, safety, security, service, and time savings), a manager should also consider the disadvantages (e.g., costs and responsibilities) and then weigh both sides. The main point is that a manager must consider the important items that affect his or her business. Making a decision in haste may result in unwanted and unprofitable actions. First and foremost, a manager should consider who should evaluate the issue: should others be involved or is the decision personal? Then the variety of variables presented in Chapter 3 should be considered; that is, those important factors that affect the manager's operation should be identified and reviewed in detail.

Cost is the second item because it usually is the main factor in the decision-making process. Aircraft and their associated requirements (pilots, spare parts, insurance, security, maintenance, etc.) are expensive. Information pertaining to the different types and costs of aircraft is presented in the next chapter. From operating a small general aviation aircraft to flying an airline type (B-737), the cost may be substantial. Deciding whether to establish an aviation flight department; to have an aircraft management firm manage an aircraft; or to time share, jointly own, or use fractional ownership involves considerable personal attention, both in time and money. Determining the extent of this cost is critical in the decision-making process.

The last C is *comparative analysis*. After considering the many variables that relate to the decision and determining the costs for each, the final step is to make a comparison of these costs to the potential methods under consideration. If air transportation is the chosen mode of travel, there needs to be a comparison among commercial airline service, charter air services, a business aircraft, or another air service. This last step can only be made after the first two steps have been *sequentially* taken. In other words, a manager must determine the factors, cost these factors, and then decide on the optimum choice. While this may appear to be a logical process, there are stories of some managers obtaining an aircraft and determining later how the aircraft will be used and costed.

Transportation Choices

Business managers have the use of five possible transportation modes: air, motor, water, rail, and pipeline. Obviously, the latter refers to moving products versus passenger travel. There is an intermodal system which combines the use of a surface mode and pipeline. Consider the "Chunnel" between France and England. A 22-mile "pipe" is used as a tunnel by rail operators to move people and goods. Using this intermodal system enables many business managers to meet their objectives more easily than by using a ferry service or aircraft across the English Channel.

Within each mode of transportation, there are various carriers to consider using. Common, contract, indirect, government, and private carriers provide transportation services. The first *(common carrier)* includes those operators who obtain a certificate of public convenience and necessity to serve the public without discrimination and at equal prices. This carrier provides a scheduled service at known service levels. A contract carrier is just that—a contracted operation. Whatever service you decide to have is priced, and negotiation skills are needed to keep costs low. *Indirect carriers* are those freight forwarders who provide consolidation and other services for small shippers of freight.

Government carriers involve military and other federal, state, and local operators who have their own transport equipment. Government carriers provide services including firefighting, law enforcement, scientific research and development, flight inspection, surveying, powerline/pipeline patrol, aerial photography, pollution control, search and rescue, drug interdiction, agricultural application, and transport of government personnel. The General Aviation Manufacturers Association (n.d.) reported that in 1990 there were 1,354 non-military government aircraft in operation.

Lastly, *private carriers* are those operators who use transportation equipment, but earn less than 50 percent of their income from direct transport operations. Corporate and business aircraft operators fall into this category when complying with FAR Part 91, General Operating and Flight Rules. Some corporate operators may operate under common carrier and private transport rules when following FAR Part 135 and FAR Part 91. FAR Part 135 involves providing air transportation services for compensation.

Determining the Transportation Service

At some point, a manager evaluates the transportation requirements for his/her company and identifies the services available from the transportation carriers. The question is, Which mode and carrier will be used? In an early NBAA guide, five major transportation determinants were presented. These same factors are applicable today; they are cost, comfort, convenience, time, and safety (NBAA, 1982). While numerous important benefits associated with each transportation mode can be identified, these five factors appear dominant in the transportation decision-making process. *Cost* is primary because a manager's resources are limited, and cost-benefit tradeoff analyses become ongoing processes. *Time* is also extremely important and is a major tradeoff concern. For example, if time is critical, using an aircraft's major advantage of speed will result in higher costs. Other tradeoffs to consider are *comfort and convenience*. When time or cost isn't an issue, then perhaps a person's comfort and convenience may become a major concern. Finally, *safety* must be paramount! A manager will pay more to ensure safety in travel. These five determinants need to be considered when selecting a mode of transportation. Most managerial concerns can be factorized within these five major determinants, no matter what size and type of company operation exist. For example, in a small corporate operation, perhaps beating the competition to a limited market (time concern) means being there absolutely first with a quality product, whereas in a larger company income may be earned through a variety of markets. A review of evaluating the type of carrier to use within the transportation mode follows. Before a manager can discern which carrier or type of air services to use, an evaluation of an executive or employee's worth must be accomplished.

Value of an Executive/Employee's Worth

Since time means money to a businessperson, it is logical to think that saving time means more opportunity to earn more money. How a person's worth is related to time savings becomes a major concern for evaluators, but determining a person's worth is an imprecise and very subjective action. Using a straight time savings to salary ratio is impractical. For example, a salesperson who has three extra hours in a marketplace may more than triple his or her hourly salary by gaining increased sales revenue. Time and sales are related; therefore, a Poisson probability distribution may be considered. This means that the probability of a sale is the same for any time interval of equal length and that the occurrence or nonoccurrence of a sale in any interval is independent of the occurrence or nonoccurrence in any other interval. What differs is the value of the sale made.

In other words, sales can be realized at any time with equal probability and not every sale is of equal value. When an opportunity results in a major contract or substantial income for a company, what is the salesperson's value? Any *extra* sales during a given period means additional income, and as long as the value of such sales exceeds the expense involved, a manager should take a positive view. Of course, not every moment of time is involved in direct sales; however, the long-term effect may be increased income not otherwise possible.

The basic method in evaluating a person's time is to use the *straight salary approach*. If an executive works 2,500 hours per year and earns $100,000, then the person's hourly value is $40.00. This does not seem logical to use because a six-figure earner may contribute substantially towards major business contracts worth millions of dollars. PRC Aviation (1995) identified alternate methods to consider.

Using a *multiplier or factor number* times a person's hourly salary is one method. For example, a senior executive is assigned a multiplier of 5.7, and mid-level managers and professional people assigned 3.8. Still other methods to consider include the five times salary, stock coverage, human resource accounting, ten percent rule of thumb, replacement cost, loss of excess earnings, social psychological, group-human value, opportunity cost, goodwill

Table 4-1. Outline of PRC Aviation Valuation Methods

Method	Remarks
Five Times Salary	Uses total compensation as multiplier. Insurance companies use as basis for determining a cost advantage. Replacements can be trained and 20% capitalization rate over five years could be obtained.
Ten Percent	Used in life insurance. Correspond to useful life of employee. Company buys as much life insurance as can be purchased with 10% of the salary.
Replacement Cost	Based on costs of inducing experienced executive to leave present firm. Employees expect 10 to 20% increases in compensation when leaving a firm for another. Various techniques are used to determine replacement costs.
Loss of Excess Earnings	Uses proportionate loss of excess earnings that firm would have gained if executive worked until retirement.

PRC Aviation. (1995). *Business aircraft operations: Financial benefits and intangible advantages.* Tucson, AZ: Author. pp. 6-17 to 6-20. Used with permission.

approach, absolute cost approach, earnings objective, and excess return differential approach methods. Each of these methods involves arbitrary techniques to equate resources and intangible values. Contact PRC Aviation at its office in Tucson, Arizona, phone 520-299-0410 (1997 data), for specific details about the above or to determine other suitable methods, but an outline of some of these methods is provided in Table 4-1. Perhaps developing a personal value system based on a synthesizing approach may be appropriate. PRC Aviation also presented an overview of various employee value factors based on multipliers of base salary (Table 4-2).

As indicated, there is a variety of factors to consider when valuing a company employee. PRC Aviation researched extensively human resource accounting and reported on many of these factors. The premise generally accepted was that a person's worth is more than total compensation received. Perhaps some top executives hear many critical comments after they receive pay increases and stock options, but they must believe their value to the company is still less than what they earn.

Table 4-2. Summary of Salary Multipliers for Determining Employee's Worth

Method	Senior Executives	Middle Management and Professionals	All Employees
Service Industries	5.0–7.0	2.5–5.0	2.5
Five Times Salary	10.9–12.2	4.8–8.0	—
Ten Percent Rule			
Whole Life	4.9	—	—
Term Insurance	20.0	—	—
Benefit Term Ins	5.0–7.0	3.0	—
Replacement Cost			
Present Value of Term of Replacement	5.0–5.3	3.5	—
Present Value of of Replacement plus Inefficiencies	6.4–6.7	4.8	—
Loss of Excess Earnings	13.1	—	—
Conference Board	—	—	2.3
PRC Aviation Evaluation	5.7	3.8	—

Source: PRC Aviation. (1995). *Business aircraft operations: Financial benefits and intangible advantages.* p. 9. Used with permission.

Service organizations are reported to have the most standardized approach for valuing a professional's time. The factors typically used by the managers were reported by PRC Aviation (1995) to include:

1. Annual wages or salaries;
2. Cost of fringe benefits;
3. Bonuses and other short-term incentive compensation;
4. Long-term incentive or contractual compensations;

5. Other contractual employment costs;
6. Operating costs of the direct area of responsibility;
7. Allocated overhead of the responsibility area;
8. Profits of the responsibility area;
9. Provisional costs for employee replacement; and
10. Potential loss or revenues, profits, and return on equity in the event of employee loss.

Another acceptable approach recommended earlier by the NBAA is the *management accountability factor,* or **MAF approach** (NBAA, 1982). This method involves determining the level of responsibility for each person and prorating the percentage of company sales to the person's number of work hours. For example, if Department A within a firm is responsible for 27.5 percent of the company sales, and total company sales equated to $400 million, then the director of Department A is assigned a value of $110 million for accountability. Say the firm made $400 million in sales in 1997 and that the director of Department A worked 2,750 hours. Divide $110 million by 2,750 hours, and the manager's hourly MAF value is $40,000. Obviously, the higher in the organizational hierarchy one is, the more responsible the person should be and the higher the person's hourly value will be; however, top managers are not the only ones assigned a high responsibility factor. Employees whose service is vital to the company; for example, the firm's lawyer or computer repair technician may be assigned a high responsibility factor. In Table 4-3, an imaginary MAF structure is presented to indicate various levels of responsibilities and hourly values.

The MAF is based on a subjective assessment of assigning a level of responsibility to every position within a company. Every person, therefore, can have his or her responsibility level percentage multiplied by the financial amount earned for a department or company section to determine a position amount. This is then divided by the number of hours worked by the employee and an hourly MAF amount is determined. Remember that the amount is only based on a *subjective assessment* of the responsibility level and there are no real earnings per hour.

Next, the time spent traveling on a commercial airline versus a corporate aircraft can be compared to determine the net cost/value of a trip (review Table 3-6). If it is shown that the increase in time savings from traveling results in more sales for the company, then a positive correlation can be attributed to the use of a corporate aircraft. After all, if the executive didn't have the extra time, perhaps sales would be lower. Although many variables affect the outcome of company sales, it stands to reason that more time means more opportunities for productive work; ergo, more revenue.

The MAF may not be the best method in determining one's value, but what is the best method? Straight salary application definitely is unreasonable, so a subjective assessment must be made. What factors must then be considered? Do these factors include responsibility, contracts signed, number of sales, amount of income earned, or others? The value of the intangible factors is based on a subjective assessment, and the managers who use the most appropriate values usually can forecast productivity and revenue amounts quite accurately.

Table 4-3. MAF Value of Responsibility ($400,000,000 Corporate Sales)

Management Level	Accountability Factor	Company/Unit Sales	Annual Hours	Hourly MAF
CEO	100%	$400 mil	3,000	$133,333
Top Mgt. Senior Vice-Pres.				
Dept A	100%	$110 mil	2,750	$ 40,000
Dept B	100%	$ 50 mil	2,750	$ 18,182
Dept C	100%	$240 mil	2,750	$ 87,273
Accounting Director	40%	$400 mil	2,650	$ 60,377
Dept A Upper Mgt. Finance Operations	30%	$110 mil	2,750	$ 12,000
Middle Mgt. Design/Office Production	17%	$110 mil	2,650	$ 7,057
Lower Mgt. $ 3,960 Technician/Personnel	9%	$110 mil	2,500	
Labor Line Services	2%	$110 mil	2,000	$ 1,100

Certain managers revert to the value of the tangibles as the decision-making point. In the end, real income and hard, cold figures are used to establish the parameters. Top executives are earning higher and higher amounts as the millennium approaches. For example, in a *Business Week* issue ("What, me overpaid?" 1992), 10 CEOs were identified, and their average salary was $23,813,000. The top earner was Anthony O'Reilly of Heinz at $75,085,000, which included the value of all stocks and other items besides straight pay.

Using a level of responsibility is one of the best methods, but it is not the only method. Salary amounts also are imprecise. Is a maintenance employee sent to repair a critical component in the field less essential to the company than a company lawyer, a manager, or other employee? Should this maintainer be scheduled on a slow transportation carrier to save transportation costs? Of course not; therefore, a person's value cannot be determined by straight salary alone and may be changed when company priorities are affected. Transportation assignments for company employees depend on the value of the person's position, company need, company resources, and subjective assessment by those who manage the company.

Managers should establish travel priorities for their employees; changes in these priorities may be made based on time-sensitive requirements. The critical concern is *time value!* How important is it to have an employee at a destination? How much is it worth to have an employee quickly depart and return? These decisions are made at top management level, but they should be made known to everyone in a firm.

Taking a person's annual salary and multiplying it by a factor and then dividing the result by the number of hours the employee worked in the year will provide a reasonable method for determining a person's worth. The responsibility held by the employee can be accounted for in the factor used. Salary and number of hours worked are tangible, but determining the right factor to assign is the manager's important task. Now that the method of determining an employee's worth has been presented, a review of developing a travel analysis is most useful because results from this analysis are the data upon which to make company travel decisions.

Travel Analysis

Data Collection

The first step in accomplishing a travel analysis is to collect data—pertinent and useful data! Begin by talking to the top executives or managers who control the company, the CEO, executive vice-president, senior vice-presidents, marketing and planning officials, and anyone who has knowledge of company travel needs. Keep records of their requirements, real and imaginary. Determine the idiosyncrasies of those who will be traveling. Review company travel records to determine dates, times, costs, origins, destinations, mission requirements, and other pertinent factors necessary. Talk to the company's strategic managers, forecasters, or anyone who can provide information about company travel plans for the foreseeable future.

A major question is how much data to obtain? This must be decided based on the interviews conducted with top managers and the data available in company records. Using the last two years' data (eliminating any extremely good or bad years unless no changes are expected) is appropriate. Too much data can confuse analysts, make procedures too complex, and lengthen the analysis. Quality, not quantity counts. Relevant data can be tabulated, analyzed, and used to prepare other travel analysis steps.

To briefly provide an example of using data collected, the following fictitious information is provided. With this information, examples of useful travel analysis forms will be shown. The company for this example is Expando, Incorporated, located in Columbus, Ohio. Briefly, Joan Neal (CEO) requested a consultant to analyze the company's travel data and make recommendations for a suitable number and type of aircraft. Joan has her private pilot's license and at times travels alone or with two other company employees.

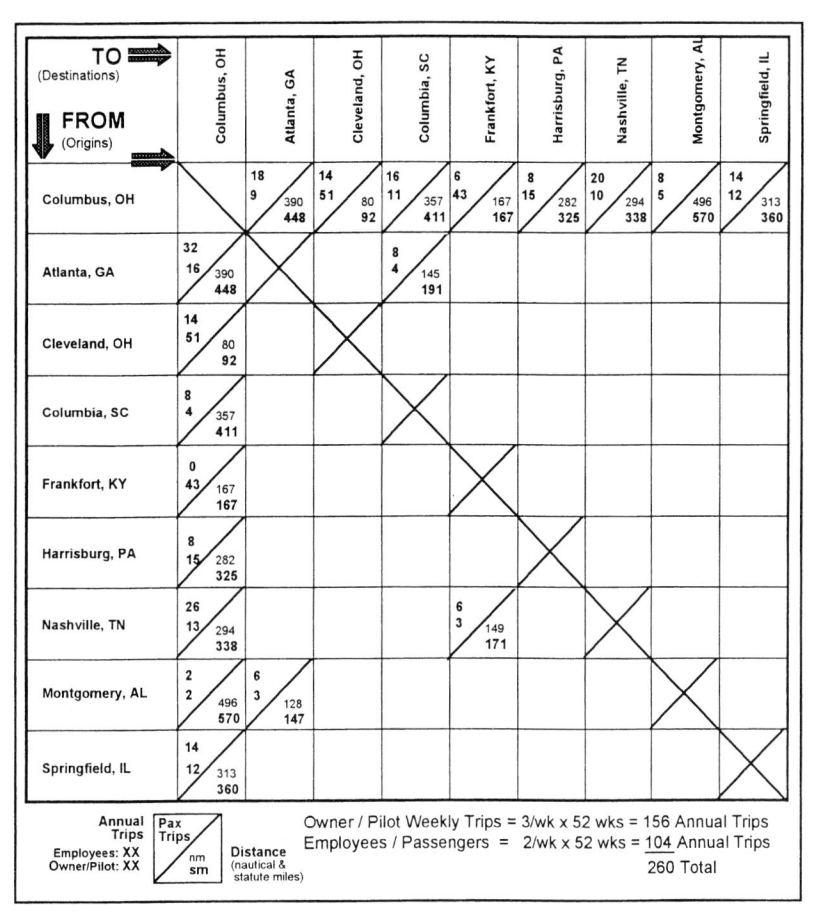

Figure 4-1. Spreadsheet. Data provided by F. W. VanCleave.

Spreadsheet

Gormley (1992) presented the use of various formats in accomplishing a travel analysis. His first recommended form is the *spreadsheet*. He simply lists the places people go across the top of a prepared form and the destinations down the side of the form. Then the number of passengers traveling and the mileage between the origins and destinations are identified in the appropriate cells. In Figure 4-1, data for Expando, Incorporated are used to prepare a spreadsheet. Statistical data provided by Fred W. VanCleave (personal communication, April 12, 1996) will be used in the examples provided.

All the trips made in the selected time period (usually a one- or two-year period) are identified. Mileage is shown in statute miles to use layman terms and in nautical miles (one nautical mile is 1.15 statute miles). The spreadsheet reads clockwise from the upper left corner; for example, 18 people traveled from Columbus, Ohio, to Atlanta, Georgia—the distance between these locations being 448 statute miles. For the purposes of this example, mileage will be approximated.

4: Corporate/Business Aviation Decision   79

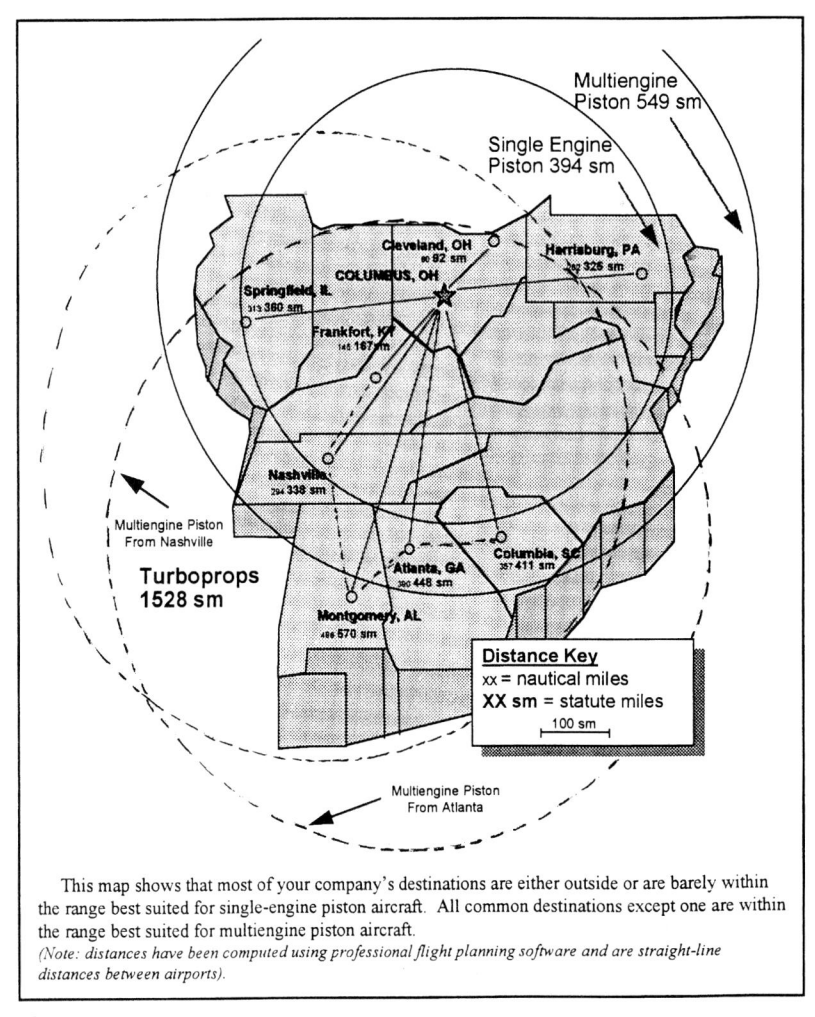

Figure 4-2. Travel Pattern Map. Extract of report by F. W. VanCleave. Permission to use.

The major outcome is the identification of the origins, destinations, number of people traveling to each location, distances involved, and total requirements. Other factors such as specific individual or corporate requirements are not identified on this form, but preparing the spreadsheet is a start in the data analysis process.

Travel Pattern Map

Next, Gormley (1992) recommends using a map to plot the trips. Figure 4-2 is an example of a travel pattern map. This allows for a quick and clear look at distances, locations, and aircraft radii. Drawing the aircraft radii from the home base or origin will allow reviewers to

TRAVEL PATTERN GRAPH

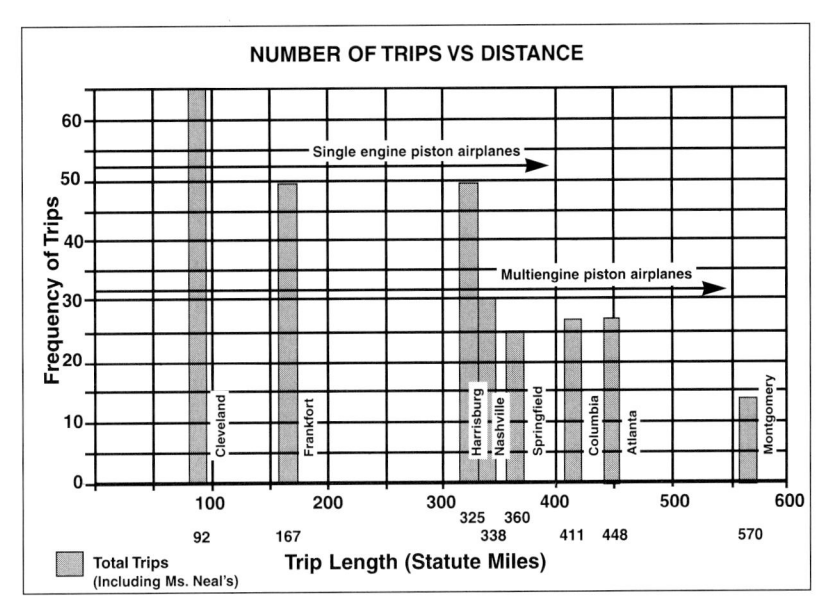

Figure 4-3. Travel Pattern Graph (Expando Incorporated).

determine the aircraft types and ranges needed; for example, a single-engine piston aircraft like the Piper Arrow PA-28R-201 has a range of about 380 statue miles (sm), but a single-engine piston, such as the Commander 1148, CDR-114B, has a range of 811 statute miles. Although a detailed analysis will be made of the aircraft recommended for the firm, at this point an analyst needs to consider the type of aircraft and average radii and speed. For the examples presented here, an average of the aircraft speeds, by type, as presented in *Business and Commercial Aviation's 1997 Planning and Purchasing Handbook* is used (handbook is published annually).

From the origin point on the map, a circle can be drawn representing a distance of 394 sm out for a single-engine piston aircraft. In Chapter 5, various aircraft types and their performance specifications will be identified, but the information presented is only used for example purposes to present the travel analysis process.

Travel Pattern Graph

Gormley's (1992) most important step is the preparation of a travel pattern graph. Figure 4-3 is an example of a travel pattern graph for Expando, Incorporated, while Figure 4-4 is an example of a more extensive transportation requirement. For convenience, the NBAA (see Table 2-4) has large, preprinted forms of a spreadsheet, travel map, and travel pattern graph for analysts to use.

Analysts need sufficient time to properly analyze the information included on a travel pattern graph. Passenger trips and trip lengths are plotted, but passenger load factors are not considered at this time. Each trip indicates one passenger and is the worst-case scenario. In Figure 4-3, the distances and frequencies of trips for specific destinations are presented. The trip to Cleveland is most frequently recorded, while the second-most-frequent distance traveled was to Frankfort, Kentucky (167 sm) and the remaining number of trips were scattered over various distances from 325 to 570 statute miles.

In Figure 4-3, the distances are divided into 50-mile blocks. By breaking the travel distances into blocks of utilization, you can select an aircraft type for each group of passengers. Mileage alone is not the basis for the type aircraft selected. Company-specific requirements that have been reported in the data collection step may guide managers to specific type aircraft. For example, facilities and equipment for handicapped individuals or travelers who require a shower, extra headroom, or other needs, might limit aircraft type to specific categories. In Figure 4-4, a more complex travel pattern graph is shown to indicate a larger business operation involving more extensive travel. To determine the number of aircraft required and the types suitable, a five-step procedure is followed:

1. Specify blocks of utilization and assign an aircraft type to each block.
2. Calculate the total cumulative passenger miles for each utilization block.
3. Divide the total cumulative mileage by the average speed of the type of aircraft chosen for each utilization block.
4. Divide the figure from step 3 by the load factor for each utilization block. The NBAA reported an average of three to four top- and mid-level managers or salespeople per flight (NBAA, 1997a).
5. Divide each utilization block by the current annual general operating average flight hour utilization for the type aircraft. For example, the NBAA reported jet aircraft averaging 419 hours, piston-powered aircraft 299 hours, turboprops 407 hours, and helicopters 348 hours per year for 1995 (NBAA, 1997a). In the example, a more generous number of 600 hours per year will be used.

For the travel data in Figure 4-4, blocks of utilization could be based on the distance of the single-engine piston aircraft and jets over 20,000 MTOW (maximum takeoff operating weight). Again, the type aircraft selected will be user-defined or based on specific requirements, either operational, financial, or other. To calculate the block utilization, each mileage block is multiplied by the number of trips; for example, in Block A, 150 miles times 140 trips equals 21,000 trip miles. Doing the math for each mileage group in Block A results in a cumulative mileage of 348,000 trip miles. For Block B, the cumulative mileage is 2,248,000 trip miles. This results from the longer distances of the trips taken. Next, in Step 3 the average aircraft speed chosen is divided into the cumulative mileage amounts. Using a single-engine piston aircraft average of 160 mph, a total of 2,175 trip miles results for Block A, and a jet speed average of 486 mph for Block B results in 4,626 trip miles. Step 4 uses the load factor to reduce the number of trips required. For Blocks A and B, the load factors used

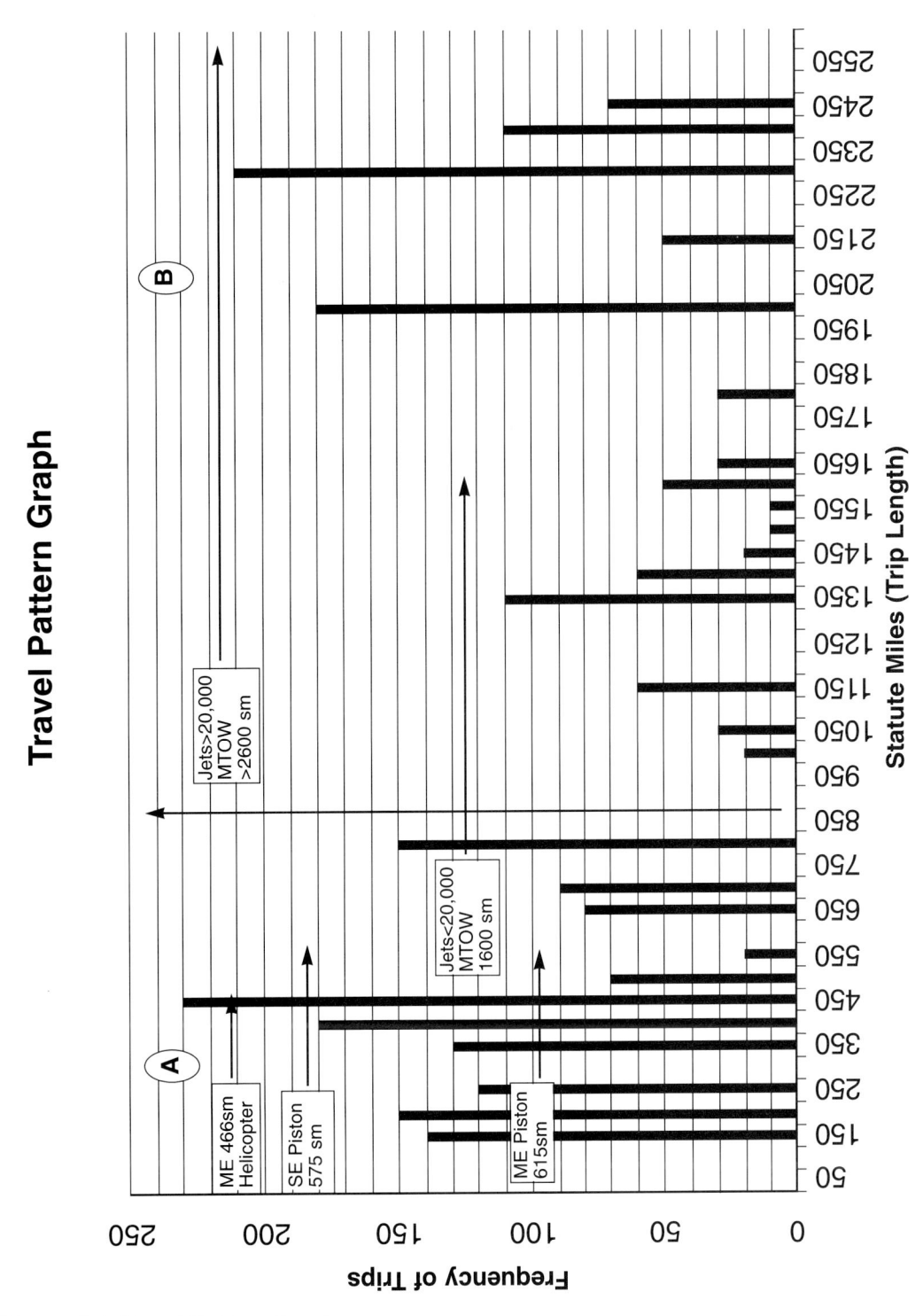

Figure 4-4. Travel Pattern Graph.

are 2.8 and 5.1, respectively. Higher load factors generally apply to longer trips—at least for this company. The results are 776.8 trip hours for Block A and 907 trip hours for Block B. Lastly, these amounts are divided by the average annual use of aircraft; in this case, 600 hours per year is used. Dividing 776.8 trip hours by 600 aircraft hours per year gives 1.3 aircraft and 907 trip hours divided by 600 hours per year gives 1.5 aircraft. The end result so far is that at least one single-engine piston aircraft and one jet over 20,000 MTOW is required.

Even if a lower aircraft flight-hour rate is used for the year's average (some managers use their aircraft more than 600 hours a year while others average much less), the aircraft requirements will probably increase. For example, dividing NBAA's jet flight-hour average for 1994 of 431 hours per year into 776.8 and 907 trip hours above gives 1.8 and 2.1 aircraft, respectively (all figures being rounded). A major item at this point is that only the one-way-trip distances have been used; what about the return trips? Do you have to double the final numbers? Unless the round-trip miles are indicated on the travel analysis worksheets, then the answer is yes!

This means that instead of 776.8 aircraft flight hours per year, approximately 1,554 flight hours are required to support company travel in Block A and 1,812 flight hours are required for Block B. Divide these amounts by 600 aircraft hours (or whatever is determined) and 2.6 and 3 aircraft are required.

Analysis of the data from the travel pattern graph is not over yet. An analyst could consider recommending a sales manager or other appropriate company employee be officed at the distant locations, thereby perhaps deleting the longer travel requirements. Commercial air services could be used for specific trips where airline service is suitable. Additionally, joint trips could be easily planned, thereby reducing the total mileage; for example, two passengers could be flown on an 800-mile trip and four others on a second leg of 1,100 miles on one flight schedule (saving flight time overall). An analyst must take time talking to the top managers, the strategic planners, and the financial officer about possible flight operations. Considerable effort is needed to analyze the data collected and determine an appropriate number of aircraft based on effective and optimum use.

A key to this procedure is to preselect an aircraft type based on the company requirements. If a firm's travel is limited to specific locations, then aircraft selection is more easily considered; that is, an aircraft to allow service to those destinations is all that is required. More complicated or unknown company trip requirements necessitate greater discussion and analyses of the data collected. Nonstop flight operations, jet or other services, seating, comfort, and numerous other considerations have to be factored into a desirable and suitable aircraft recommendation. Load factors and trip segments have to be considered. For example, if the average load factor for Block B in Figure 4-4 was 3.0 versus 5.1, then the number of one-way-trip requirements would increase to 1,542 units. Divide this number by 600 hours and five aircraft are needed (1,542 divided by 600 = 2.57 × 2 (round-trip) = 5.14 aircraft).

Again, the decision-making process is imprecise because of the number of variables that have to be considered. Making the decision is neither impractical nor impossible. The important factor is the *data collection* step. Accurate data are needed for good decisions! The process of accomplishing a travel analysis applies when management decides to evaluate air transportation services. Accomplishing a travel analysis allows for comparative analysis of commercial and private air services, resulting in good management. To do so, all suitable and available tools to make this travel analysis must be used with understanding and analytical skills. The NBAA has developed the following program, which provides a cost/benefit analysis of a firm's travel requirements, and should be considered as an excellent evaluation tool.

Travel$ense® Program

The NBAA's Travel$ense® program has already been introduced as a business management tool. In this section, more details about this useful program are provided because of its value to managers when making decisions for air travel. With the kind permission of John Olcott to use NBAA material, the following information has been extracted from the Internet (National Business Aircraft Association, 1997b).

Travel$ense® is a program that analyzes travel options and provides the user with various reports that can be interpreted by managers. Every air transportation trip is analyzed with respect to costs and benefits to a company based on user-defined parameters and assumptions. Managers can preview trip decisions or evaluate flight services taken. The program goals are to track productivity and efficiency gains. Corporate flight is not recommended for every business trip; airline usage is also favored since the program has been set to be dispassionate in its analysis.

Basically, Travel$ense® is designed for Windows 95, Windows 3.X, or Apple Macintosh (limited on the Mac). The basic Internet demands need to be met; that is, fast modem, sufficient RAM, and free disk space. Information gained through the CompuServe program and a centralized reservation system (CRS) enables users to have airline data, including fares and schedules for approximately 250 airlines (a Compuserve account is required). Sensitivity to the efficient use of time is a major program activity; input times (departure, arrival, on-site, etc.) are provided by the user. Four key reports are provided. These include:

1. Trip report—analysis of the currently reviewed trip,
2. Detailed trip report—data for chief pilots or trip analysts,
3. Summary report—provides totals for all trips within a given date range by trip status, and
4. Defaults report—estimates and assumptions for trip analysis.

The default settings for Travel$ense® values that are used to calculate the trip results are established by the NBAA, but may be changed by the user since accuracy is critical to the success of the program. These default settings affect corporate defaults, costs, times, airlines, and miscellaneous tabs. Some of these default settings are identified in Table 4-4.

Table 4-4. Travel$ense® Default Settings

Heading	Default	Remarks
Corporate	Company name	
	Aircraft	
	Business hours/weeks per year	actual time values
Costs	Aircraft setup	medium bizjet
	Time valuation	PRC Aviation averages
	Unproductive flight time	business aircraft 10 min & Airline aircraft 30 min
	Crew expense	Charge expenses
	Airline productivity	% non-bus. time 50%
		Jet 40%
		Turboprop 20%
	Airline scheduling	
	Business hours definition	
	Miscellaneous	Drive time productivity
	Trip report	Passenger time
		Passenger expenses
		Time triggered expenses
Times	Drive time to airport	
	Drive time from airport	
	Processing time—Departure	
	Processing time—Arrival	
	Leg dead time	
Airline	Enroute productivity jet airliner	40%
	Enroute productivity turboprop	20%
	Earliest airline travel time	6 AM
	Latest airline travel time	11 PM
	Minimum acceptable % on-site time	90%
	Percent productive during extra time on-site	50%
Miscellaneous	Valuation method	PRC Aviation average
	Value non-business travel	50%
	Productivity percent during drive	10%

Note: Travel$ense® tabs are only identified. For explanations, review the NBAA Travel$ense® program. Each user sets specific defaults or uses those set by the NBAA—examples of NBAA defaults are placed in the *Remarks* column. A blank area indicates no NBAA default.

Business Strategies

Business aviation has come a long way since the Knox Gelatin advertisement from Beachey's dirigible in 1906. Throughout the economic cycles, political changes, and social activities that have affected every country and business, using an aircraft as a business tool has undoubtedly helped many managers. Since corporate operations' information is scarce at best, analysts can only surmise that aircraft have been used as productive tools. When most of the top companies use aircraft, the correlation between company success and aircraft use can be assumed to be high.

NBAA's No Plane-No Gain report identifies that of the top 50 Fortune 500 companies over a decade (1982–1992), 46 of them use business aircraft. That means that 92 percent of these firms have managers who value business aircraft (NBAA & GAMA, n.d.). The No Plane-No Gain concept is a proactive approach to identify the value and benefits of using aircraft for business. It projects that planes equal gains. Time, flexibility, productivity, security, customer relations, and safety are the factors that have been seen by business managers to be affected when business aircraft are used. The end result is that stockholders benefit through better company sales, earnings, total return, and company productivity. For more information about the No Plane-No Gain concept, contact the NBAA (see Table 2-4).

Corporate managers have been affected by numerous and substantial political changes globally. Governments have changed, westernization has affected many eastern bloc countries, commercialization has resulted in a global marketplace, and competition has been demanding on company resources, especially people.

In a 1997 personal survey of 50 aviation managers, 23 different topics were identified as having an affect on corporate and business aviation activities in the millennium. These topics are listed in Table 4-5.

The main foci from the survey appear to be on global markets, increased productivity of employees, use of technological improvements, and flexibility to be where needed at the right time. When aviation services are required, commercial airlines can only provide a restricted service based on set schedules for their users. They provide limited services to business people when these people travel to locations where the airlines have good services. When air services are infrequent or non-existent, then business managers have to find other air service providers, either through aircraft charter flights, air management services, or their own aircraft operations.

Competition in the marketplace will increase as more sources, domestic and international, provide products and services. Therefore, business managers cannot be complacent; they must continually review their strategic plans and grand strategies. For those managers who have created excellent products, obtaining a larger share of sales through market development, mergers, and alliances may be suitable. Other managers are looking toward diversification, joint ventures, and integration. Whatever the strategy used, a business manager must consider the competition and beat it to the marketplace, while being cost efficient and productive. The use of a business aircraft has provided increased value and benefits for many top firms, whether it will for others depends on the evaluation of its application. The key point is that an evaluation should be made.

Table 4-5. Issues Affecting Business in the Millennium

More alliances	Strength of foreign currencies
Rightsizing	Service economy
Mega-carriers in aviation	Use of human resources
Flexibility required	Market globalization
Congested airspace	Pacific rim expansion
Just-in-time manufacturing	Just-in-time inventory
Bring customer to you	Technological improvements
Time savings	From executives to line users
Need for more research analysis	Sharing costs (aircraft operations)
Essential Air Service program ends	Central/South America trade increase
Multi-national businesses	Centralization of management
Meeting customer needs	

Summary

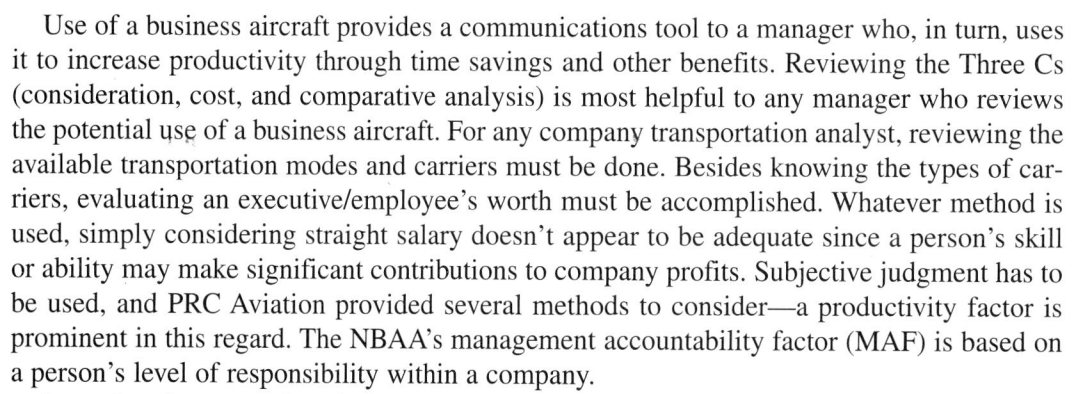

Use of a business aircraft provides a communications tool to a manager who, in turn, uses it to increase productivity through time savings and other benefits. Reviewing the Three Cs (consideration, cost, and comparative analysis) is most helpful to any manager who reviews the potential use of a business aircraft. For any company transportation analyst, reviewing the available transportation modes and carriers must be done. Besides knowing the types of carriers, evaluating an executive/employee's worth must be accomplished. Whatever method is used, simply considering straight salary doesn't appear to be adequate since a person's skill or ability may make significant contributions to company profits. Subjective judgment has to be used, and PRC Aviation provided several methods to consider—a productivity factor is prominent in this regard. The NBAA's management accountability factor (MAF) is based on a person's level of responsibility within a company.

In performing a travel analysis, several key forms are introduced. These forms, developed by David Almy and the NBAA, include a spreadsheet, travel pattern map, and travel pattern graph. Extremely important and the key to the use of these forms are the data collected. The first step in the aircraft decision-making process, then, is to collect accurate data. Using the data to complete the forms results in a recommended number and type of aircraft to obtain. Which aircraft and how it should be obtained will be covered in the next chapter.

One of the major helpful programs is NBAA's Travel$ense® program, and its purpose and scope are presented. A table provides a listing of the program defaults, but readers are recommended to use the Internet to obtain a copy of Travel$ense® from the NBAA. This tool will help managers evaluate the use of air transportation services and highlight the most productive option. Lastly, various business strategies for the millennium are identified. The main foci are global markets, increased productivity, technological improvements, and flexibility to respond to business demands.

Activities

1. Review available texts and professional publications to outline the advantages and disadvantages of the five transportation modes. Make a comparison chart outlining at least five major characteristics and compare the five modes to them.
2. Use the Internet and obtain a copy of the Travel$ense® program. Review its scope, reports generated, and default settings. Consider its use for your company, if applicable.
3. Discuss valuing an executive or employee's worth with business managers. How is this accomplished in your company?
4. Obtain travel analysis forms from the NBAA or prepare your own. Accomplish a mini review of a firm's transportation requirements or discuss with available company officials how decisions are made to establish transportation policies for your company.
5. Discuss with co-workers or others the issues that will affect business managers in the millennium.

Chapter Questions

1. List the five transportation modes and identify two major advantages and two disadvantages of each mode.
2. In choosing a transportation mode, various determinants are considered. Identify three determinants and explain why each is important.
3. Explain what a travel analysis is and its importance to a manager.
4. Discuss data collection methods for company travel requirements and explain why their accuracy is vital.
5. Why should statute miles be used in presenting company travel requirements?
6. Explain why costs and time are major determinants in choosing a transportation mode and carrier.
7. Discuss the management accountability factor and how it is applied in a company travel analysis.
8. Explain the purpose and method of the NBAA's Travel$ense® program.
9. Identify and explain the Three Cs in the corporate and business aviation decision-making process.
10. List four major business strategies for the millennium and explain your top one.
11. Explain what is meant by optimizing a firm's resources.
12. Why is the Three Cs a sequential process?
13. Company managers use various transportation carriers within the modes. Identify these types and explain their services.
14. Choose or develop a method for evaluating an executive's worth besides straight salary. Explain your choice and provide an example of its application.

15. When would traveling on a commercial airline be more advantageous to a firm than using its own aircraft? Explain your example.
16. Explain the development and use of a spreadsheet, a travel pattern map, and a travel pattern graph.
17. Why is the travel pattern graph the most complicated form to complete and use in performing a travel analysis?
18. Explain the significance of quality, not quantity for data collection.
19. Discuss the use of the characteristics of aircraft speed and range in a travel analysis.
20. Is consideration of a cost-time benefit analysis the most important activity to be accomplished in a travel analysis? Explain your answer.

CHAPTER 5

Business Aviation Aircraft

Overview

In this chapter, various methods used by corporate managers to select and obtain business aircraft are presented. Buying, leasing, chartering, time sharing, fractional ownership, and other means are reviewed. Numerous figures and statistics are used to indicate the types and numbers of general aviation aircraft, available airports, and pilots. Then the aircraft team selection process is presented through a project management approach. Finally, information on obtaining or operating aircraft is supplemented with specific aircraft manufacturer's aircraft charts and performance details.

Goals

- Understand the difference between general aviation aircraft and commercial airline aircraft.
- Become familiar with aircraft types, airports, and pilots available for corporate aviation.
- Understand the team aircraft selection process.
- Identify various types of general aviation aircraft and their performance capabilities.
- Discuss the advantages of buying, leasing, and chartering business aircraft.
- Discuss fractional ownership in business aviation.
- Differentiate between fixed and variable aircraft costs.
- Explain costs of aircraft acquisition and possession.

Introduction

When business managers first started using aircraft to support their companies, either they used their own designs (e.g., Henry Ford's trimotor, Clyde Cessna's Bobcat, Beech 18, etc.) or they mainly used military derivatives (B-24s, B-26s, T-39s, and many others). It was not until the development of the Gulfstream I, or the Grumman Model 159, in May, 1959 that any aircraft was designed specifically for corporate aviation.

The Second World War had given corporate aviation its biggest boost. Priority had been given to military travelers during the war while corporate and business managers used the limited commercial air service. Managerial decisions to obtain and use business aircraft resulted in the establishment of corporate flight departments and the use of private aircraft. After the war, it was only natural to internalize these operations and maintain a corporate aviation department that could use the readily available military aircraft (converted into corporate designs).

Today, business managers are using aircraft not just because they are available. Aircraft are business tools used to save time or provide flexible air transportation services not readily available. As seen in Chapter 3, many tangible and intangible benefits are realized through the use of business aircraft. *General aviation* includes those activities and aircraft that are not military or commercial. Presented are some of the pistons, turboprops, turbojets, and helicopters available for business use. The *millennium factors* for corporate and business aviation include aircraft, pilots, and airports, and they must be considered and managed if business aviation is to continue to provide vital support for managers.

General Aviation

In the general aviation category, the Federal Aviation Administration's Statistics and Forecast Branch conducts annual aircraft surveys. For 1994, it identified approximately 35,300 aircraft being flown 5.553 million hours by corporate and business operators (see Figures 5-1 and 5-2). Aircraft types being used included fixed-wing piston and turboprop aircraft with one or two engines, or those identified as "other." Aircraft with experimental airworthiness certificates, lighter-than-air, homebuilt, and so forth are grouped into the "other" category. Rotorcraft were labeled piston, turbine, single-engine, multi-engine, or other (Office of Aviation Policy and Plans, 1996).

In 1994, the number of hours flown by corporate and business aviation operators represented 23.3 percent of the total of 23.866 million hours in the general aviation category, which was almost a 2.0 percent decline from 1993's total of 24.34 million hours. FAA statisticians accounted for the lower numbers by the decline of U.S. general aviation aircraft manufacturers from eleven to nine during this period, and a decline in shipments of single piston-engine aircraft. Additional variables that the FAA reported affecting the decline were less flight instructors, deteriorating physical facilities of FBOs, changes in people's disposable income, increases in airspace restrictions applied to VFR aircraft, and shifts in personal preferences (Office of Aviation Policy and Plans, 1996).

GENERAL AVIATION ACTIVE AIRCRAFT BY PRIMARY USE
(In Thousands)

Use Category	1994	1993	1992	1991	1990
Corporate	9.7	9.9	9.4	10.0	10.1
Business	25.6	27.8	28.9	31.6	33.1
Personal	100.8	102.1	108.7	115.1	112.6
Instructional	14.6	15.6	16.0	17.9	18.6
Aerial Application	4.2	5.0	5.1	7.0	6.2
Aerial Observation	4.9	4.8	5.6	5.1	4.9
Sight Seeing	1.3	1.6	N/A	N/A	N/A
External Load	0.1	0.1	N/A	N/A	N/A
Other Work	1.2	1.0	1.7	1.7	1.4
Air Taxi	3.9	3.8	4.6	5.5	5.8
Other	4.2	4.2	3.5	3.9	4.1
TOTAL	170.6	175.9	183.6	198.5	198.0

SOURCE: 1990-1994 General Aviation Activity and Avionics Surveys.

N/A = Not applicable. Sight Seeing and External Load added in 1993 as new use categories. Prior to 1993 these aircraft were included in one of the other nine use categories, as appropriate.

Notes:
Commuter aircraft were excluded from survey beginning in 1993. Commuter aircraft in 1990 - 1992 were as follows: 1990 = 1,200; 1991 = 700; and 1992 = 800.

Columns may not add to totals due to rounding and estimation procedures.

Figure 5-1. General Aviation Active Aircraft by Primary Use. Source: Statistics and Forecast Branch. (1996, March). *FAA Aviation Forecasts—Fiscal Years 1996–2007.* Washington, DC: Office of Aviation Policy and Plans.

The General Aviation Manufacturers Association reported 35,940 general aviation aircraft flown 6.29 million hours in 1995 by corporate and business operators (see Figures 5-3 and 5-4). This represented 19.8 percent of the number of aircraft by type and 24.7 percent of the number of hours flown by these aircraft (General Aviation Manufacturers Association [GAMA], 1997).

GENERAL AVIATION HOURS FLOWN BY PRIMARY USE
(In Thousands)

Use Category	1994	1993	1992	1991	1990
Corporate	2,548	2,659	2,262	2,617	2,913
Business	3,005	3,345	3,537	4,154	4,417
Personal	8,116	7,938	8,592	9,685	9,276
Instructional	4,156	4,680	5,340	6,141	7,244
Aerial Application	1,210	1,167	1,296	1,911	1,872
Aerial Observation	1,750	1,750	1,730	1,797	1,745
Sight Seeing	323	412	N/A	N/A	N/A
External Load	172	105	N/A	N/A	N/A
Other Work	226	175	343	471	572
Air Taxi	1,670	1452	2,009	2,241	2,249
Other	640	656	358	473	475
TOTAL	23,866	24,340	25,800	29,497	30,763

SOURCE: 1990-1994 General Aviation Activity and Avionics Surveys.

N/A = Not applicable. Sight Seeing and External Load added in 1993 as new use categories. Prior to 1993 these aircraft were included in one of the other nine use categories, as appropriate.

Notes:
Commuter aircraft were excluded from survey beginning in 1993. Total hours for commuter aircraft in 1990 - 1992 were as follows: 1990 = 1,333,000; 1991 = 570,000; and 1992 = 693,000.

Columns may not add to totals due to rounding and estimation procedures.

Figure 5-2. General Aviation Hours Flown by Primary Use. Source: Statistics and Forecast Branch. (1996, March). *FAA Aviation Forecasts—Fiscal Years 1996–2007*. Washington, DC: Office of Aviation Policy and Plans.

U.S. FLEET AND FLIGHT HOURS BY TYPE AND USE

Number of Active General Aviation Aircraft By Type and Primary Use—1995
(Excluding Commuters)

Aircraft Type	Active GA Aircraft	Corporate	Business	Personal	Instructional	Aerial Application	Aerial Observation	Sight-Seeing	External Load	Other Work	Air Taxi	Other
ALL AIRCRAFT—TOTAL	181,341	9,944	25,996	108,492	14,389	4,924	4,536	990	186	1,174	4,273	6,430
PISTON—Total	145,454	3,011	24,269	90,401	13,063	3,979	3,185	356	0	820	2,617	3,748
One-Engine	128,804	997	18,775	85,346	12,139	3,724	2,864	340	0	794	844	2,975
Two-Engine	16,594	2,014	5,493	5,054	923	251	317	15	0	26	1,772	725
Other Piston	55	0	0	0	0	4	3	0	0	0	0	47
TURBOPROP—Total	4,530	2,345	560	266	27	297	26	0	0	59	709	238
One-Engine	746	66	92	41	14	274	4	0	0	0	241	11
Two-Engine	3,754	2,253	467	225	13	22	21	0	0	59	468	223
Other Turboprop	28	24	0	0	0	0	0	0	0	0	0	3
TURBOJET—Total	4,577	3,444	230	119	13	0	2	3	0	0	326	436
Two-Engine	4,256	3,215	230	115	13	0	2	3	0	0	324	349
Other Turbojet	320	228	0	4	0	0	0	0	0	0	1	86
ROTORCRAFT—Total	5,117	893	255	602	528	603	1,096	94	186	38	458	359
Piston	1,474	10	136	345	356	313	197	42	15	16	0	38
Turbine	3,643	882	119	257	172	289	898	52	170	21	458	320
One-Engine	2,773	633	112	250	166	218	788	48	110	21	201	222
Two-Engine	870	249	6	7	5	71	110	3	60	0	257	98
GLIDERS—Total	1,905	0	33	1,481	210	0	1	120	0	6	0	51
LIGHTER-THAN-AIR—Total	3,374	0	29	2,284	219	1	30	379	0	208	55	165
EXPERIMENTAL—Total	16,382	250	617	13,335	326	42	194	35	0	41	104	1,431
Homebuilt	10,964	0	270	9,531	210	0	100	0	0	0	0	850
Exhibition	1,841	0	41	1,340	47	0	14	0	0	10	2	386
Other	3,576	250	305	2,464	69	42	79	35	0	31	102	194

Note: Row and column summation may differ from printed totals due to estimation procedures, or because some active aircraft did not report use. Source: FAA

Figure 5-3. Number of Active General Aviation Aircraft by Type and Primary Use—1995 (excluding Commuters). Source: FAA/GAMA.

Statistical Analysis

An understanding of the type and number of aircraft can be gained from statistics; for example, seven types of aircraft are identified in Figures 5-3 and 5-4. For 1996, GAMA reported the total units of new general aviation aircraft at 1,130, up 4.9 percent from 1995's number of 1,077. Total 1996 dollar billings rose 11 percent to $3.127 billion. GAMA (1997) reported that in 1996, compared to 1995, total new piston-engine aircraft shipments increased to 600 from 576 (4.2%); single-engine piston aircraft increased to 530 from 515 (2.9%), and multi-engine piston shipments increased 14.8 percent to 70 from 61 units. Turboprop aircraft shipments increased to 289 from 255 (13.3%), and turbojets held about equal (241 in 1996 to 246 in 1995). Billings from these sales of aircraft in 1996 were $2.247 million for the turbojets, $734 million for the turboprops, and $146 million for the pistons, or a total of $3.127 billion for the 1,130 new aircraft.

General Aviation Hours Flown By Aircraft Type and Primary Use—1995
(In Thousands of Hours)

Aircraft Type	Active GA Aircraft	Corporate	Business	Personal	Instruc- tional	Aerial Applica- tion	Aerial Observa- tion	Sight- Seeing	External Load	Other Work	Air Taxi	Other
ALL AIRCRAFT—TOTAL	25,447	3,007	3,283	9,129	3,788	1,349	1,467	219	156	261	1,691	1,097
PISTON—Total	18,886	610	3,088	8,105	3,483	994	825	100	0	219	872	590
One-Engine	16,246	200	2,384	7,622	3,253	958	753	98	0	214	308	457
Two-Engine	2,636	410	704	483	230	35	72	3	0	5	564	130
Other Piston	3	0	0	0	0	0	0	0	0	0	0	3
TURBOPROP—Total	1,356	664	76	40	9	138	6	0	0	16	271	136
One-Engine	273	11	15	7	6	134	1	0	0	0	98	3
Two-Engine	1,069	640	62	33	3	4	5	0	0	16	172	133
Other Turboprop	14	13	0	0	0	0	0	0	0	0	0	0
TURBOJET—Total	1,392	1,134	26	23	7	0	0	1	0	0	118	84
Two-Engine	1,326	1,081	26	23	7	0	0	1	0	0	117	73
Other Turbojet	65	53	0	0	0	0	0	0	0	0	1	11
ROTORCRAFT—Total	2,333	465	26	59	210	192	608	49	156	11	381	175
Piston	341	1	10	21	126	75	86	10	4	4	1	4
Turbine	1,992	464	16	38	84	117	522	39	152	7	380	171
One-Engine	1,450	316	15	37	81	86	483	38	43	7	198	145
Two-Engine	541	148	0	1	3	31	40	1	109	0	182	26
GLIDERS—Total	185	0	3	98	56	0	0	24	0	1	0	3
LIGHTER-THAN-AIR—Total	134	0	1	73	6	0	1	22	0	9	2	20
EXPERIMENTAL—Total	1,162	133	62	730	18	25	26	24	0	5	48	89
Homebuilt	578	0	17	510	10	0	12	0	0	0	0	30
Exhibition	125	0	3	71	2	0	1	0	0	1	0	47
Other	459	133	43	150	6	25	13	24	0	4	48	13

Note: Row and column summation may differ from printed totals due to estimation procedures, or because some active aircraft did not report use.   Source: FAA

Figure 5-4. General Aviation Hours Flown by Aircraft Type and Primary Use—1995 (in thousands of hours). Source: FAA/GAMA.

Figure 5-5 shows the billings for the number and type of aircraft sold in 1996. Although only 53 more aircraft were sold in 1996 versus 1995, the value of the aircraft sold differed by $285 million. Raytheon (formerly Beech) aircraft were bought most in 1995, with Cessna and Piper purchases next in order. The number of aircraft imports into the United States (Figure 5-6) was also interesting. For 1995 and 1996, total imports increased by 28, 259 to 287 shipments, respectively. Billings for these aircraft were $1.448.8 million for 1995 and $2.136.2 million for 1996 (GAMA, 1997).

The number of rotorcraft used in 1994 is indicated in Figure 5-7. Approximately 18.5 percent of corporate and business aircraft operators used helicopters in 1994. Only 175 piston-powered helicopters were used, and of those, only two were used by corporate operators. However, corporate operators used 450 of the 638 turbine-powered helicopters. Aastad (1996–97) reported that a total of 51 new piston and 71 new turbine helicopters, and 834 used helicopters (455 turbine and 379 piston) were sold in the first nine months of 1996. For 1996, he reported a forecast of 100 new turbine, 600 used turbine, 70 new piston, and 500 used piston helicopter sales. For a copy of the Helicopter Transaction Report, contact the Aastad Company at 610-459-9086 or fax 610-459-0509. Civil helicopter shipments for 1991 through 1995, including the number, type, and billings are shown in Figure 5-7. For the 292 helicopters sold in 1995, the average price was approximately $664,500, while the average price of the 1,077 fixed wing aircraft sold in 1995 was approximately $2.64 million. For 1996, the

GENERAL AVIATION AIRCRAFT SHIPMENTS By Selected Manufacturers Calendar Years 1991-1995					
	1991	1992	1993	1994	1995
NUMBER OF AIRCRAFT SHIPPED	1,021	899	964	928	1,077
Single-engine, Piston	564	510	516	444	515
Multi-Engine, Piston	49	41	39	55	61
Turboprop	222	177	211	207	255
Turbojet	186	171	198	222	246
VALUE OF SHIPMENTS[a]					
(Millions of Dollars)	$1,968	$1,836	$2,144	$2,357	$2,842
Piston	$ 93	$ 92	$ 76	$ 94	$ 123
Turboprop	527	460	595	595	653
Turbojet	1,348	1,284	1,473	1,681	2,066
Number of Aircraft By Selected Manufacturer					
American Champion	NA	NA	38	22	46
American General	82	51	30		
Aviat	71	63	56	47	42
Bellanca	1	3	4	2	1
Cessna	176	140	173	172	200
Classic	8	9	7	4	7
Commander	NA	25	31	22	25
Fairchild	10	14	20	16	7
Gulfstream	29	25	26	22	26
Lake	11	9	3		
Learjet	25	23	38	36	43
Maule	25	23	38	36	43
Mooney	88	69	64	71	84
Piper	41	85	99	132	165
Raytheon[b]	402	348	305	317	363
Taylorcraft	11	2			

Source: General Aviation Manufacturers' Association.
[a]Manufacturers' net billing price.
[b]Formerly reported as Beech.
NA Not available.

Figure 5-5. General Aviation Aircraft Shipments.

U.S. Civil Aircraft Imports from the World—1992-1996
Units and Dollar Value ($Millions)

	1992		1993		1994		1995		1996	
	Units	Dollars	Units	Dollars	Units	Dollars	Units	Dollars	Units	Dollars
Single-Engine	67	$24.6	96	$28.6	105	$65.9	117	$48.5	100	$57.8
Multi-Engine–under 4,400 lbs	7	$3.1	N/A	N/A	8	$2.8	5	$0.3	0	$0
Multi-Engine–4,400-10,000 lbs	18	$75.7	6	$14.8	2	$2.4	2	$3.0	1	$8
Multi-Engine–Turbojet/Turbofan 10,000-33,000 lbs.	52	$612.0	66	$792.3	82	$1,030.4	72	$902.4	96	$1,286.6
Muti-Engine–Other —including Turboshaft 10,000-33,000 lbs	72	$659.5	44	$402.1	64	$609.4	63	$494.6	90	$783.9
TOTAL	216	$1,374.9	212	$1,237.8	261	$1,710.9	259	$1,448.8	287	$2,136.2

Source: Aerospace Industries Association

Figure 5-6. U.S. Civil Aircraft Imports from the World—1992–1996 Units and Dollar Value ($millions). Source: Aerospace Industries Association.

average price of the 1,130 fixed-wing aircraft was $2.77 million. Wide differences existed based on the type and model purchased, but for the cost-conscious buyer, average price might be a starting point in aircraft selection. From there, consideration of the average price for specific types of rotorcraft and fixed-wing aircraft may be the next step. Although not every detail of the aircraft selection process needs to be written down, numerous thought-provoking items have to be considered before formulating a specific analysis outline in the aircraft selection process. Some of these quick decisions can be applied immediately when preparing the travel pattern map and travel pattern graph referenced earlier.

What these statistics and others presented by GAMA, the FAA, and industry officials mean is that general aviation is vital to our individual, societal, and political needs. General aviation is a multi-billion dollar industry ($15 billion in 1996), and exports exceed imports. Over 130 million passengers use general aviation aircraft, and the aircraft are flown twice as much as commercial aircraft (over 23 million hours). Also, more than 5,000 communities depend on air transportation services provided with general aviation aircraft (GAMA, 1997). Corporate and business aviation users have a significant share in these contributions.

NBAA Fleet

NBAA members operated more than 6,000 aircraft in 1996, with jets weighing less than 30,000 pounds being the most popular (see Figure 5-8). Other details of the NBAA fleet were provided in the NBAA's *Business Aviation Fact Book 1997* (NBAA, 1997a). The Cessna Citation was the most popular jet aircraft under 30,000 pounds, followed by Learjet aircraft, Israel Aircraft Industries' Westwinds and Astras, and Raytheon's Hawkers and Beechjets. Of the "heavy iron" turbofan aircraft weighing over 30,000 pounds, nearly half were Gulfstream aircraft. Dassault's Falcon jets, Canadair's Challenger, and Raytheon's Hawker 1000 were other heavy jets used.

CIVIL HELICOPTER SHIPMENTS[a]
Calendar Years 1991-1995

Company and Model	1991	1992	1993	1994	1995
CIVIL SHIPMENTS	**571**	**324**	**258**	**308**	**292**
Value (Millions of Dollars)	$211	$142	$113	$185	$194
Bell TOTAL	4	1	2		
212					
214 series		1	2		
412	4				
Enstrom TOTAL	17	6	10	17	11
F-28	8	3	(b)	(b)	(b)
280 series	9	3	8[b]	13[b]	3[b]
480 series			2	4	8
Hiller TOTAL	2	3			1
UH12E	2	3			1
Kaman TOTAL			5	6	
McDonnell Douglas TOTAL . .	50	51	26	36	34
500 series	42	23	5	3	12
520N series	3	17	21	9	10
530 series	5	11		22	
900 series				212	
Robinson TOTAL	402	212	166	195	179
R22	402	212	135	89	83
R44			31	106	96
Schweizer TOTAL	78	39	45	40	47
300C	78	39	40	35	22
300CB					21
330			5	5	4
Sikorsky TOTAL	18	12	9	15	14
S-76	18	12	9	15	14

Source: Aerospace Industries Associations, based on company reports
NOTE: All data exclude production by foreign licenses.
[a] Domestic and export helicopter shipments for non-military use. Helicopters in military configuration exported to foreign governments and purchased under commercial contract are reported elsewhere. Models which may be shipped in either a civil or a military configuration appear in both tables.
[b] Reporting of F-28 and 280 series combined.
[c] Formerly reported as Rogerson.

Figure 5-7. Civil Helicopter Shipments.

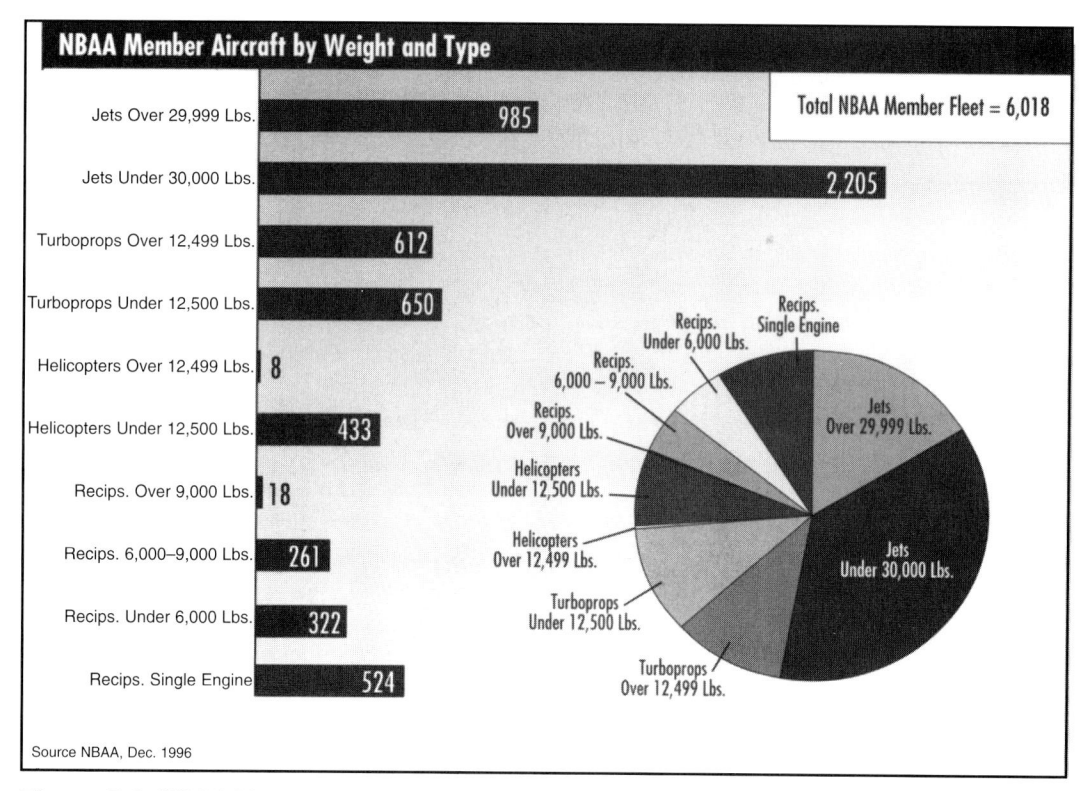

Figure 5-8. NBAA Member Aircraft by Weight and Type. Courtesy of NBAA.

Turboprop aircraft used were somewhat divided at the 12,500-pound category with 650 under and 612 over this weight, or 11 percent for the lighter and 10 percent for the heavier aircraft of the total NBAA fleet. Turboprop manufacturers included Raytheon (Beech King Airs), Cessna, Fairchild, Piaggio, and Piper (NBAA, 1997a).

Nearly 8 percent of the NBAA fleet were helicopters (441), and only eight of these were over 12,499 pounds. Bell Helicopter Textron, Sikorsky, McDonnell Douglas, and Eurocopter were providers of the turbine-powered helicopters for NBAA members. Beech, Cessna, and Piper provided most of the 1,125 piston-powered aircraft; 46.6 percent single-engine and 53.4 percent multi-engine aircraft (NBAA, 1997a).

In Table 5-1, hours of utilization for the various type aircraft used by NBAA members are indicated and are based on survey data collected by the NBAA from January through March of 1995. Using the mean to indicate hours of aircraft use does not provide for those high and low users. It should allow managers who are becoming involved in corporate aviation for the first time and are beginning their analysis of the number of aircraft needed to consider a good estimate (remember the travel pattern graph). Overall, for 1995, average annual aircraft usage totaled 419 hours for jets, 407 hours for turboprops, 299 for pistons, and 348 hours for helicopters.

Table 5-1. NBAA Average Annual Flight Hours by Aircraft Type

Aircraft Type	Average Annual Flight Hours 1993	1994	1995
Jets	406	415	419
Turboprops	419	421	407
Pistons	313	307	299
Helicopters	404	357	348

Source: NBAA. (1997). *Business aviation fact book 1997.* Washington, DC. Author. Courtesy of NBAA. Note: Data provided by 1,018 flight departments using 2,336 aircraft; 42.4% response rate.

Before presenting various aircraft manufacturers and specific corporate aircraft, two other items need to be considered by managers when deciding on the use of business aircraft. These include the airports available for use and the pilots needed.

Airports and Pilots

Why are airports important to corporate and business aircraft operators besides the obvious answer? Over 18,200 airports, heliports, stolports, and seaplane bases in the U.S. enable the corporate and business aircraft operators to go where the airlines do not go or cannot go to generate business in locations perhaps not otherwise feasible, and to increase management's productivity. In Figure 5-9, GAMA (1997) reported 5,129 public airports available for use. Additional private facilities allow the corporate and business flyers to use their vital business tool—the aircraft. In some instances, corporate operators have developed their own airport or heliport to provide for increased productivity.

GAMA (1997) also presented the number of pilots by region and state, as well as by category and age (see Figures 5-10 and 5-11). Reviewing these statistics in conjunction with those on the number of aircraft being flown should indicate to business managers the extent and potential for using an aircraft as a business tool. A private pilot may fly his or her aircraft in the furtherance of a business, but when professional pilots fly aircraft for a corporation, at least a commercial pilot's certificate is required. The NBAA requires its members to use pilots with an air transport certificate.

The number of pilots available includes over 32,500 commercial and airline pilots who are over age 60. The small growth in newly qualifying pilots causes a major concern for the operation of a corporate flight department. The FAA (Office of Aviation Policy and Plans, 1996) forecasted only a very small 10-year average growth rate for total pilots, only 0.76 percent until 2007. The largest rate was for airline transport pilots at 1.5 percent, but the others fell

AIRPORTS BY GEOGRAPHIC AREA

U.S. Civil and Joint Use Airports, Heliports, Stolports, and Seaplane Bases On Record
By Type Of Ownership—1996

FAA Region and State	Total Facilities	Total Facilities, by Ownership - Public	Total Facilities, by Ownership - Private	Paved Airports[1] - Lighted	Paved Airports[1] - Unlighted	Unpaved Airports[1] - Lighted	Unpaved Airports[1] - Unlighted	Total Airports
GRAND TOTAL	18,292	5,129	13,163	3,609	294	399	802	5,104
UNITED STATES—Total*	18,216	5,079	13,137	3,586	291	399	789	5,065
ALASKAN—Total	546	383	163	45	5	99	156	305
Alaska	546	383	163	45	5	99	156	305
CENTRAL—Total	1,493	499	994	386	17	41	54	498
Iowa	305	136	169	96	1	13	14	124
Kansas	384	132	252	101	9	17	21	148
Missouri	507	137	370	113	5	6	10	134
Nebraska	297	94	203	76	2	5	9	92
EASTERN—Total	2,359	349	2,010	332	26	48	82	488
Delaware	35	4	31	5	0	3	1	9
District of Columbia	17	8	9	2	0	0	0	2
Maryland	200	21	179	28	1	3	3	35
New Jersey	348	47	301	36	2	4	7	49
New York	535	91	444	83	13	22	31	149
Pennsylvania	755	74	681	89	5	14	28	136
Virginia	365	73	292	63	1	1	3	68
West Virginia	104	31	73	26	4	1	9	40
GREAT LAKES—Total	4,259	905	3,354	744	21	145	156	1,066
Illinois	896	122	774	92	0	19	5	116
Indiana	603	87	516	78	4	5	21	108
Michigan	471	134	337	124	7	43	56	230
Minnesota	475	150	325	107	0	23	16	146
North Dakota	434	96	338	66	5	12	11	94
Ohio	739	136	603	124	2	17	24	167
South Dakota	156	77	79	54	1	14	5	74
Wisconsin	485	103	382	99	2	12	18	131
NEW ENGLAND—Total	705	142	563	116	17	4	27	164
Connecticut	136	16	120	17	2	0	4	23
Maine	150	46	104	30	7	2	7	46
Massachusetts	226	35	191	36	4	1	5	46
New Hampshire	94	16	78	16	3	1	6	26
Rhode Island	26	9	17	7	0	0	0	7
Vermont	73	20	53	10	1	0	5	16
N.W. MOUNTAIN—Total	1,938	683	1,255	407	51	21	144	623
Colorado	391	92	299	64	4	3	8	79
Idaho	227	131	96	44	9	1	64	118
Montana	243	123	120	73	7	8	31	119
Oregon	411	103	308	58	17	4	18	97
Utah	125	58	67	41	5	0	1	47
Washington	437	127	310	93	8	4	17	122
Wyoming	104	49	55	34	1	1	5	41
SOUTHERN—Total	2,634	840	1,794	633	49	25	52	759
Alabama	240	103	137	86	5	4	6	101
Florida	769	163	606	100	5	8	15	127
Georgia	401	136	265	97	9	2	2	110
Kentucky	179	73	106	53	9	0	3	65
Mississippi	221	90	131	70	9	2	2	83
North Carolina	358	95	263	87	4	5	16	112
Puerto Rico	32	17	15	10	1	0	0	11
South Carolina	167	69	98	57	1	4	6	68
Tennessee	258	88	170	71	6	0	2	79
Virgin Islands	9	6	3	2	0	0	0	2
SOUTHWEST—Total	2,949	816	2,133	637	47	12	74	770
Arkansas	265	115	150	90	5	0	5	100
Louisiana	426	110	316	71	2	1	7	81
New Mexico	164	73	91	43	9	0	10	62
Oklahoma	410	160	250	112	13	6	17	148
Texas	1,684	358	1,326	321	18	5	35	379
WESTERN—Total	1,409	512	897	309	61	4	57	431
Arizona	276	88	188	54	8	0	14	76
California	933	319	614	207	42	1	11	261
Hawaii	46	18	28	11	2	0	0	13
Nevada	119	60	59	26	7	3	19	55
South Pacific**	35	27	8	11	2	0	13	26

*Excludes Puerto Rico, Virgin Islands, Northern Mariana Islands, and South Pacific.
**American Samoa, Guam, and Trust Territories.

Source: FAA

Figure 5-9. U.S. Civil and Joint Use Airports, Heliports, Stolports, and Seaplane Bases on Record by Type of Ownership—1996. Courtesy of GAMA.

PILOTS BY GEOGRAPHIC AREA
Estimated Active Pilots and Flight Instructors By FAA Region and State—1996

FAA Region and State	Total Pilots	Student[1]	Private	Commercial	Airline Transport	Misc.[3]	Flight Instructor[4]
GRAND TOTAL[5]	**622,261**	**94,947**	**254,002**	**129,187**	**127,486**	**16,639**	**78,251**
UNITED STATES—Total	592,874	90,610	247,859	119,755	119,624	15,026	76,126
ALASKAN—Total	8,781	988	3,834	2,250	1,623	86	1,027
Alaska	8,781	988	3,834	2,250	1,623	86	1,027
CENTRAL—Total	30,154	4,496	14,661	6,299	4,341	357	3,836
Iowa	6,217	914	3,353	1,273	624	53	767
Kansas	8,508	1,324	4,116	1,821	1,138	109	1,083
Missouri	11,233	1,635	5,097	2,216	2,121	164	1,531
Nebraska	4,196	623	2,095	989	458	31	455
EASTERN—Total	76,606	13,410	31,852	14,504	14,208	2,632	10,396
Delaware	1,453	246	591	259	322	35	220
District of Columbia	476	99	202	97	42	36	53
Maryland	8,556	1,425	3,639	1,667	1,536	289	1,185
New Jersey	11,893	2,064	5,087	2,062	2,253	427	1,536
New York	19,035	3,814	8,282	3,522	2,515	902	2,478
Pennsylvania	18,425	3,039	7,807	3,377	3,679	523	2,638
Virginia	14,037	2,195	5,072	2,908	3,511	351	1,959
West Virginia	2,198	407	1,033	459	258	41	263
Armed Forces Europe[6]	533	121	139	153	92	28	64
GREAT LAKES—Total	102,351	16,649	47,771	19,905	16,371	1,655	13,387
Illinois	22,177	3,630	9,658	4,208	4,283	398	3,101
Indiana	11,719	2,023	5,713	2,214	1,602	167	1,434
Michigan	18,219	3,239	8,880	3,291	2,438	371	2,284
Minnesota	15,018	2,112	6,671	3,143	2,930	162	2,027
North Dakota	2,682	373	1,186	907	203	13	313
Ohio	18,991	3,048	8,866	3,598	3,055	424	2,572
South Dakota	2,258	362	1,053	537	292	14	253
Wisconsin	11,287	1,862	5,744	2,007	1,568	106	1,403
NEW ENGLAND—Total	27,028	4,198	11,917	4,901	5,174	838	3,407
Connecticut	6,727	957	2,801	1,112	1,635	222	866
Maine	3,239	551	1,515	675	464	34	371
Massachusetts	10,040	1,708	4,766	1,802	1,372	392	1,227
New Hampshire	4,149	517	1,539	746	1,246	101	596
Rhode Island	1,317	229	601	255	203	29	151
Vermont	1,556	236	695	311	254	60	196
NORTHWEST MOUNTAIN—Total	63,278	9,086	26,426	12,809	13,236	1,721	7,969
Colorado	16,633	2,152	6,035	3,369	4,397	680	2,455
Idaho	4,461	644	2,134	983	629	71	507
Montana	3,622	494	1,734	922	422	50	429
Oregon	10,006	1,637	5,016	2,026	1,101	266	1,144
Utah	5,845	1,005	2,327	1,078	1,321	114	667
Washington	20,873	2,900	8,265	4,055	5,098	555	2,553
Wyoming	1,838	254	915	376	268	25	214
SOUTHERN—Total	112,477	16,504	41,598	23,680	28,514	2,181	13,936
Alabama	7,639	1,268	3,108	1,893	996	374	972
Florida	45,218	6,279	16,017	10,190	11,974	758	6,205
Georgia	17,240	2,377	5,769	3,208	5,587	299	1,890
Kentucky	6,389	1,028	2,432	1,061	1,778	90	840
Mississippi	4,075	691	1,535	1,162	611	76	478
North Carolina	13,919	2,152	5,725	2,621	3,172	249	1,613
South Carolina	6,131	998	2,614	1,373	1,036	110	405
Tennessee	11,789	1,704	4,373	2,146	3,345	221	1,522
Armed Forces Europe[6]	77	7	25	26	15	4	11
SOUTHWEST—Total	71,037	10,346	27,198	15,337	16,716	1,440	9,499
Arkansas	5,293	891	2,269	1,386	695	52	651
Louisiana	6,086	891	2,266	1,604	1,103	222	790
New Mexico	4,444	667	1,884	1,018	749	126	605
Oklahoma	8,981	1,574	4,025	2,063	1,234	85	1,280
Texas	46,233	6,323	16,754	9,266	12,935	955	6,173
WESTERN—Total	101,119	14,932	42,600	20,070	19,439	4,078	12,669
Arizona	16,298	2,691	5,970	3,351	3,759	527	2,374
California	75,801	11,098	33,779	14,935	12,845	3,144	9,031
Hawaii	2,561	311	566	567	907	210	378
Nevada	5,892	640	2,151	1,059	1,870	172	837
Armed Forces Pacific[6]	567	192	134	158	58	25	49
UNKNOWN	43	1	2	0	2	38	0
OUTSIDE U.S.—Total[7]	29,387	4,337	6,143	9,432	7,862	1,613	2,125
OTHER U.S. AREAS—Total	2,130	564	570	413	521	62	253
American Samoa	12	0	1	5	5	1	3
Guam	216	22	23	36	131	4	22
Puerto Rico	1,624	498	450	306	317	53	192
Virgin Islands	241	42	90	50	55	4	27
U.S. Territories	37	2	6	16	13	0	9

[1] Category of medical certificate unknown.
[2] Includes pilots with an airplane-only certificate. Also includes those with an airplane and a helicopter and/or glider certificate.
[3] Includes helicopter, glider, and recreational.
[4] Not included in total.
[5] Includes Outside U.S.
[6] Military personnel holding civilian certificates stationed in foreign country.
[7] Outside U.S. includes areas of the U.S. outside of the 50 states and foreign countries.

Source: FAA

Figure 5-10. Estimated Active Pilots and Flight Instructors by FAA Region and State—1996. Courtesy of GAMA.

Estimated FAA Active Pilot Certificates Held by Category and Age Group of Holder—1996

Age Group	Total	Student[1]	Recreational[1]	Private	Commercial	Airline Transport	Rotorcraft (Only)	Glider (Only)[3]	Flight Instructor[4]
TOTAL	**622,261**	**94,947**	**265**	**254,002**	**129,187**	**127,486**	**6,961**	**9413**	**78,551**
14-15	220	220	0	0	0	0	0	0	0
16-19	14,087	10,495	3	3,234	243	0	32	80	57
20-24	36,205	13,427	5	13,430	8,523	313	304	203	4,359
25-29	54,300	13,799	13	16,480	17,122	5,688	904	294	11,516
30-34	68,330	12,964	13	22,866	15,133	15,460	1,289	605	10,462
35-39	82,494	12,914	34	32,006	13,679	21,578	1,241	1,042	9,536
40-44	86,772	10,829	38	39,048	13,464	20,536	1,065	1,792	9,260
45-49	83,012	8,101	30	36,590	14,227	21,144	1,159	1,761	9,590
50-54	70,017	5,265	35	29,929	14,397	18,417	630	1,344	8,594
55-59	48,794	2,994	25	21,060	10,448	13,205	219	843	5,788
60-64	33,362	1,809	17	15,547	8,734	6,538	60	657	4,235
65-69	24,054	1,296	32	13,378	5,982	2,543	41	782	2,573
70-74	13,697	572	16	7,382	4,433	1,274	15	5	1,540
74-79	5,522	188	3	2,416	2,258	652	2	3	818
80 and over	1,395	74	1	636	544	138	0	2	223

[1] Category of certificate unknown.
[2] Includes pilots with an airplane only certificate. Also includes those with an airplane and a helicopter and/or glider certificate.
[3] Glider pilots are not required to have a medical examination; however, totals above represent pilots who received a medical examination.
[4] Not included in total active pilots.

Source: FAA

Figure 5-11. Estimated FAA Active Pilot Certificates Held by Category and Age Group of Holder—1996. Courtesy of GAMA.

below 1 percent. Overall, not only is there a need for aircraft, but there is a vital need for qualified pilots and readily available airports so that business managers can increase their opportunities for company survival, profitability, and expansion.

Aircraft Manufacturers

Some of the aircraft manufacturers are Augusta, Bell, Boeing, Bombardier, Cessna, Dassault, Gulfstream, Learjet, Raytheon, Sikorsky, and Sino-Swearingen. For a detailed listing, review Appendix G. Additionally, several manufacturers have been indicated in the chapter figures. Not all corporate and business aircraft are general aviation type. Boeing provides corporate and government aircraft in the form of 60 Boeing 737 versions (Boeing, 1995). Specific examples of the B-737 for corporate operators are provided later in this chapter, along with examples of several of the other corporate models.

An excellent source of information for aircraft prices, specifications, and performance can be obtained from the individual aircraft manufacturers and from reports in *Business and Commercial Aviation* (B&CA). The latter's annual *Planning & Purchasing Handbook* is an excellent source of unbiased information. For example, information about the Learjet 45, Cessna Citation Excel, Raytheon Hawker 800XP, Cessna Citation X, Dassault Falcon 50EX and 900EX, Canadair Challenger and Global Express, and Gulfstream G-V can be reviewed. Monthly issues of the periodical will provide you with current aircraft changes and analyses. An extract from B&CA's May 1996 issue is provided in Figure 5-12, but review current issues for more details and new aircraft.

Business Airplanes

JETS 20,000 lbs MTOW or Greater

Manufacturer Model			Dassault Falcon 2000 DA-2000	Canadair Challenger 604 CL-604	Canadair Corporate Jetliner CL-601RJ	Dassault Falcon 900B DA-900B	Gulfstream Gulfstream IV-SP G-IVSP	Dassault Falcon 900EX DA-900EX
B/CA Equipped Price			$18,150,000	$19,450,000	$21,000,000	$24,950,000	$27,000,000	$27,500,000
Characteristics		Seating	2 + 8/19	2 + 9/19	2 + 18/30	2 + 13/19	2 + 14/19	2 + 13/19
		Wing Loading	67.9	96.7	91.2	86.3	78.5	91.6
		Power Loading	3.13	2.73	2.92	3.19	2.69	3.22
		Noise (EPNdB)	79.4/86.4	80.9/90.3	78.6/92.1	79.8/91.7	76.8/91.0	NA/NA
Dimensions (ft)	External	Length	66.3	68.4	87.8	66.3	88.3	66.3
		Height	23.2	20.7	20.4	24.8	24.4	24.8
		Span	63.4	64.3	69.6	63.4	77.8	63.4
	Internal	Length	23.3	22.0	41.0	30.1	31.4	30.1
		Height	6.2	6.1	6.1	6.2	6.2	6.2
		*Width	7.7/6.3	8.2/7.2	8.2/7.2	7.7/6.3	7.3/5.5	7.7/6.3
Power		Engines	2 CFE CFE 738-1-1B	2 GE CF34-3B	2 GE CF34-3B1	3 ASE TFE731-5BR-1C	2 RR Tay MK 611-8	3 ASE TFE731-60
		Thrust	5,725 lbs ea.	8,729 lbs ea.	8,729 lbs ea.	4,750 lbs ea.	13,850 lbs ea.	5,000 lbs ea.
		TBO	OC	OC	OC	OC	7,000	OC
Weights (lbs)		Max Ramp	36,000	47,700	51,250	45,700	75,000	48,500
		Max Takeoff	35,800	47,600	51,000	45,500	74,600	48,300
		Max Landing	33,000	38,000	47,000	42,000	66,000	42,000
		Zero Fuel	28,660c	32,000c	44,000c	28,220c	49,000c	30,865c
		BOW	21,450	26,630	31,860	24,700	42,500	24,700
		Max Payload	7,210	5,370	12,140	3,520	6,500	6,165
		Useful Load	14,550	21,070	19,390	21,000	32,500	23,800
		Executive Payload	1,600	1,800	3,600	2,600	2,800	2,600
		Max Fuel	12,155	19,852	14,305	19,165	29,500	20,825
		Payload – Max Fuel	2,395	1,218	5,085	1,835	3,000	2,975
		Fuel – Max Payload	7,340	15,700	7,250	17,480	26,000	17,635
		Fuel – Exec. Payload	12,155	19,270	14,305	18,400	29,500	20,825
Limits		M_{MO}	0.870	0.850	0.850	0.870	0.880	0.870
		FL/V_{MO}	FL 250/370	FL 222/348	FL 254/335	FL 250/370	FL 280/340	FL 250/370
		V_{FE} (app.)	180	231	230	180	220	180
		PSI	9.3	9.2	8.3	9.3	9.5	9.3
Airport Performance		TOFL (SL ISA)	5,440	5,665	5,640	4,950	5,450	5,290
		TOFL (5,000 ft @ 25°C)	8,015	9,085	9,050	7,510	8,925	7,595
		Mission Weight	35,800	47,600	51,000	45,500	74,000	48,300
		NBAA/IFR	3,195	4,069	2,535	3,920	4,220	4,595
		V2	125	146	143	130	150	134
		V_{REF}	107	116	121	107	125	107
		LD	2,500	2,300	3,920	2,320	2,658	2,320
Climb		All-Engine Rate (fpm)	3,425	4,345	3,555	4,000	4,122	4,000
		Engine-Out Rate (fpm)	850	1,175	1,070	2,000	1,054	2,000
		Engine-Out Grad. (ft/nm)	305	400	346	486	352	486
Ceilings (ft)		Certificated	47,000	41,000	41,000	51,000	45,000	51,000
		All-Engine Service	43,000	37,500	37,230	41,000	43,000	41,000
		Engine-Out Service	31,000	22,500	21,600	31,000	31,800	31,000
		Sea-Level Cabin	25,300	23,200	21,100	25,300	22,000	25,300
Cruise	Long Range	TAS	430	424	424	430	459	438
		Fuel Flow	1,318	1,681	1,820	1,842	2,713	1,765
		Altitude	FL 450	FL 410	FL 410	FL 450	FL 450	FL 430
		Specific Range	0.326	0.252	0.233	0.233	0.169	0.248
	High Speed	TAS	479	459	459	482	480	474
		Fuel Flow	2,018	1,945	2,166	2,420	3,429	2,295
		Altitude	FL 390	FL 410	FL 410	FL 430	FL 410	FL 390
		Specific Range	0.237	0.236	0.212	0.199	0.140	0.207
NBAA IFR Ranges (200 nm alt.)	Max Payload	Nautical Miles	1,380	2,895	775	3,329	3,338	3,410
		Average Speed	427	415	390	421	437	428
		Trip Fuel	5,525	13,655	4,659	15,325	22,900	15,560
		Specific Range/Altitude	0.250/FL 430	0.212/FL 410	0.166/FL 370	0.217/FL 450	0.146/FL 450	0.219/FL 450
	Max Fuel**	Nautical Miles	2,930	4,029	2,316	3,827	4,033	4,325
		Average Speed	441	418	412	422	443	430
		Trip Fuel	10,340	17,963	12,072	17,100	26,180	18,750
		Specific Range/Altitude	0.283/FL 470	0.224/FL 410	0.192/FL 390	0.224/FL 450	0.154/FL 450	0.231/FL 470
	Ferry	Nautical Miles	3,175	4,142	2,581	4,000	4,243	4,690
		Average Speed	443	419	414	422	443	430
		Trip Fuel	10,435	18,007	12,256	17,100	26,180	18,835
		Specific Range/Altitude	0.304/FL 470	0.230/FL 410	0.211/FL 410	0.234/FL 450	0.162/FL 450	0.249/FL 470
Missions (4 pax)	300 nm	Runway	3,365	3,270	2,935	2,685	3,200	3,000
		Flight Time	0 + 48	0 + 49	0 + 50	0 + 46	0 + 45	0 + 48
		Fuel Used	1,450	1,545	1,862	1,820	2,822	1,585
		PSR/Altitude	0.828/FL 450	0.777/FL 410	0.644/FL 390	0.659/FL 450	0.425/FL 450	0.757/FL 450
	600 nm	Runway	3,390	3,300	3,120	2,685	3,250	3,000
		Flight Time	1 + 28	1 + 32	1 + 33	1 + 26	1 + 25	1 + 28
		Fuel Used	2,335	2,667	3,132	2,918	4,387	2,595
		PSR/Altitude	1.028/FL 450	0.900/FL 410	0.766/FL 410	0.822/FL 450	0.547/FL 450	0.925/FL 450
	1,000 nm	Runway	3,410	3,400	3,370	2,700	3,300	3,000
		Flight Time	2 + 21	2 + 28	2 + 30	2 + 19	2 + 17	2 + 20
		Fuel Used	3,545	4,201	4,886	4,448	6,512	3,990
		PSR/Altitude	1.128/FL 450	0.952/FL 410	0.819/FL 410	0.899/FL 450	0.614/FL 450	1.003/FL 450
Remarks		Certification Basis	FAR 25 1995 Optional 36,700-lb max ramp weight, 36,500-lb MTOW.	FAR 25, 1980/83/87/1995 Optional 48,300-lb max ramp weight, 47,600-lb MTOW.	FAR/JAR 25, 1992 Also available: 3,137-nm range, nine passenger Canadair SE for $23,500,000.	FAR 25, 1979/91 optional 46,700-lb max ramp weight, 46,500-lb MTOW, 30,870-lb MZFW.	FAR 25, 1983/93 Price B/CA estimate.	FAR 25 pending Preliminary data.

*Width at center cabin/floor. **Max fuel with available payload.

Figure 5-12. Extract of B&CA's Planning and Purchasing Handbook. From *Business & Commercial Aviation,* May 1996, p. 84. Permission to use by B&CA.

Aircraft Selection

With the many types and number of aircraft available, how does a manager select the right aircraft for business use? Is there only one type of aircraft suitable for a firm's use? Theoretically, the answer to the second question is, no. A corporate operator who has to use a helicopter to serve an oil rig offshore probably could not use another type. It would not be impossible, however, to use a lighter-than-air or microlight aircraft. The answer to the first question above is, with considerable effort. The aircraft selection issue is more complicated when a multitude of choices is available, especially for a manager who has to provide for mid-range company air trips of 300–1,000 miles. The NBAA and schools like Embry-Riddle Aeronautical University offer special seminars to enable managers to understand aircraft selection processes.

If a CEO or other top manager decides to obtain a helicopter, then the type is known, but not necessarily the model. Perhaps, the same CEO only wants a two-engine helicopter for safety concerns, then again the choice is narrowed down. If, however, a boss asks for a recommendation and provides no other information, then the data from a travel analysis are used at the start of a complex aircraft selection process.

Preliminary decisions to consider a helicopter for short-range trips (under 350 miles) and a mid-size business jet for trips of 500–1,200 miles may have been made. If not, then the aircraft performance specifications provided by aircraft manufacturers, the NBAA, business aircraft operators, or other sources may contribute to the aircraft selection process. What must be determined are the suitable makes and models after the type aircraft has been identified. If a jet is desired, would a Falcon, Gulfstream, Learjet, or other make be suitable?

Research into the various needs of the company travelers, the origins/destinations, cargo and mail requirements, inflight services, executive preferences, and passenger needs has to be accomplished—probably in conjunction with consideration of a targeted budget for the overall company mission.

For those managers who personally select the make and model of aircraft for business or company use, the decision may have been based on a few key factors. Cost may not even be one of these factors; however, for those managers wanting to select the most appropriate aircraft based on the company needs, then the aircraft selection process becomes more complicated.

Some executives hire an aircraft management firm to analyze the transportation requirements and to recommend the number and type of aircraft for the firm. Other managers use aviation consultants to provide critical information, but the aircraft review is mainly accomplished in house.

Project Management

The best method to select an appropriate company aircraft for a large company is to employ the team concept using project management. The first step in this team approach is the appointment of a team leader, and this may not necessarily be an aviation manager or chief pilot. It should be a person who understands project management and the team approach itself, and has the authority and interests to provide direction, control, and wisdom to the team. What should the size and composition of the team be? These initial questions are answered by the team leader, who should decide the team composition based on experience and knowledge of the employees and the task at hand. For example, a finance officer can be vital in budgeting; a pilot to the aircraft performance requirements; a flight department manager to the scheduling, dispatching, and many other functions; and other key people for transportation and support requirements. The process can be divided into phases wherein different people are brought in to present requirements and provide important information at different times in the overall selection process.

For a smaller company, a team comprised of only a few key people might be all that is required to collect the data and accomplish the analysis. A key person familiar with flight operations can review all requirements, make comparisons, and provide a final recommendation without involvement by many other people. The main point is that the aircraft selection process is specific to a company and to the people within the firm. The major emphasis is that a project management approach should result in a more appropriate solution to the air transportation needs of the company. Whatever method is used, accurate data are needed, along with a person familiar with aviation. Outside aviation consultants may be useful for some companies, but most aviation managers have the capability to conduct this analysis. However, some firms may wish to have the entire process accomplished by a contracted aircraft management firm. An example of such an analysis is provided in the Appendices, as well as situations that can be used for conducting an original study.

Aircraft Selection Processes

The following aircraft selection suggestions have been provided by NBAA and Embry-Riddle Aeronautical University officials in their professional seminars. Their four goals were to

1. Use a model for project management,
2. Determine the customers' needs,
3. Define the mission profile of customers' air travel needs, and
4. Build awareness of important decisions in aircraft selection, purchase, and outfitting (Center for Professional Development, n.d.).

For the first goal, emphasis was placed on people's interaction. Spirit, sense of mission, active participation, skills, enthusiasm, and unity of effort were some of the important team characteristics desired. The four steps of a project management model included:

1. Plan,
2. Organize,
3. Implement, and
4. Control. Specific functions addressed for each are presented in Table 5-2.

Table 5-2. Project Management Steps

Step	Activities
Plan	Analyze facts of the situation Set objectives Develop all reasonable courses of action Identify negatives of each action Develop strategies Determine measures of progress
Organize	Identify various job tasks Define relationships, responsibilities and authority Establish qualifications for positions Determine allocation of resources
Implement	Find qualified people Train and develop for new roles Develop agreed upon performance objectives Assign responsibility, accountability, and authority
Control	Measure project progress to goals Measure individual performance Take corrective action on project Deliver consequences for individuals

Additional Key Point:

Groups make better decisions than individuals through group dynamics, interpersonal skills, and teamwork.

Source: Abstracted from NBAA and Embry-Riddle Aeronautical University *Seminar guide for aircraft selection*, (n.d.). Courtesy of Center for Professional Development, Embry-Riddle.

Factors of air travel value were reviewed relative to measuring tangible and intangible items. These factors were confidentiality, cost efficiency, safety/security, service, and time efficiency. The importance of each factor has to be company specific, but after a firm's needs and expectations are identified, the corporate strategic mission profile must be defined. Typical descriptive factors of this profile included

    a. business alliances,
    b. business strategy,
    c. culture,
    d. financial policy,
    e. growth strategy,

f. travel policy, and
g. mission definition.

Following the strategic mission profile, the tactical mission profile must be identified. In this, answers to the *Who, What, When, Why, and Where* questions will provide operational requirements. Some of the various key items which have to be addressed in determining the tactical mission profile have been identified in Table 5-3. For the aircraft selection process, many of the factors and items of concern have to be grouped into major categories. For example, pilots can best identify aircraft performances in speed, range, engine power, avionics, and other key items, while a transportation or logistics official would consider number of passengers, baggage, cargo, mail, special equipment, and spare parts. A team should group the many factors into a manageable number of items for evaluation. Important group issues can be capacity, performance, operations, and economics.

Aircraft Costs

One of the major issues in obtaining an aircraft is the cost of an aircraft; however, this is not a single item. Costs involve *costs of acquisition* and *costs of possession.* The sales price of an aircraft can be obtained from the aircraft manufacturers, aircraft management firms, sales representatives, B&CA's *Planning and Purchasing Handbook,* or a variety of other sources. Perhaps an evaluator can call another company that has a similar aircraft or contact a company like Executive Jet Aviation of Columbus, Ohio, or Jet Aviation Business Jets Incorporated, of Teterboro, New Jersey, or one of the other aircraft sales companies. Still another source of aircraft costs can be found in various publications like *Aviation International News, Executive Controller, Business Air Today, Professional Pilot,* or other informative publications.

Undoubtedly, an aircraft price is negotiable, depending on the services and equipment required. A budget target price for a selected aircraft type should be identified and used in the selection process.

The second area of costs is *costs of possession.* These costs vary greatly depending on the financial program and the operation/use of the aircraft. A number of factors has to be considered, but typically a manager has to identify the fixed and variable costs. *Fixed costs* must be paid regardless of whether the aircraft is flown or not. These costs include salaries of the aircrew, maintenance, and administrative personnel; employee benefits; hangar rental; insurance; training; and miscellaneous others. *Variable costs* depend on the use of the aircraft; for example, fuel and oil costs, landing fees, travel charges, maintenance per flight hour, spare parts, servicing fees, catering, and other items that result from flying an aircraft. Figure 5-13 is a sample aircraft cost form that the NBAA uses to identify costs. The same items can be used to convert the form into an annual aircraft costs form, and other items such as annual maintenance, fuel, engine costs, and user-defined items can be identified with a specific aircraft.

Table 5-3. Tactical Mission Profile	
Item	**Characteristics**
Information Resources	Aircraft operators
	Airframe/engine technical representative
	Airframe manufacturer
	Airworthiness directives
	Service bulletins
	Completion centers
	Consultants
	Flight crews
	Maintenance personnel
	NBAA MOBs
	NBAA technical committee (aircraft type)
	Repair and support facilities
	Training facilities
	Users/passengers
	Vendors
	Historic data
Destination and Route Issues	Airport and airway capabilities
	Distances and geography
	Frequencies of travel needs
Key Passenger Profile	Frequency of travel
	Number of pax
	Strategic value
	Tactical value
	Time cost
	Trends and changes in travel patterns
	Baggage
Alternative Resources Review	Airlines' abilities to meet travel needs
	Charter services available to meet needs

Source: Abstracted from NBAA and Embry-Riddle Aeronautical University *Seminar guide for aircraft selection.* (n.d.). Courtesy of Center for Professional Development, Embry-Riddle.

Aircraft Selection Matrix

Except when an aircraft is specifically favored or chosen for personal or other reasons, the aircraft selection matrix is recommended. The matrix is simply a tabulation of chosen categories to weighing factors for specific aircraft (see Figure 5-14). After a team or individual

SAMPLE

Aircraft Operating Costs Form

Date _____

VARIABLE EXPENSE
  Fuel, oil, additives, etc.
  Travel, flight crew
  Maintenance total
    Labor, outside general
    Parts maintenance, replacement, etc.
    Overhaul, engine major
    Overhaul, airframe major
  Miscellaneous (catering, landing fees, enroute storage, customs, foreign permits, etc.)
  Total Variable

FIXED EXPENSE, NORMAL
  Salaries total
    Flight crew (No. ___)
    Maintenance (No. ___)
    Administrative/clerical (No. ___)
  Employee benefits
  Base hangar rent or cost
  Hangar and shop equipment cost and supplies
  Contract training
  Miscellaneous (office supplies, publications, manuals, memberships, telephone, etc.)
  Total Fixed, Normal
  Total Variable plus Fixed Normal

FIXED EXPENSE, ADDITIONAL
  Insurance total
    Liability
    Hull (Rate ___%)
  Major nonrecurring
  Depreciation (Based upon ___ years with ___% residual)
  Total Fixed, Additional
  Total Operation Expense

OPERATIONAL DATA
For Period _____ through _____
Hours Flown _____ Miles Flown _____   Total Sorties _____

	COST PER HOUR		COST PER MILE	
	Variable	_____	Variable	_____
	Fixed normal	_____	Fixed normal	_____
	Subtotal	_____	**Subtotal**	_____
	Fixed additional	_____	Fixed additional	_____
	TOTAL	_____	**TOTAL**	_____

Figure 5-13. Aircraft Operating Costs Form. Source: NBAA.

Item	Weight Factor (1–10) (1–10)	Aircraft A I.S F.V.	Aircraft B I.S. F.V.	Aircraft C I.S. F.V.
Safety	10	100 1,000	100 1,000	100 1,000
Costs	10	100 1,000	100 1,000	100 1,000
Aircraft Systems	9	100 900	100 1,000	100 1,000
Capacity (Pax/Cargo)	10	90 900	95 950	90 900
Range	9.5	92 874	90 855	96 912
Maintenance	9.5	100 950	90 855	90 855
Fuel Use	8	95 760	90 720	98 784
Environmental Concerns	9	100 900	95 855	90 810
Runway Requirements	9.5	100 950	100 950	98 931
Interior Design	8	95 760	98 784	90 720
Totals		972 9,094	958 8,969	942 8,812

Note: I.S. = Item score; F.V. = Factor value

Figure 5-14. Aircraft Selection Matrix.

short lists specific aircraft of a desired type, a comparison of the aircraft is made by weighing each factor to each aircraft and totaling the score. For example, if Aircraft A, B, and C relate to a Gulfstream, Falcon, and Canadair jet, respectively, and the first category was safety with a weighing factor of ten, each aircraft would be evaluated for its safety features. In the example, the aircraft were considered very safe and assigned a score of 100. This score times the weight factor of ten gives a total factor value of 1,000. While 100 was used in the example, any number range could be used. As for the categories, factorizing many items into specific categories that will result in a comparison is highly recommended; for example, performance factors could be used.

The key to using the aircraft selection matrix is *subjective assessment.* A team of evaluators should be given tasks, as several people working together usually produce better decisions. The knowledge each person brings to a team may be different. One member may know the type of business conducted during the flight and at the various destinations; another member may have knowledge of the aircraft specifications and performance. How many people and which ones depends on the team manager. Groups no larger than five people should work together; beyond this number results in poor intercommunications. For large companies that have a major undertaking in selecting aircraft, more than one team may be required. Overall, the team manager is to recommend to the CEO and/or Board of Directors a specific type, model, and number of aircraft. The NBAA Education and Safety Foundation conducts a seminar involving corporate aircraft selection. During the seminar, the team concept is expanded even further with the recommendation of a five-part effort. These parts include needs analysis, aircraft selection, design, completion, and operational start up.

Prior to developing the matrix, the team should study the aircraft available and narrow its choices down to two or three aircraft makes and models. An aviation flight department manager and chief pilot will have a major role in the discussion and choice. Manufacturers' material, along with any other pertinent information about the aircraft in question, needs to be thoroughly reviewed. In some instances, test flights may be obtained from the aircraft sellers, not only for the team, but for transporting company passengers on company business—a try-before-you-buy approach. In the end, the team should be satisfied with the aircraft selected and able to answer any question from company managers.

What if two or more aircraft have been scored fairly equally? That is when a priority of factors becomes important. One aircraft may score higher in capacity, but lower in range performance. The team has to review all close scores before recommending the best aircraft for the company. Best does not always mean the cheapest; best means the aircraft that is most suitable for the company based on priorities and capabilities. For example, in Figure 5-14, is Aircraft A better than B just because it scored 125 points higher? What would a maintenance manager say? A pilot?

Various aircraft, their interiors, performance specifications, and other details are displayed in the figures at the end of this chapter in order to provide a brief look at selected aircraft. Contact any of the many aircraft manufacturers, dealers, brokers, management firms, or other sales representatives of new and/or used aircraft for more types, models, and details. Before reviewing these aircraft, however, how to obtain aircraft is presented.

Obtaining Aircraft

There are various ways to obtain an aircraft. You can buy a new or used aircraft, lease, charter, time share, joint own, interchange, and/or fractionally own. Each of these methods will be briefly covered.

Buying

Purchasing a new or used aircraft outright may be a favorable option for firms with ready cash. When you buy an aircraft, you have an asset and direct control; however, you will also have the full responsibility of ownership. Warranties may be considered a reason for buying new aircraft, but limited warranties may still exist with used aircraft. Attractive financial terms may be offered to a buyer, and depreciation and tax considerations may be desirable. Additionally, for a manager who buys an aircraft, FAR Part 135 operations may generate revenue for the firm when company personnel are not using the aircraft. With an asset, the aircraft has a resale value that could provide revenue at a later date, and pride of ownership may be a factor. Finally, some managers want to buy a used aircraft to start their corporate flight operations, and then decide to buy new or obtain aircraft using another method after gaining experience with operations.

Leasing

Leasing an aircraft may be desirable when a manager is not sure about keeping the aircraft or wants to use it before making a final decision on buying. In other words, try before you buy. There are many potential financial and tax benefits to leasing, and lease assets may not indicate to stockholders the use of a "high-priced" asset. Some of these stockholders may not consider that the value of the intangibles outweigh the high costs of a private aircraft operation. (This is the job of the aviation department manager—to develop the value of the company aircraft.) Even if it can be shown that business aircraft are cost effective, some people will never accept logic when it comes to private aircraft operations.

New aircraft operators may want to trade up at a later date as managers develop their firms' businesses. Not only are aircraft leased, but facilities to store and maintain the aircraft, as well as administrative office space, may be leased. Other concerns may be market volatility, managerial changes, and changing company missions.

Dry leases refer to obtaining an aircraft only, while wet leases relate to having an aircraft with other resources and support; for example, maintenance, fuel, and aircrew. Accountants classify leases as operating leases and capital leases. The former involves time periods (e.g., year-to-year) and the aircraft is not a company asset. There is no charge against the company's working capital, but the disadvantages mean that there is no ownership and finance rates are usually higher.

Capital leases may appear as a sale over time when ownership may transfer at the end of the lease period. The interest expense may be deductible for tax purposes, but the lease obligation reduction is not. In other words, a lease aircraft may be obtained for 75 percent of the aircraft value, and at the end of the lease, an option to buy is made. Pseudo leases are not authorized; they are those leases where an item is valued at extremely low value for outright purchase at the end of a lease period. No one expects to buy an aircraft for $100, so the government tax officials watch these transactions quite closely.

A manager who leases an aircraft may be experimenting with its use and suitability for his or her company. The terms of the lease contract have to be followed or else penalty payments may occur. Such terms may involve refurbishing actions, deductions for wear and tear, or other items resulting in monetary payments.

Chartering

Chartering offers a manager flexibility of aircraft use. Individual trips may be chartered or aircraft used over a specific limited time based on a company's travel needs. The disadvantage is the relatively high cost of chartering. In the short term, chartering an aircraft for peak travel requirements may be cost effective and provide ready transportation for some company managers, especially for trips to locations where the airlines do not service or service infrequently. Disadvantages include loss of depreciation, lack of management control, and costs.

Whether new or used aircraft are concerned, weighing the advantages and disadvantages of buying, leasing, and chartering should be accomplished by a firm's management. Contacting an air charter broker may help one to identify services available. Time sharing, interchange, joint ownership, and fractional ownership can also be reviewed as possible methods of using a business aircraft.

Time Sharing, Joint Ownership, and Interchange

FAR Part 91.501(c) defines *time sharing* as "an arrangement whereby a person leases his airplane with flight crew to another person and no charge is made for the flights conducted under that arrangement other than those specified in paragraph (d) of this section." *Interchange* " means an arrangement whereby a person leases his airplane to another person in exchange for equal time, when needed, on the other person's airplane, and no charge, assessment, or fee is made, except that a charge may be made not to exceed the difference between the cost of owning, operating, and maintaining the two airplanes." *Joint ownership* "means an arrangement whereby one of the registered joint owners of an airplane employs and furnishes the flight crew for that airplane and each of the registered owners pays a share of the charge specified in the agreement" (Federal Aviation Administration, 1997). Under FAR Part 91.501(d), the following may be charged as expenses to a specific flight as authorized under time sharing:

1. Fuel, oil, lubricants, and other additives;
2. Travel expenses of the crew, including food, lodging, and ground transportation;
3. Hanger and tiedown away from the aircraft's base of operation;
4. Insurance obtained for the specific flight;
5. Landing fees, airport taxes, and similar assessments;
6. Customs, foreign permit, and similar fees directly related to the flight;
7. In-flight food and beverages;
8. Passenger ground transportation;
9. Flight planning and weather contract services; and
10. An additional charge equal to 100 percent of the expenses listed in (1) above.

Various benefits arise from these methods, namely flexibility of using an aircraft, economic use, tax advantages, and prestige arising from use. Using an aircraft when needed and sharing costs results in lower payments. A spin-off from the shared ownership concept has been termed fractional ownership.

ional Ownership

About 1986, fractional ownership became prevalent when the business jet was recognized as a business tool. Sharing costs was desired by corporate managers because of the high costs of ownership. Two major fractional ownership programs, NetJets managed by Executive Jet Aviation, and FlexJet managed by Business Jet Solutions, an executive aviation affiliate of American Airlines, are the biggest names in this service program.

Before identifying their program requirements and advantages, an overview of fractional ownership is provided. Contracts vary among these type programs, but, in general, individuals or corporate managers purchase an undivided interest in a specific aircraft from the company owning the aircraft. This purchase enables a manager to be registered as an owner. Generally, at least a one-eighth share of the aircraft is obtained, and in addition to the purchase price, monthly management fees and hourly rates are paid for the time specified. In return for the payments, the buyer is entitled to a specific number of flight hours per year. Normally 800 hours is considered full use; therefore, a buyer of a one-eighth share obtains 100 flight hours.

Importantly, the owner is guaranteed aircraft use within four to eight hours of demand, based on distance. Aircraft operations are conducted under FAR Part 91, while the legal basis for this program is the interchange agreement. Shared ownership of corporate aircraft has been increasing, especially for managers who do not need a high-priced jet throughout the year. Companies like Executive Jet Aviation, Executive Jet International, Business Jet Solutions, JetCo, and Jet Shares provide specialized services for corporate managers. Following are data provided by Business Jet Solutions and Executive Jet Aviation.

Business Jet Solutions

The FlexJet program allows a manager to purchase a share of one or more of various aircraft tailored to the needs of the company. All aspects of the aircraft operations are included in the payment made, which covers maintenance, hangar security, aircrew, insurance, catering, flight planning, and more. The aircraft is available 24 hours-a-day, 365 days-a-year on as little as four hours notice. Major aircraft available include the Learjet 31A, Learjet 60, and Challenger 604. Figures 5-15 through 5-17 are extracts from Business Jet Solutions' literature that provide specific details. The company is at 8001 Lemmon Avenue, Dallas, Texas, and communication numbers are 1-800-590-JETS or fax 214-956-1709.

Executive Jet Aviation

The NetJets program was founded in 1987 by Richard T. Santulli, CEO and chairman of Executive Jet Aviation. The program allows customers to purchase an undivided interest in aircraft like the Citation S/II, Citation V Ultra, Citation VII, Citation X, or Gulfstream IV-SP. The operating costs are defined over a fixed five-year period, and like FlexJet, a monthly management fee covers all other costs. However, unlike FlexJet, only actual flight hours used by the owner are paid, not the deadhead or other unproductive flights. Ownership begins at one-eighth shares (using 800 hours as the annual basis). Managers who need aircraft less than 400 hours a year, especially managers who only average 50–200 flight hours annually need to review the program options. Important to the buyer is that Executive Jet will repurchase the manager's share based on fair market value. Figures 5-18 through 5-22 are extracts from Executive Jet Aviation's brochures. The company's NetJets program is out of Columbus, Ohio (800-821-2299).

Aircraft Examples

While an extensive number of aircraft are available for review and analysis of their specifications, the following aircraft have been chosen to represent only a few makes and models for comparison, especially to provide information to those who have not previously studied this subject or experienced traveling in the aircraft. By reviewing the following figures, details of the aircraft performance factors, capacity, and other information will be useful in realizing the extent of service availability. Contact with the aircraft manufacturers or other aircraft servicing companies will provide readers with more current and additional information.

Summary

The Gulfstream I was developed for corporate and business aviation in 1959, and since then numerous aircraft makes and models have been made available. General aviation is shown to be an important and extensive part of the aviation industry and involves the use of over 35,000 general aviation aircraft for business. Specific details of the number and types of aircraft, airports, and aircrew members support the overall finding that this industry is a multi-billion dollar activity. The NBAA fleet of aircraft indicates specific types and numbers operated by NBAA members.

Manufacturers include, among others, Gulfstream, Dassault, Canadair, Cessna, Piper, Raytheon, Sikorsky, Bell, Boeing, Bombardier, Sino-Swearingen, and Learjet. To select one of the makes and models, a manager can use project management's team approach. The aircraft selection matrix enables a quick comparison of listed aircraft for company use, based on strategic and tactical needs and analysis. Costs of acquisition and possession must be considered prior to any purchase, and while tangible items can be measured, oftentimes the intangible values help managers to decide on an appropriate aircraft.

After deciding business aircraft are needed, various methods of obtaining an aircraft need to be considered; for example, aircraft management firms, buying, leasing, chartering, time sharing, and others. Ownership programs of Executive Jet Aviation and Business Jet Solutions provide a new approach to aircraft ownership. Various aircraft specifications and performance indicators help to provide readers knowledge of the capabilities and resources available for the business flyer.

FlexJet Benefits

- *Flex*Jet IS A FRACTIONAL OWNERSHIP PROGRAM PROVIDING AFFORDABLE ACCESS TO BUSINESS JETS FOR THOSE REQUIRING LOW TO MODERATE USE (SEE THE FOLLOWING "COMPARATIVE 5 YR. COST OF OWNERSHIP" EXAMPLE FOR LIGHT JETS).

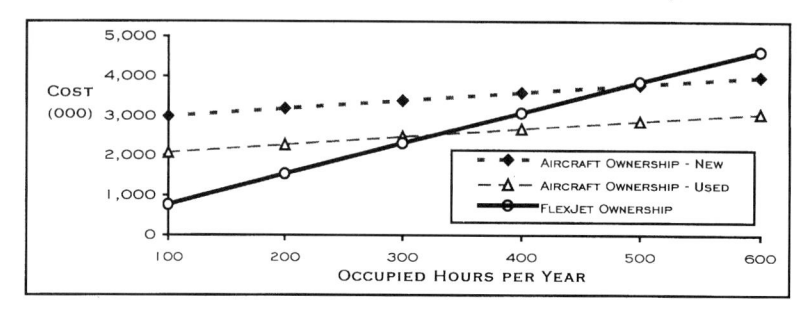

- *Flex*Jet IS A TROUBLE-FREE PROGRAM FOR NON AIRCRAFT OPERATORS:
  - CREW, MAINTENANCE, HANGARS, ETC., ARE PROFESSIONALLY MANAGED BY JET SOLUTIONS L.L.C.

- *Flex*Jet PROVIDES EXISTING BUSINESS JET OPERATORS THE FLEXIBILITY OF SUPPLEMENTING THEIR FLEET:
  - DURING OPERATORS' AIRCRAFT DOWNTIME.
  - WHEN SIMULTANEOUS USE OF AIRCRAFT IS NEEDED.
  - ALLOWING REDUCTION OR ELIMINATION OF DEAD-HEADINGS.

- *Flex*Jet IS OFFERED BY POWERFUL AND FINANCIALLY SECURE PARTNERS.

- *Flex*Jet IS FLEXIBLE:
  - AIRCRAFT AVAILABILITY YEAR ROUND, 24 HOURS PER DAY.
  - ACCESS TO MULTIPLE AIRCRAFT SIMULTANEOUSLY.
  - GUARANTEED RESPONSE WITH 4-10 HOUR NOTICE.
  - EXIT PROGRAM AT ANY TIME.

- HIGHEST LEVELS OF SAFETY STANDARDS (GPWS AND TCAS II ON ALL OWNERS AIRCRAFT).

- CUSTOMERS DEAL DIRECTLY WITH ALL THE RESOURCES OF THE AIRCRAFT MANUFACTURERS. ALL AIRCRAFT ARE SERVICED AND MAINTAINED TO THE FAA AND MANUFACTURER STANDARDS.

- ADVANTAGES OF USING BUSINESS AIRCRAFT:
  - PERSONNEL TIME SAVINGS (UP TO 5 HOURS SAVED PER ROUND TRIP).
  - ACCESS TO OVER 5,000 AIRPORTS IN THE US VS. 500 FOR COMMERCIAL AIRLINES.
  - DIRECT FLIGHTS ANYWHERE ACCORDING TO YOUR SCHEDULE.
  - REDUCTION IN OVERNIGHT STAYS.
  - ENHANCEMENT OF MENTAL ALERTNESS.
  - FULL PRIVACY, ABILITY TO CONDUCT MEETINGS EN ROUTE.
  - PERSONNEL SECURITY.

JANUARY 96

Figure 5-15. FlexJet Benefits. Courtesy of FlexJet.

FlexJet Options

FlexJet IS A FRACTIONAL OWNERSHIP PROGRAM, GIVING YOU THE FLEXIBILITY OF SELECTING FROM A VARIETY OF AIRCRAFT TYPES AND SHARE SIZES IN ORDER TO TAILOR SERVICES AND COSTS TO YOUR EXACT TRAVEL REQUIREMENTS.

AIRCRAFT TYPE OPTIONS:

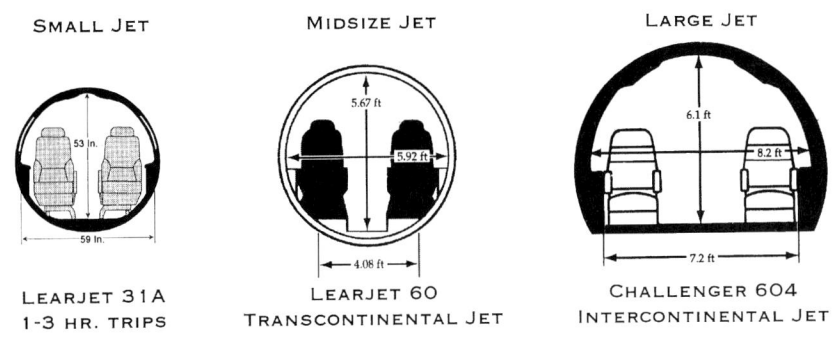

SMALL JET
LEARJET 31A
1-3 HR. TRIPS

MIDSIZE JET
LEARJET 60
TRANSCONTINENTAL JET

LARGE JET
CHALLENGER 604
INTERCONTINENTAL JET

OWNERSHIP OPTIONS:

YOU CAN PURCHASE A FRACTION OF A BUSINESS JET THAT CORRESPONDS TO FLIGHT TIME PER YEAR:

- A MINIMUM OF 100 HOURS/YEAR PER AIRCRAFT IS NEEDED TO ENTER THE FlexJet PROGRAM. 100 HOURS/YEAR CORRESPONDS TO A SHARE OF 1/8 (12.5%). THEN, MULTIPLES OF 50 HOURS ARE AVAILABLE.
- FlexJet ALLOWS YOU TO SELECT VARIOUS SHARE SIZES (HOUR COMBINATIONS) WITH DIFFERENT AIRCRAFT TYPE WITH A MINIMUM OF 1/8 OWNERSHIP PER AIRCRAFT TYPE. EXAMPLE:
  - IF 350 HOURS ARE NEEDED, IT IS POSSIBLE TO BUY 3/16 (150 HR.) ON A LEARJET 31A, 1/8 (100 HR.) ON A LEARJET 60 AND 1/8 (100 HR.) ON A CHALLENGER.
  - IF 250 HOURS ARE NEEDED, IT IS POSSIBLE TO BUY 3/16 (150 HR.) ON A LEARJET 31A AND 1/8 (100 HR.) ON A LEARJET 60.
- FlexJet ALSO OFFERS THE ABILITY TO DOWNGRADE OR UPGRADE AIRCRAFT TYPE. EXAMPLE: IF YOU HAVE OWNERSHIP OF A LEARJET 60, IT IS POSSIBLE TO TRADE UP TO THE CHALLENGER OR DOWN TO THE LEARJET 31A.
- NOTE: HOURS ALWAYS CORRESPOND TO OCCUPIED HOURS ON THE AIRCRAFT FROM SIX MINUTES BEFORE TAKEOFF TO SIX MINUTES AFTER LANDING. A FLIGHT HOUR IS APPROXIMATELY 375 S.MILES (EXAMPLE: ONE FLIGHT HOUR = NEW YORK CITY TO PITTSBURGH OR SAN FRANCISCO TO LOS ANGELES).

	LEARJET 31A	LEARJET 60	CHALLENGER 604
PURCHASE PRICE-1/8	$700,000	$1.3 M	$2.7 M
BUY-BACK	FAIR MARKET VALUE	FAIR MARKET VALUE	FAIR MARKET VALUE
NUMBER OF HR./YR.	100	100	100
FIXED MONTHLY FEE*	$5,640	$7,690	$12,980
FIXED CHARGE/HR.*	$1,170	$1,475	$2,010

*OPERATION COSTS SUBJECT TO TRANSPORTATION TAX OF 10%.

JANUARY 96

Figure 5-16. FlexJet Options. Courtesy of FlexJet.

5: Business Aviation Aircraft **119**

FlexJet
Program Details

1. LEARJET 31A

	1/8	**3/16**	**1/4**	**5/16**	**3/8**	**7/16**	**1/2**
	12.50%	18.75%	25.00%	31.25%	37.50%	43.75%	50.00%
Purchase Price	$700,000	$1,050,000	$1,400,000	$1,750,000	$2,100,000	$2,450,000	$2,800,000
Allocated Hours/year	100	150	200	250	300	350	400
Excess Hours	25	37.5	50	62.5	75	87.5	100
Available Hours	125	187.5	250	312.5	375	437.5	500
Mgmt. Fee per month*	$5,640	$8,460	$11,280	$14,100	$16,920	$19,740	$22,560
Variable Rate per hr.*	$1,170	$1,170	$1,170	$1,170	$1,170	$1,170	$1,170
Response Time	8 hrs	8 hrs	4 hrs	4 hrs	4 hrs	4 hrs	4 hrs
Differential Learjet 60 Challenger	1.6 2.7	1.6 2.7	1.6 2.7	1.6 2.7	1.6 2.7	1.6 2.7	1.6 2.7
Maximum Cancel.Fee	$2,340	$2,340	$2,340	$2,340	$2,340	$2,340	$2,340
Simult. Aircraft	None	None	2	2	2	2	4
Catering	Normal	Normal	Normal	Normal	Normal	Normal	Normal

* Operation costs subject to transportation tax of 10%.

Notes: -Pricing and service levels are subject to revision.
-Allocated hours per year correspond to occupied hours only (no charge for aircraft positioning)

January 96

Figure 5-17. FlexJet Program Details. Courtesy of FlexJet.

NETJETS® INVESTMENT SUMMARY

	CITATION S/II QUARTER SHARE	CITATION V *ULTRA* QUARTER SHARE	HAWKER 1000 QUARTER SHARE	GULFSTREAM IV-SP QUARTER SHARE
ACQUISITION COST*	$660,000	$1,510,000	$3,216,000	$6,650,000
OPERATING COSTS:				
· Indirect Costs (Pilots, Insurance, Training, Hangaring & Administrative)	$12,148/Month**	$12,148/Month**	$18,950/Month**	$29,000/Month**
· Direct Costs (Fuel, Maintenance, Engine Reserves, Catering, etc.)	$1,149/ Occupied Flight Hour**	$1,149/ Occupied Flight Hour**	$1,570/ Occupied Flight Hour**	$2,480/ Occupied Flight Hour**
USAGE FLEXIBILITY:				
· Average Annual Utilization***	200 Occupied Flight Hrs.	200 Occupied Flight Hrs.	200 Occupied Flight Hrs.	200 Occupied Flight Hrs.
· Five Year Utilization	1,000 Occupied Flight Hrs.	1,000 Occupied Flight Hrs.	1,000 Occupied Flight Hrs.	1,000 Occupied Flight Hrs.
· Guaranteed Response Time	4 Hours	4 Hours	6 Hours	6 Hours (Domestic)
· Simultaneous Aircraft Availability Per Day	2 Citation S/II	2 Citation V *Ultra*	1 Hawker 1000	1 GIV-SP

ADDITIONAL INFORMATION:

*Owner purchases an undivided interest in a specific Citation S/II, Citation V *Ultra* or Gulfstream IV-SP Aircraft with a guaranteed buy back option from Executive Jet. Or Owner purchases undivided interest in specific Hawker 1000 aircraft with a buy back guarantee of $2,520,000 from the Raytheon Company after five (5) years. Acquisition costs for Citation S/II Aircraft are subject to change based on current market conditions and subject to verification at contract closing.

**The Indirect and Direct Costs will be adjusted annually on January 1 of each year commencing January 1, 1997, based on the CPI-U for the previous fiscal year ending November 30th.

***Average annual utilization for an quarter (1/4) share is based upon 1000 Occupied Flight Hours per five (5) year period. However, for Owner's flexibility, a maximum of 250 Occupied Flight Hours can be flown in any one (1) year. An excess charge will apply for Occupied Flight Hours above 250 hours per year or 1000 hours over the five year term.

All prices are subject to change and deliveries of aircraft are subject to availability. Above pricing in effect as of December 20, 1995, for 1996.

ACTUAL TERMS AND CONDITIONS ARE SUBJECT TO DEFINITIVE AGREEMENTS.

c\kr\nvsum14.895
Rev. 12/20/95

Figure 5-18. NetJets Investment Summary. Executive Jet extracts. Used with permission by Kevin R. Russell.

Executive Jet
Financing and Leasing Program
Provided by AT&T Capital Corporation
As of 2/15/96

	Cessna Citation V Ultra	Hawker 1000	Gulfstream IV-SP
LOAN PROGRAM			
Loan Rates (1)	8.00%	8.00%	8.00%
Monthly Loan Payments for:			
One-Eighth Shares	$ N/A	$12,395	$ 26,978
One-Quarter Shares	11,639	24,789	51,259
Three-Eighth Shares	17,459	37,185	76,889
One-Half Shares	23,278	49,578	102,518
Down Payment	15%	15%	15%
Balloon Payment	70%	70%	70%
LEASE PROGRAM			
Lease Factor:			
Shares over One-Eighth	1.165%	1.165%	1.165%
One-Eighth Shares	N/A	1.165%	1.212%
Monthly Lease Payment for:			
One-Eighth Shares	$ N/A	$18,733	$ 42,420
One-Quarter Shares	17,592	37,466	77,472
Three-Eighth Shares	26,387	56,199	116,209
One-Half Shares	35,184	74,932	154,944
Deposit			$625,000 per one-eighth (2)

Notes:

(1) Loan Rate is based on 90 day LIBOR plus 2.75%; the monthly loan payment amounts listed above are for illustrative purposes - the actual monthly loan payment amount will be adjusted periodically to reflect changes in LIBOR.

(2) May be financed through lease payment.

Loan and Lease Payments listed above assume a five year term. Lease is cancelable after 36 months without penalty.

Loan and Lease Rates are as of the Date Listed and are Subject to Change.

Lease Factors based on Four-year U.S. Treasury Rates of 5.07%. Actual lease factor is fixed at Closing.

AT&T3.RTE

Figure 5-19. Executive Jet Financing and Leasing Program. Executive Jet extracts. Used with permission by Kevin R. Russell.

The Citation S/II

CESSNA CITATION S/II

- 430 mph cruise speed
- 37,000 ft. typical cruise altitude (43,000' max.)
- 1,500 miles non-stop range*
- 850 lbs. external baggage capacity
- 7 individual, fully adjustable passenger seats
- Private rear lavatory
- Full service refreshment center
- Outstanding short runway capability

* Range may vary due to weather conditions, altitude and aircraft issues.

EXECUTIVE JET
800-848-6436

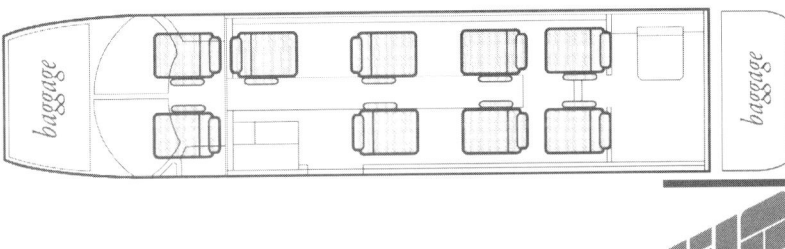

Figure 5-20. The Citation S/II. Executive Jet extracts. Used with permission by Kevin R. Russell.

Figure 5-21. The Citation V Ultra. Executive Jet extracts. Used with permission by Kevin R. Russell.

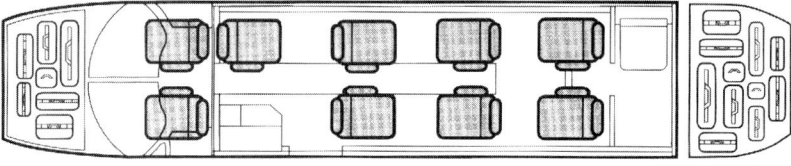

124 *Chapter 5*

The Hawker 1000

Hawker 1000

- 515 mph cruise speed
- 41,000 ft. typical cruise altitude (43,000' max.)
- 3,200 miles non-stop range*
- 850 lbs. baggage capability
- 6 individual passenger seats plus 3 place divan
- Large private rear lavatory
- Full service refreshment center
- Elegant stand-up cabin
- Designed for in-flight PC compatibility
- Advanced engine and avionics systems
- Full intercontinental capabilities

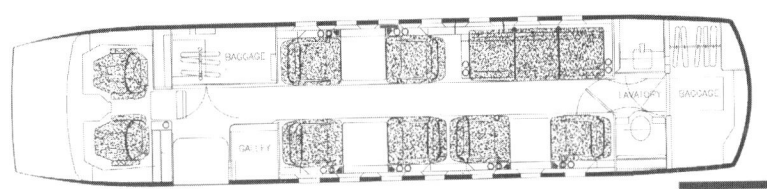

* NBAA IFR 3,200 nmi. Range may vary due to weather conditions, altitude and aircraft load.

Figure 5-22. The Hawker 1000. Executive Jet extracts. Used with permission by Kevin R. Russell.

5: Business Aviation Aircraft 125

Figure 5-23. Super King Air B200. Photo courtesy of Beech Aircraft Corporation.

Figure 5-24. Beechjet 400A. Courtesy of Beech Aircraft Corporation.

Caravan Specifications

DESCRIPTION	CARAVAN	GRAND CARAVAN
INTERIOR DIMENSIONS		
Cabin (Aft of Pilot Area)		
Length FT / M	12.7 / 3.8	16.7 / 5.1
Height FT / M	4.3 / 1.3	4.3 / 1.3
Width FT / M	5.2 / 1.6	5.2 / 1.6
Volume CU FT / CU M	254 / 7.2	340 / 9.6
Max. Seating FAR 23 / Other	10 / 14	11 / 14
EXTERIOR DIMENSIONS		
Overall		
Length FT / M	37.6 / 11.5	41.6 / 12.7
Span FT / M	52.1 / 15.9	52.1 / 15.9
Height FT / M	14.8 / 4.5	14.8 / 4.5
Cargo Pod (External)	OPTIONAL	OPTIONAL
Weight Limit LBS / KG	820 / 372	1090 / 494.4
Volume CU FT / CU M	83.7 / 2.4	111.5 / 3.2
ENGINE	PT6A-114	PT6A-114A
SHP	600	675
PERFORMANCE		
Cruise Speed (10,000 ft) KTS / KM / HR	184 / 341	184 / 341
Range (10,000 ft) NM / KM	1085 / 2009	1026 / 1900
Includes takeoff, climb, cruise, descent and 45-min reserve at max. range power		
S.L. Rate of Climb FPM / MPM	1050 / 320	975 / 297
Service Ceiling (Max. Gross)		
FT / M	25,500 / 7772	23,700 / 7224
Takeoff S.L. ISA		
Ground Roll FT / M	1205 / 367	1365 / 416
50 Ft. Obs. FT / M	2210 / 674	2420 / 738
Stall Speed (ldg) KTS / KM / Hr	61 / 113	61 / 113
Landing S.L. ISA		
Ground Roll FT / M	745 / 227	950 / 290
50 Ft. Obs. FT / M	1655 / 504	1795 / 547
Maximum Weight LB / KG		
Ramp	8035 / 3645	8785 / 3985
Takeoff	8000 / 3629	8750 / 3969
Landing	7800 / 3538	8500 / 3856
Standard Empty Weight LBS / KG	3925 / 1780	4237 / 1922
Maximum Useful Load LBS / KG	4110 / 1864	4548 / 2063

Note: Caravan Floatplane interior dimensions are the same as Caravan shown above. Super Cargomaster dimensions are the same as Grand Caravan shown above. Other specifications will vary slightly for both models not shown, and specific information is available upon request.

CESSNA CARAVAN

Count On It!

Cessna
A Textron Company

Cessna Aircraft Company, Caravan Marketing, P.O. Box 7704, Wichita, Kansas 67277, Telephone 316-941-6081, Fax 316-941-7850

SPA95019-15

Figure 5-25. Cessna Caravan. Courtesy of Cessna Aircraft Company.

The Ultimate in Affordable Mission Flexibility

The world's most versatile aircraft... period.

For passenger and cargo transport.

For air cargo transport.

For commuter transport.

For special missions.

For floatplane operation.

Passenger and Cargo Interior.

10-Seat Commuter Interior.

14-Seat Commuter Interior.

Executive Interior.

Cargo Interior.

Four Models To Choose From

The Caravan family consists of four distinct models providing complete versatility based on customer needs.

- The Caravan—Multi use aircraft ideal for passenger, cargo, special missions or a combination of both.
- The Grand Caravan—The same multi use applications combined with top of the line performance with increased horsepower, payload and cabin volume. Also ideal for commuter transport.
- Super Cargomaster—The largest cargo-dedicated single engine turboprop manufactured today with 452 cubic feet of cargo space.
- Caravan Floatplane—The largest single engine floatplane manufactured today, available on seaplane or amphibious floats for transportation anywhere.

Figure 5-26. Caravan Models. Courtesy of Cessna Aircraft Company.

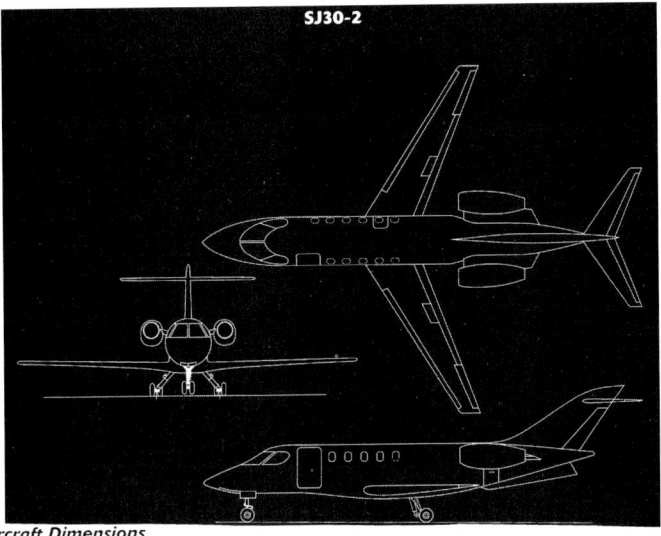

Aircraft Dimensions

SJ30-2 General Specifications
Aircraft Dimensions

External:
Length	46.97 ft
Height	14.29 ft
Wing Span	36.33 ft
Wing Area	165 ft²
Sweep	30°
Aspect Ratio	8

Design Data:
Equipped Weight Empty	6,800 lbs
Fuel Capacity (gals/lbs)	716/4,800
Max. Take-Off Wt.	12,300 lbs
Certificated Altitude	49,000 ft

Engines:
Williams-Rolls (2)	FJ44-2C
Take-Off Thrust Rating, each	2,300 lbs

Performance:
Max. Operating Mach No.	0.83
High Speed Cruise (Mach/kts/mph)	0.83/476/548
Long Range Cruise (Mach/kts/mph)	0.78/447/515
VMO	320 kcas
VFR Range (45 min. reserve fuel)	3,000 nm
NBAA IFR Range (1 pilot, 3 pax)	2,500 nm
FAA Take-Off Balanced Field Length	3,800 ft
FAA Landing Distance	2,960 ft
Stall Speed	91 kcas
Two-Engine Rate-Of-Climb	4,100 fpm
Pressurization	12.0 psi
Sea Level Cabin Altitude Maintained At	41,000 ft

Internal:
Length (between pressure bulkheads)	17.26 ft
Cabin Length	12.0 ft
Cabin Height	4.3 ft
Cabin Width	4.7 ft
Cabin Volume	316.22 ft³
Cabin Door Width	2.67 ft

Weights:
Max. Ramp Wt.	12,400 lbs
Max. Take-Off Wt.	12,300 lbs
Max. Landing Wt.	11,690 lbs
Max. Zero Fuel Wt.	8,400 lbs
Equipped Weight Empty	6,800 lbs
Operating Weight Empty	7,000 lbs
Max. Fuel	4,800 lbs
Max. Payload	1,400 lbs

Design Speeds:
VMO, Up To 29,000 ft.	320 kcas
MMO, Above 29,000 ft.	0.83 Mach
Speed Brake Operation	No Limit
VFE (approach flaps)	200 kcas
VFE (landing flaps)	160 kcas
Landing Gear Operation	225 kcas
Stall Speed @ Max. Landing Wt.	91 kcas

Figure 5-27. SJ30-2. Courtesy of Sino Swearingen.

Figure 5-28. Falcon 900B. Courtesy of Dassault Falcon Jet.

Figure 5-29. Falcon 900B Interior. Courtesy of Dassault Falcon Jet.

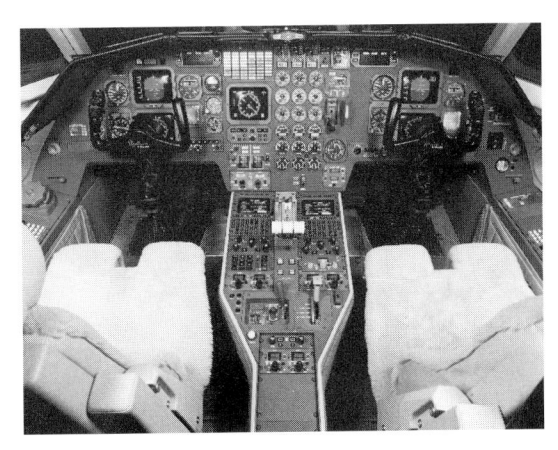

Figure 5-30. Falcon 900B Cockpit. Courtesy of Dassault Falcon Jet and Bud Shannon Photo, Inc.

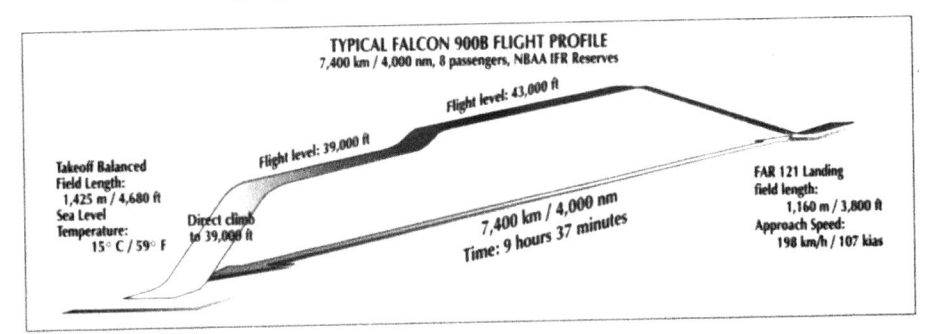

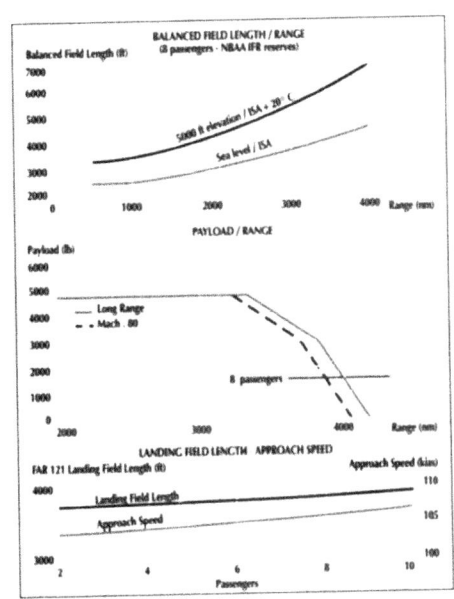

Figure 5-31. Falcon 900B Specifications.

5: Business Aviation Aircraft

Figure 5-32. Falcon 900EX. Courtesy of Dassault Falcon Jet.

BAGGAGE

The baggage compartment is pressurized and accessible in flight. With a full 3.60 cubic meters (127 cubic feet) of space, it is sized to handle baggage requirements for long flights with maximum passenger load.

LONG-RANGE COMFORT

For a business aircraft that routinely flies non-stop for ten hours or more, comfort takes on a whole new meaning.

The Falcon 900EX's standard layout offers a cabin divided into three independent sections, allowing passengers to work or rest at their own pace. Excellent natural lighting is provided by a large number of evenly distributed windows.

With a lavatory at each end of the cabin, the rear section, fitted with a convertible sofa, can be arranged as a private lounge. But this section also offers all the high-tech conveniences of a modern business office. Located well ahead of the engines, and further insulated by the lavatory and baggage bay, this lounge offers a particularly low level of noise, equivalent to the general passenger cabin.

The balanced proportions of the Falcon 900EX cabin, free of any avionics equipment, enables many different interior arrangements.

All galley equipment is sized to meet the demands of long-duration flights.

CABIN DIMENSIONS

Length .... 11.90 m/468 in
Width:
- maximum .... 2.34 m/92 in
- floor .... 1.91 m/75 in
Height .... 1.87 m/74 in

VOLUME

Flight deck .... 3.75 m³/132 cu. ft
Cabin .... 35.9 m³/1,267 cu. ft
Baggage compartment 3.6 m³/127 cu. ft

Figure 5-33. Falcon 900EX Interior. Courtesy of Dassault Falcon Jet.

5: Business Aviation Aircraft **133**

Figure 5-34. Falcon 900EX Cockpit. Courtesy of Dassault Falcon Jet.

TYPICAL FALCON 900EX FLIGHT PROFILE
8,335 km / 4,500 nm, 8 passengers, NBAA IFR Reserves

Takeoff Balanced Field Length:
1,535 m / 5,035 ft
Sea Level Temperature:
15°C / 59°F

Direct climb to 39,000 ft
39,000 ft
43,000 ft
47,000 ft

8,335 km / 4,500 nm
Time: 10 hours 46 minutes

FAR 121 landing field length:
1,210 m / 3,975 ft
Approach Speed:
202 km/h / 109 kias

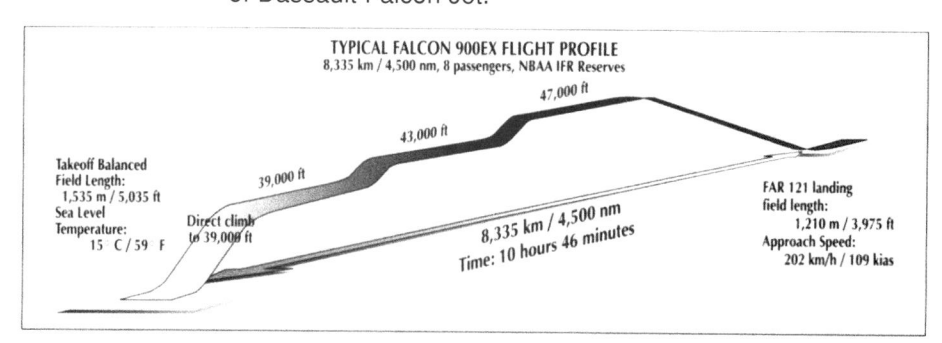

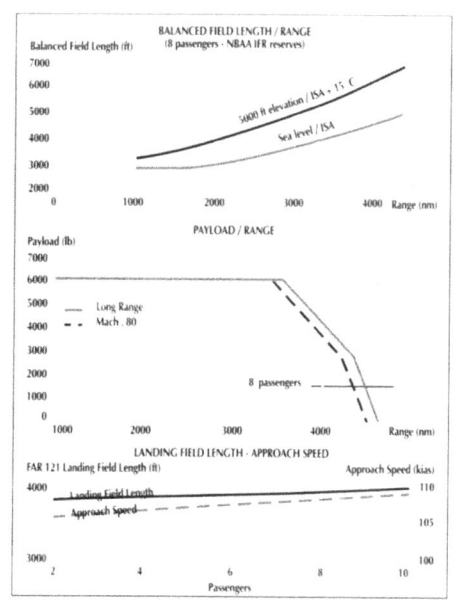

DIMENSIONS

Overall length	20.21 m	66 ft 4 in
Overall height	7.55 m	24 ft 9 in
Wing span	19.33 m	63 ft 5 in
Wing area	49 m²	527 sq.ft

WEIGHTS

Equipped empty weight	10,830 kg	23,875 lb
Maximum payload	2,795 kg	6,165 lb
Maximum payload with max fuel	1,310 kg	2,890 lb
Maximum zero fuel weight	14,000 kg	30,865 lb
Maximum fuel	9,445 kg	20,825 lb
Maximum take-off weight	21,910 kg	48,300 lb
Maximum landing weight	19,050 kg	42,000 lb

PERFORMANCE

VMO		350 - 370 kias
		650 - 685 km/h
MMO		Mach .87 - .84
Range with 8 passengers and NBAA IFR reserves		
- in Long Range cruise	8,335 km	4,500 nm
- at Mach .80	8,020 km	4,330 nm
FAR 25 balanced field length (SL, ISA)		
- at maximum take-off weight	1,610 m	5,290 ft
- with 8 passengers and max. fuel	1,535 m	5,035 ft
FAR 121 landing field length		
- at maximum landing weight	1,790 m	5,870 ft
- with 8 pax., NBAA IFR reserves	1,210 m	3,975 ft
Approach speed		
(8 passengers, NBAA IFR reserves)	202 km/h	109 kias

Figure 5-35. Falcon 900EX Specifications. Courtesy of Dassault Falcon Jet.

Chapter 5

Figure 5-36. Falcon 2000. Courtesy of Dassault Falcon Jet.

Figure 5-37. Falcon 2000 Interior. Courtesy of Dassault Falcon Jet and Bud Shannon Photo, Inc.

Figure 5-38. Falcon 2000 Cockpit. Courtesy of Dassault Falcon Jet and Bud Shannon Photo, Inc.

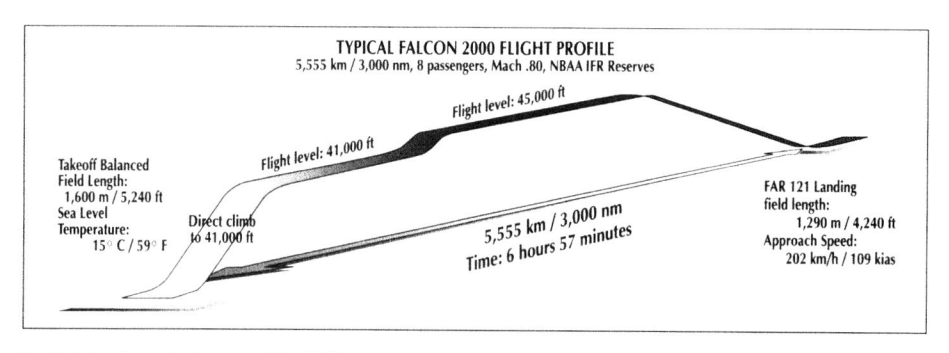

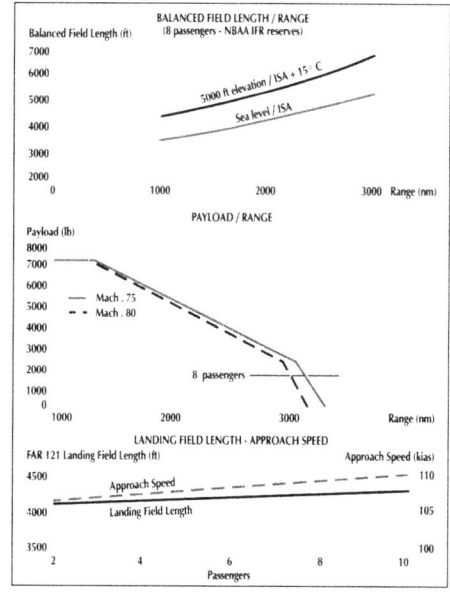

Figure 5-39. Falcon 2000 Specifications. Courtesy of Dassault Falcon Jet.

Figure 5-40. Challenger 601 3R. Courtesy of Bombardier Inc.

Figure 5-41. Challenger 601 3R Interior A. Courtesy of Bombardier Inc.

Figure 5-42. Challenger 601 3R Interior B. Courtesy of Bombardier Inc.

Figure 5-43. Challenger 604. Courtesy of Bombardier Inc.

CHALLENGER 604 SPECIFICATIONS

GENERAL

Passengers (Business/High Density Interior)	8/19
Crew (Business/High Density)	2/3

INTERIOR DIMENSIONS Cabin & Accommodations

Cabin length (from cockpit divider to end of pressurized compartment)	28 ft 4 in	8.6 m
Cabin height	6 ft 1 in	1.9 m
Cabin width (centerline)	8 ft 2 in	2.5 m
Cabin width (floorline)	7 ft 2 in	2.2 m
Floor area	202 ft^2	18.75 m^2
Total volume (from cockpit divider to end of pressurized compartment)	1,150 ft^3	32.56 m^3
Baggage volume	115 ft^3	3.25 m^3

*PERFORMANCE

Takeoff distance (SL, ISA, MTOW)	6,160 ft	1,878 m
Maximum range NBAA IFR Reserve ISA with 5 pax/2 crew and max fuel		
(Mach 0.74)	4,000 nm	7,408 km
(Mach 0.80)	3,750 nm	6,945 km

*CRUISE SPEED

Initial cruise altitude (ISA, MTOW)	37,000 ft	11,280 m
Maximum operating altitude	41,000 ft	12,497 m
Long range cruise	M 0.74	
Normal cruise speed	M 0.80	
High speed cruise	M 0.83	

ENGINES

General Electric CF34-3B	

WEIGHTS

Max takeoff weight	47,700 lbs	21,636 kg
Max landing weight	38,000 lbs	17,237 kg
Payload (with max fuel)	5 Pax	

*All performance figures and claims in this brochure are based on preliminary engineering data and are subject to change without prior notice.

Figure 5-44. Challenger 604 Specifications. Courtesy of Bombardier Inc.

Gulfstream II

Gulfstream IV-SP

Gulfstream IV

Gulfstream I

Gulfstream Fleet

Gulfstream III

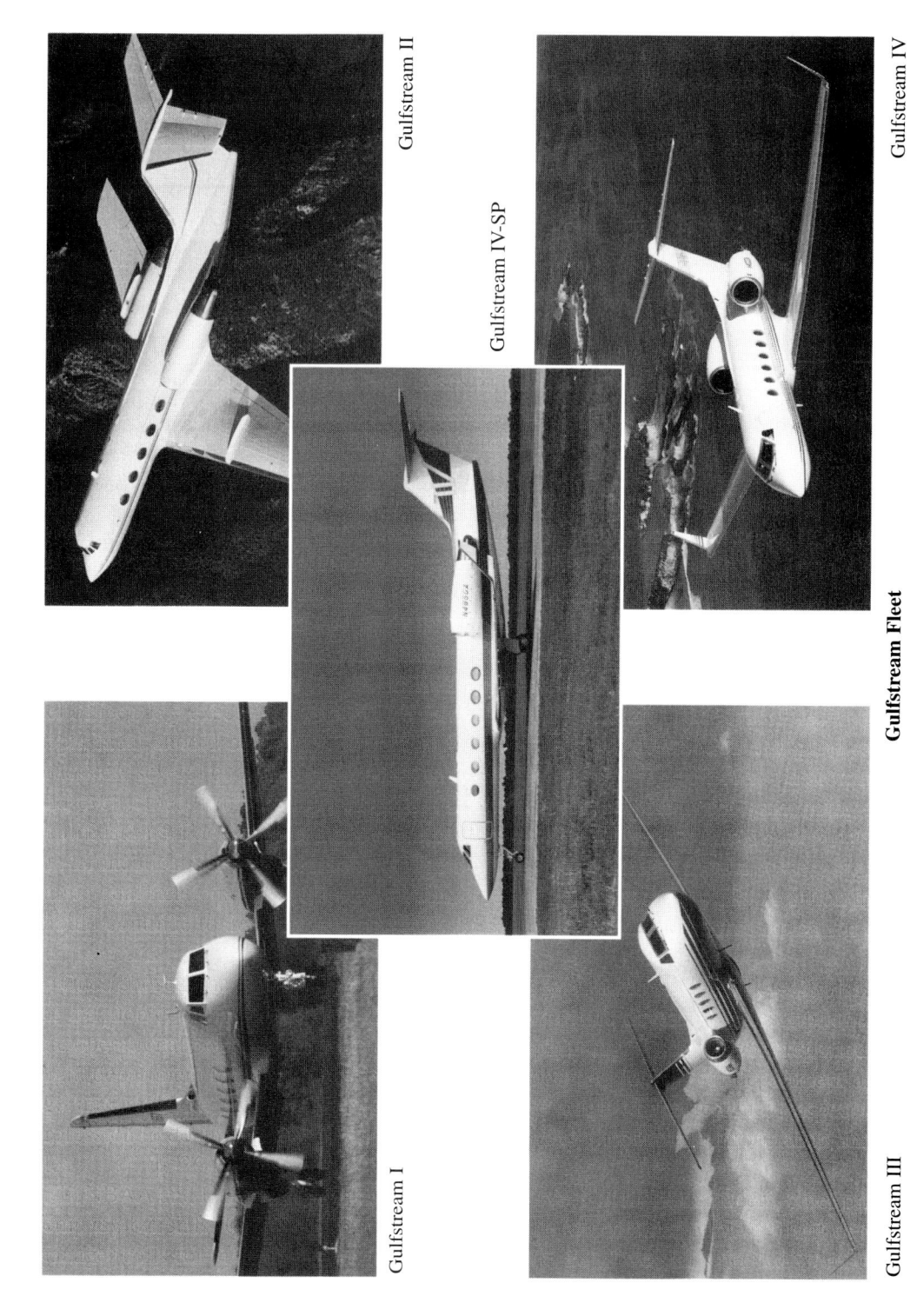

Figure 5-45. Gulfstream Fleet. Courtesy of Gulfstream.

140   *Chapter 5*

Figure 5-46. Gulfstream's Service Center (200,000 sq ft at Savannah, Georgia). Courtesy of Gulfstream.

5: Business Aviation Aircraft **141**

Figure 5-47. Gulfstream IV-SP. Courtesy of Gulfstream.

Figure 5-48. Gulfstream IV-SP Cockpit. Courtesy of Gulfstream.

Figure 5-49. Gulfstream V (First Flight, November 28, 1995). Courtesy of Gulfstream.

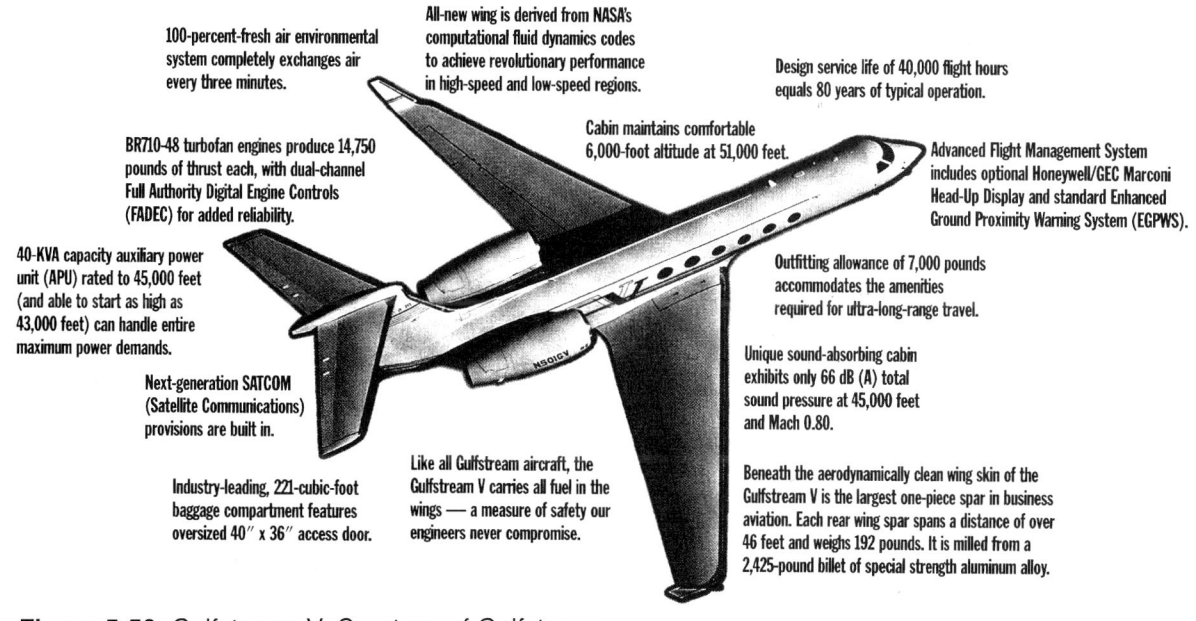

Figure 5-50. Gulfstream V. Courtesy of Gulfstream.

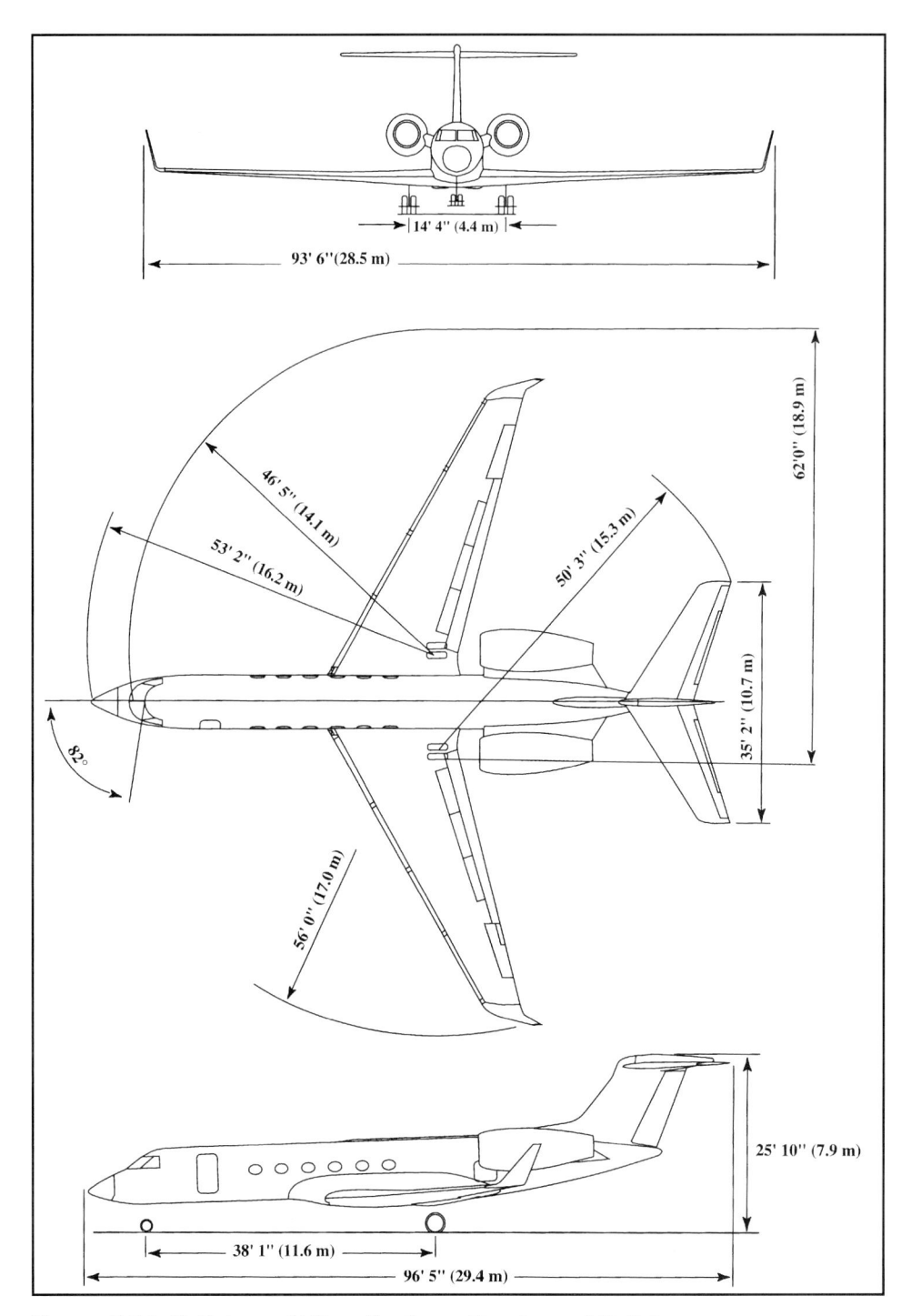

Figure 5-51. Gulfstream V Specifications. Courtesy of Gulfstream.

CABIN AND ACCOMMODATIONS

Cabin Length[1]	50'1"	15.3 m
Cabin Height	6'2"	1.9 m
Cabin Width	7'4"	2.2 m
Cabin Volume[2]	1669 cu ft	47.2 cu m
Baggage Volume	221 cu ft	6.4 cu m
Flight Deck Volume	178 cu ft	5.0 cu m
Maximum Seating Capacity	19	
Typical Business Seating	13 - 15	

PRESSURIZATION

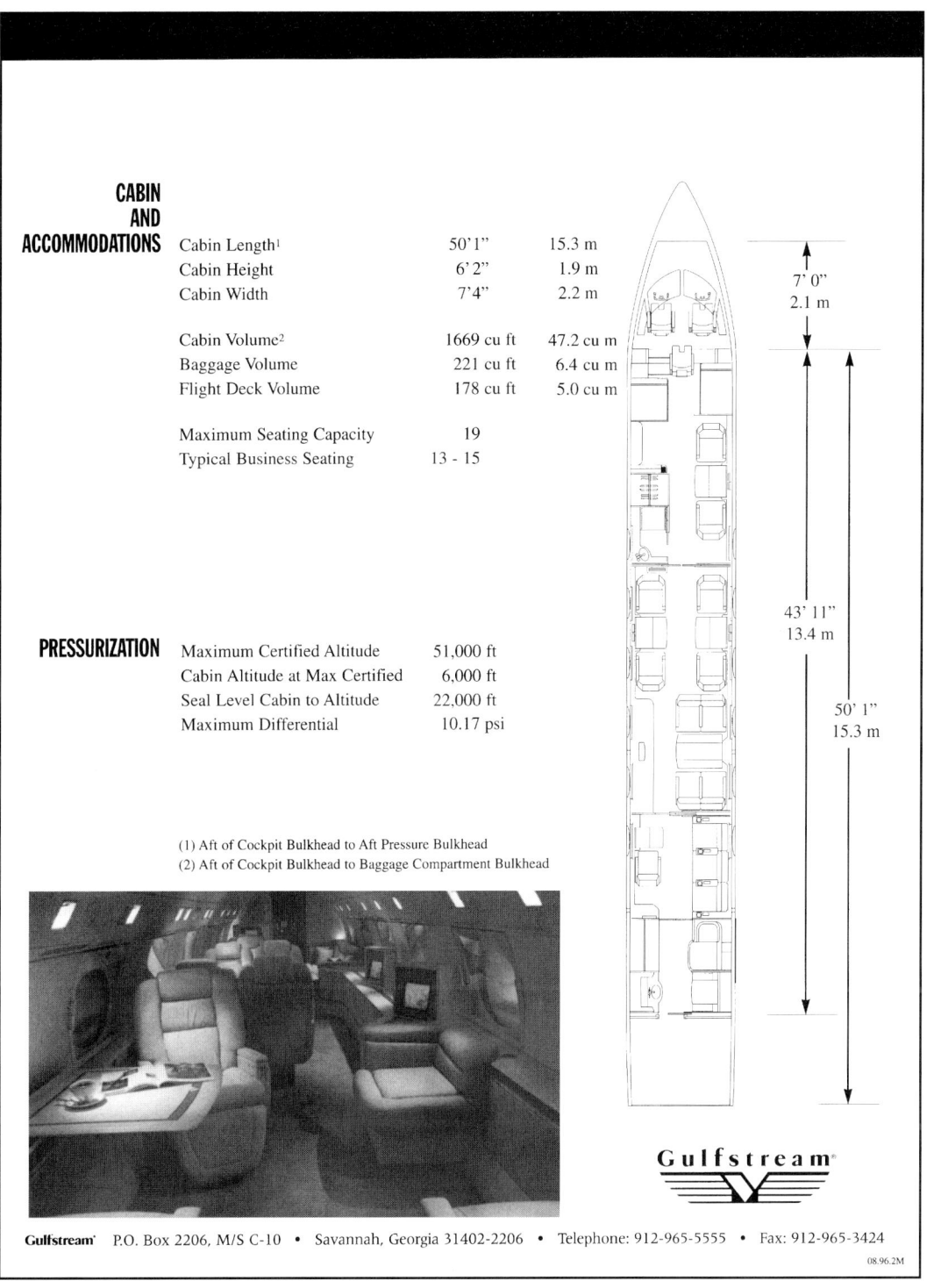

Maximum Certified Altitude	51,000 ft
Cabin Altitude at Max Certified	6,000 ft
Seal Level Cabin to Altitude	22,000 ft
Maximum Differential	10.17 psi

(1) Aft of Cockpit Bulkhead to Aft Pressure Bulkhead
(2) Aft of Cockpit Bulkhead to Baggage Compartment Bulkhead

Gulfstream P.O. Box 2206, M/S C-10 • Savannah, Georgia 31402-2206 • Telephone: 912-965-5555 • Fax: 912-965-3424

Figure 5-52. Gulfstream V Cabin. Courtesy of Gulfstream.

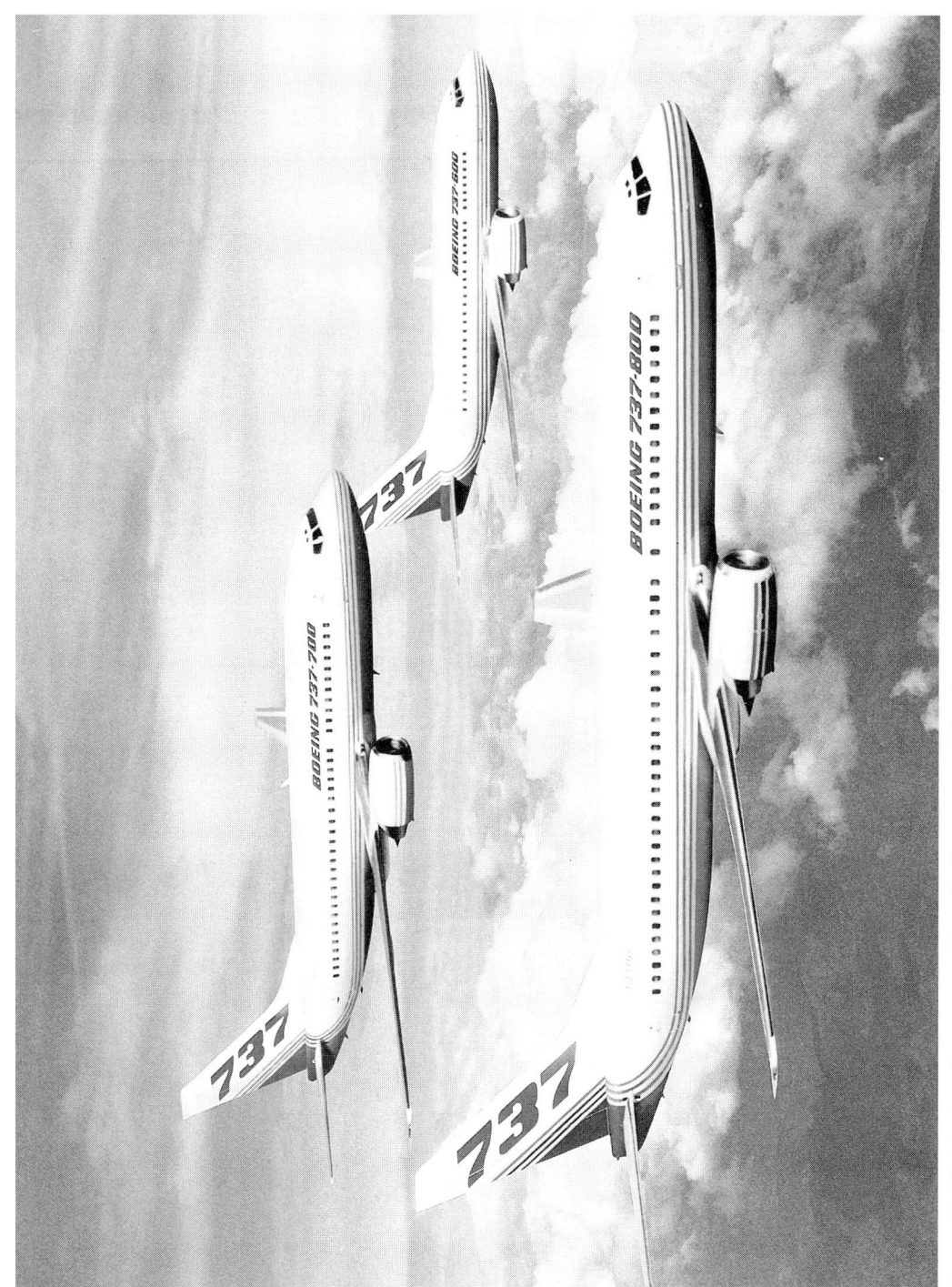

Figure 5-53. Boeing Corporate Aircraft. Courtesy of Boeing.

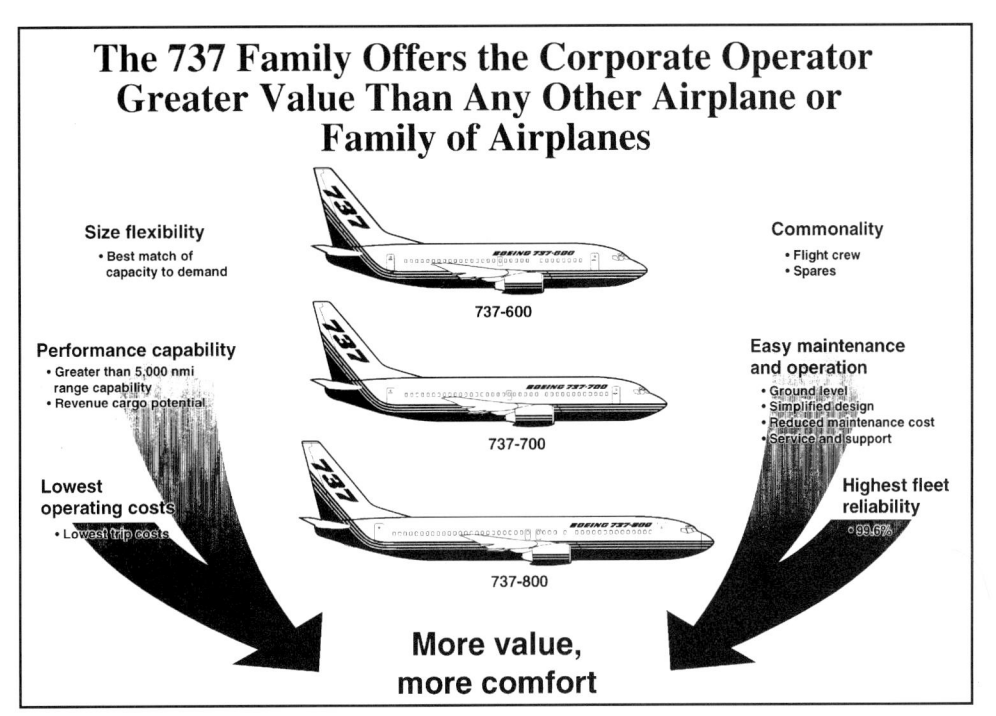

Figure 5-54. Boeing 737s. Courtesy of Boeing.

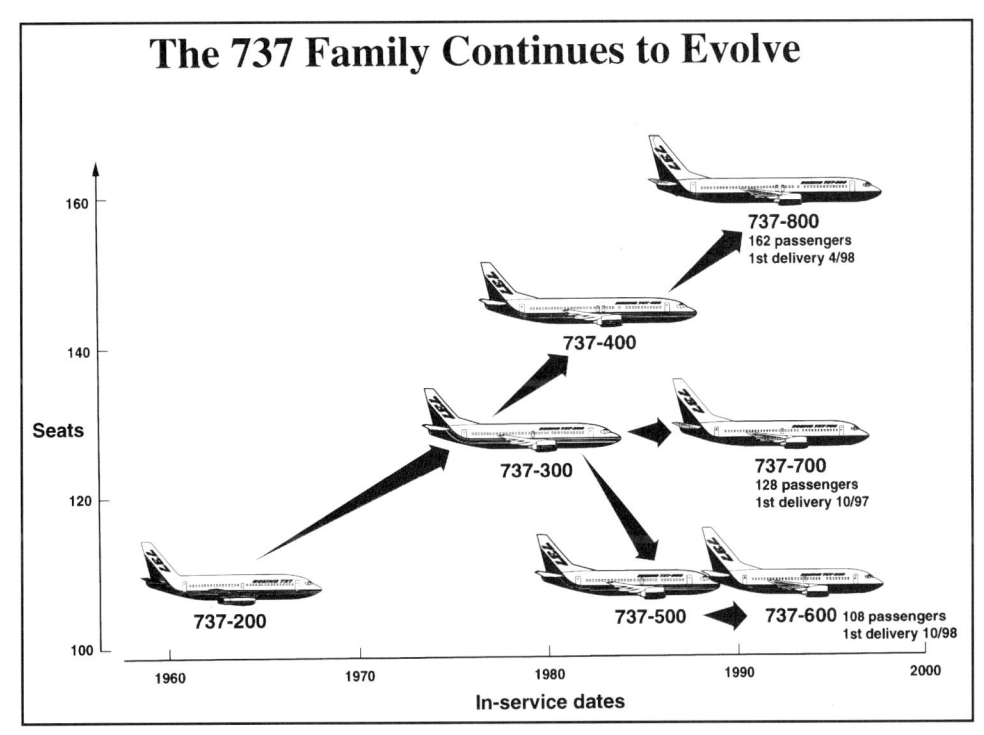

Figure 5-55. Boeing 737 Family. Courtesy of Boeing.

5: *Business Aviation Aircraft* 147

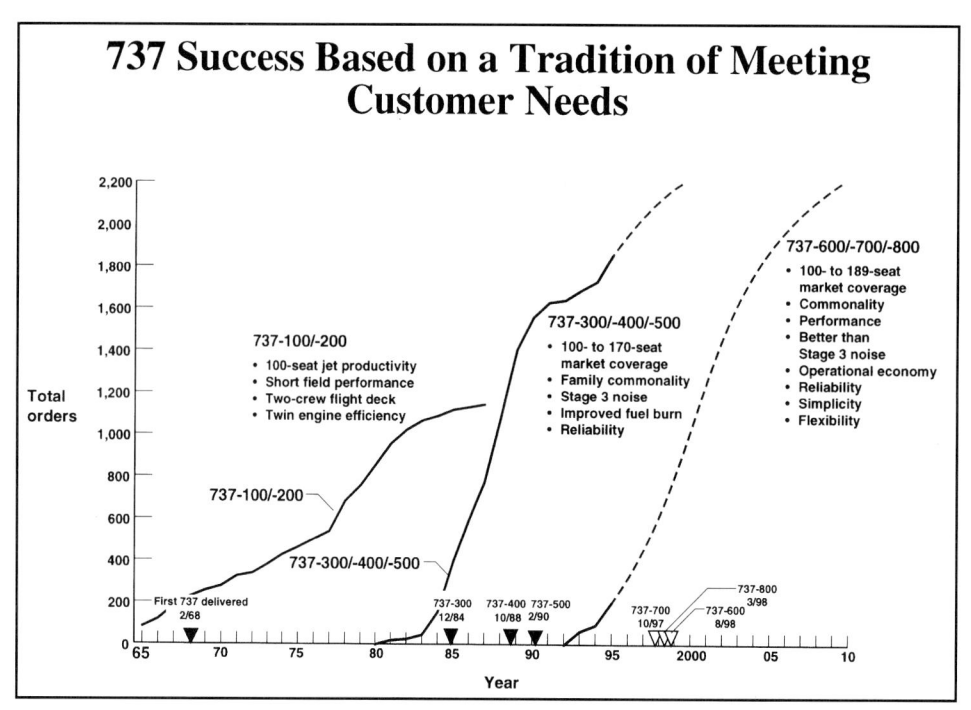

Figure 5-56. Boeing Total Orders. Courtesy of Boeing.

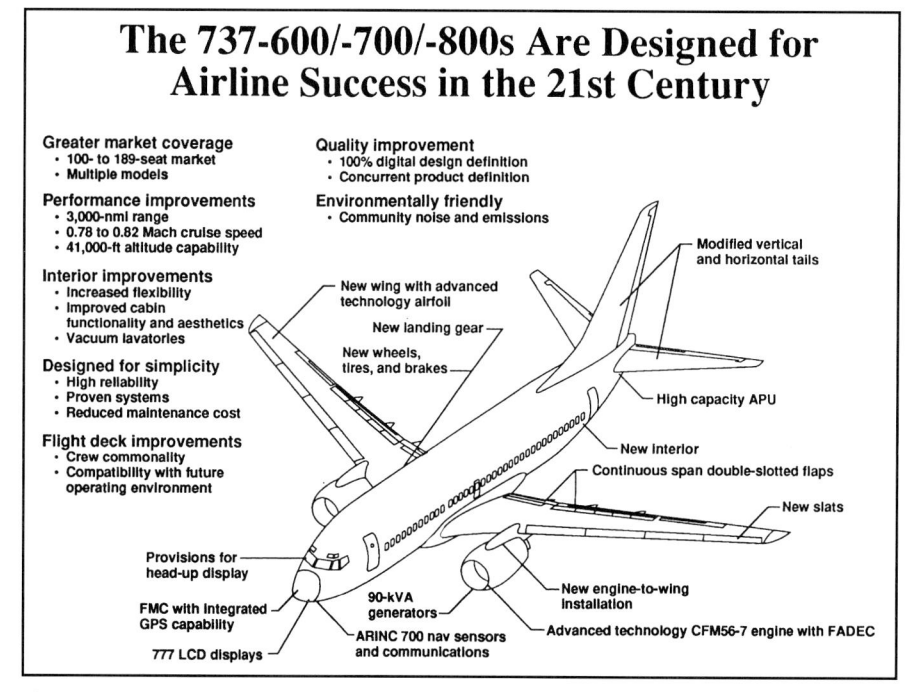

Figure 5-57. Boeing Aircraft Design. Courtesy of Boeing.

148 *Chapter 5*

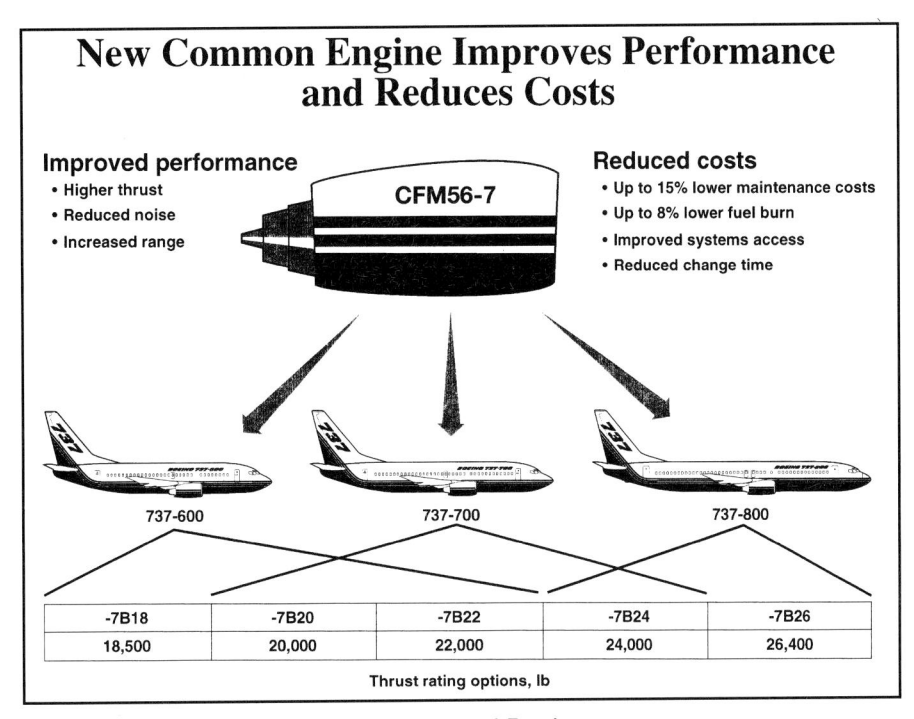

Figure 5-58. Boeing Engine. Courtesy of Boeing.

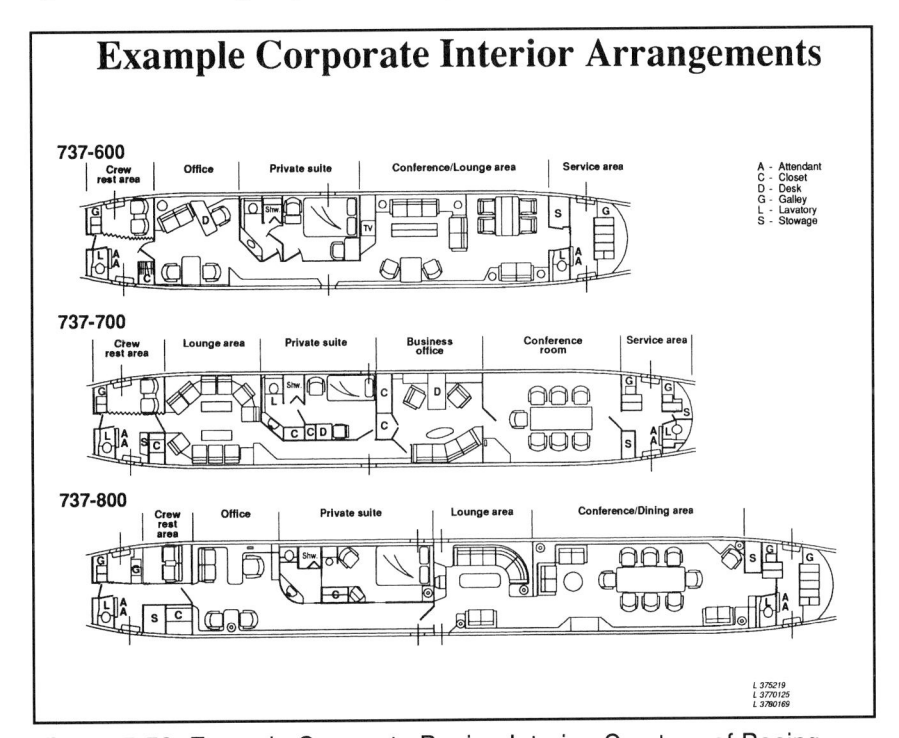

Figure 5-59. Example Corporate Boeing Interior. Courtesy of Boeing.

5: Business Aviation Aircraft **149**

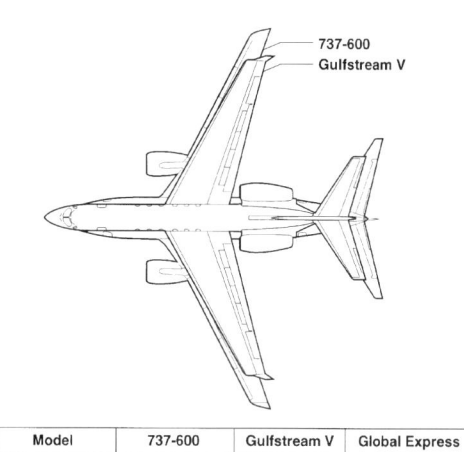

Corporate Cabin Comparison

Dimensions	737-600	737-700	737-800	Global Express	Gulfstream V
Height	7 ft 1 in	7 ft 1 in	7 ft 1 in	6 ft 3 in	6 ft 2 in
Width (center line)	11 ft 7 in	11 ft 7 in	11 ft 7 in	8 ft 2 in	7 ft 4 in
Width (floor line)	10 ft 8 in	10 ft 8 in	10 ft 8 in	6 ft 11 in	5 ft 6 in
Length	71 ft 4 in	79 ft 2 in	99 ft 1 in	48 ft	51 ft 1 in
Floor area	727 ft^2	807 ft^2	1,010 ft^2	335 ft^2	281 ft^2
Volume	4,695 ft^3	5,250 ft^3	5,555 ft^3	2,077 ft^3	1,905 ft^3

The 737 has more than twice the cabin volume of its corporate competitors.

Model	737-600	Gulfstream V	Global Express
Length	102 ft 6 in	96 ft 6 in	99 ft 0 in
Wing Span	112 ft 7 in	90 ft 10 in	91 ft 10.5 in
Height	41 ft 3 in	24 ft 6 in	24 ft 4 in

Figure 5-60. Overall Size Comparison. Boeing 737-600 Versus Gulfstream and Global Express. Courtesy of Boeing.

Figure 5-61. Corporate Cabin Comparison. Courtesy of Boeing.

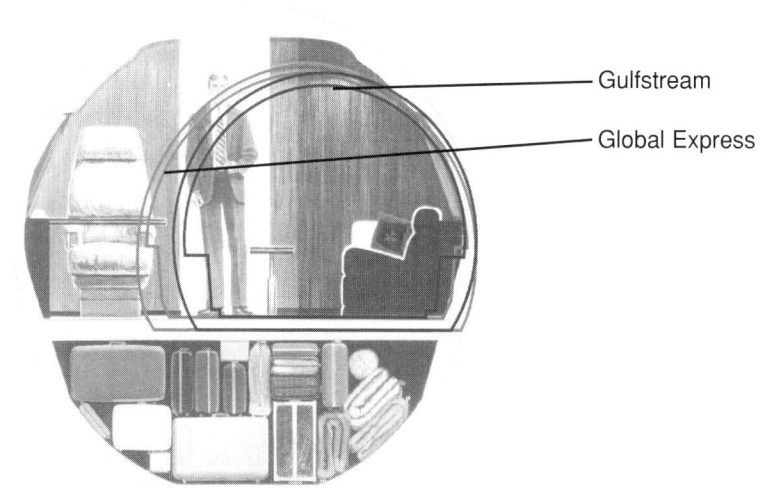

Figure 5-62. Interior Size Comparison. Boeing 737 Versus Gulfstream and Global Express. Courtesy of Boeing.

Figure 5-63. Boeing Corporate Jet Interior. Courtesy of Boeing.

Figure 5-64. Boeing Corporate Jet Interior. Courtesy of Boeing.

737-600/-700/-800 Lower Hold Volumes

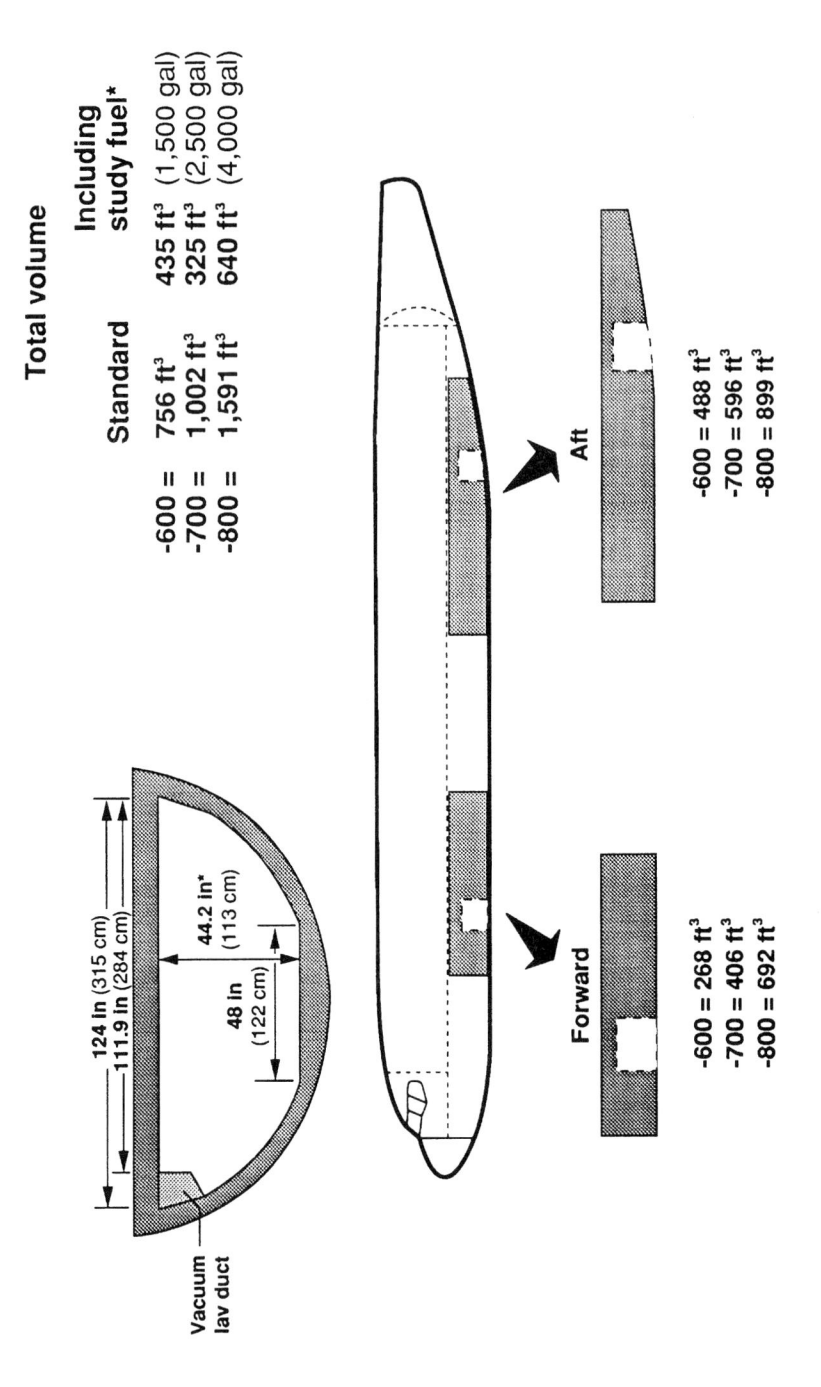

Total volume

	Standard	Including study fuel*
-600 =	756 ft^3	435 ft^3 (1,500 gal)
-700 =	1,002 ft^3	325 ft^3 (2,500 gal)
-800 =	1,591 ft^3	640 ft^3 (4,000 gal)

Aft
-600 = 488 ft^3
-700 = 596 ft^3
-800 = 899 ft^3

Forward
-600 = 268 ft^3
-700 = 406 ft^3
-800 = 692 ft^3

124 in (315 cm)
111.9 in (284 cm)
44.2 in* (113 cm)
48 in (122 cm)

Vacuum lav duct

* Estimated volumes.

Figure 5-65. Boeing 737 Volume. Courtesy of Boeing.

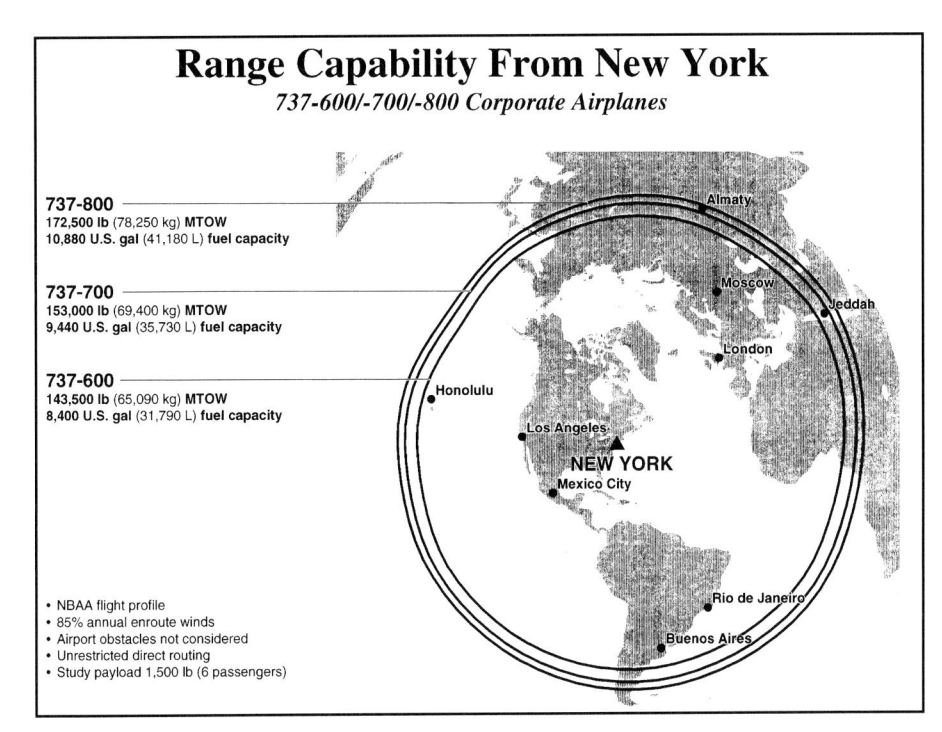

Figure 5-66. Range Capability from New York. Courtesy of Boeing.

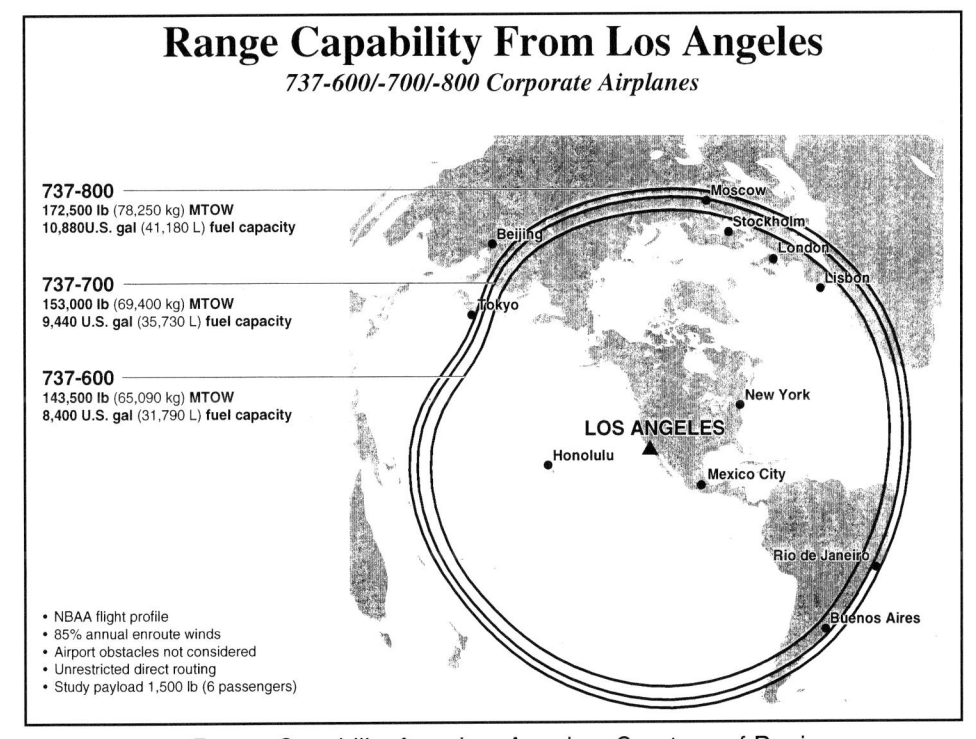

Figure 5-67. Range Capability from Los Angeles. Courtesy of Boeing.

5: Business Aviation Aircraft 153

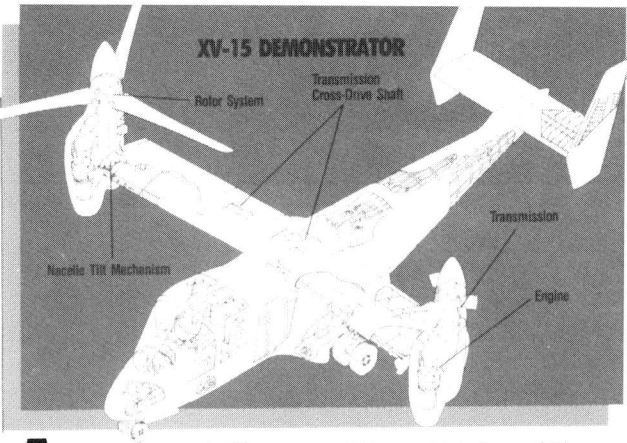

HOW DO TILTROTOR AIRCRAFT OPERATE?

At the end of each wing of a TiltRotor aircraft is a nacelle. Each nacelle houses:
- A modern fuel-efficient turboshaft engine.
- A transmission (proprotor gearbox) that takes the power output of the engine and reduces the engine rpm to the slower rpm needed to drive the proprotor system.
- A rotor system that drives and controls the main rotor blades/propellers (proprotors).
- A hydraulic screw drive system that can move the entire nacelle from vertical to horizontal to vertical.

In addition, the transmission in each nacelle is connected by a drive shaft running through the wing. This insures that one engine alone can provide power for safe operation of both rotor systems.

HELICOPTER MODE.
When the nacelles are vertical the proprotor blades turn in a horizontal plane — just like helicopter main rotor blades. In this mode, control is accomplished with conventional helicopter flight control procedures. The TiltRotor can hover or accelerate forward through 100 knots (115 MPH/185 KPH) or move off sideways or backwards at more than 35 knots (40 MPH/65 KPH). The nacelle can also be tilted backward about seven degrees to increase maneuverability.

TRANSITION MODE.
Once airborne, the pilot simply turns a switch and the nacelles begin to tilt forward. This movement may be continuous until the nacelles are horizontal, or the pilot may stop the tilting action and fly in the transition mode. During conversion, the flight controls automatically transition smoothly from the helicopter system to the airplane control system. The nacelles can move from vertical to horizontal in as little as 12 seconds.

The conversion corridor between helicopter and airplane mode is very broad and allows safe, easy-to-operate transition flight.

AIRPLANE MODE.
As transition between helicopter and airplane flight occurs, the TiltRotor aircraft automatically accelerates. When in the airplane mode, the pilot controls the normal airplane flaperons, rudders and elevators. As the TiltRotor goes forward, the wings begin to provide lift and engine speed is automatically reduced, proprotors operate at the most efficient speed for airplane flight, and fuel consumption is reduced.

Unlike the fuselage of a fixed-wing aircraft that must pitch up when taking off and down when landing, the TiltRotor fuselage remains level.

For more information contact

Bell Boeing
THE TILTROTOR TEAM

Bell Helicopter Textron Inc.
P.O. Box 482 ■ Fort Worth, TX 76101

©1991 Bell Helicopter Textron Inc.

Figure 5-68. How Do TiltRotor Aircraft Operate? Courtesy of Bell Boeing.

Figure 5-69. TiltRotor Aircraft. Courtesy of Bell Boeing.

VIP TRANSPORTER -- More Fortune 100 CEOs fly in Sikorsky S-76 helicopters than all others combined, and corporate operators remain the largest customer for the S-76B helicopter. Of the 75 S-76B helicopters delivered to date, more than 75 percent are operated in the VIP/executive transport mission. The twin Pratt & Whitney PT6B-36 engines of the S-76B helicopter allow for additional performance and payload capabilities in the hot day and high altitude applications often encountered in cities and in city-center-to-city-center operations.

###

93-#61 CL

Figure 5-70. VIP Transporter.

Figure 5-71. Sikorsky S-76C. Courtesy of United Technologies Sikorsky Aircraft.

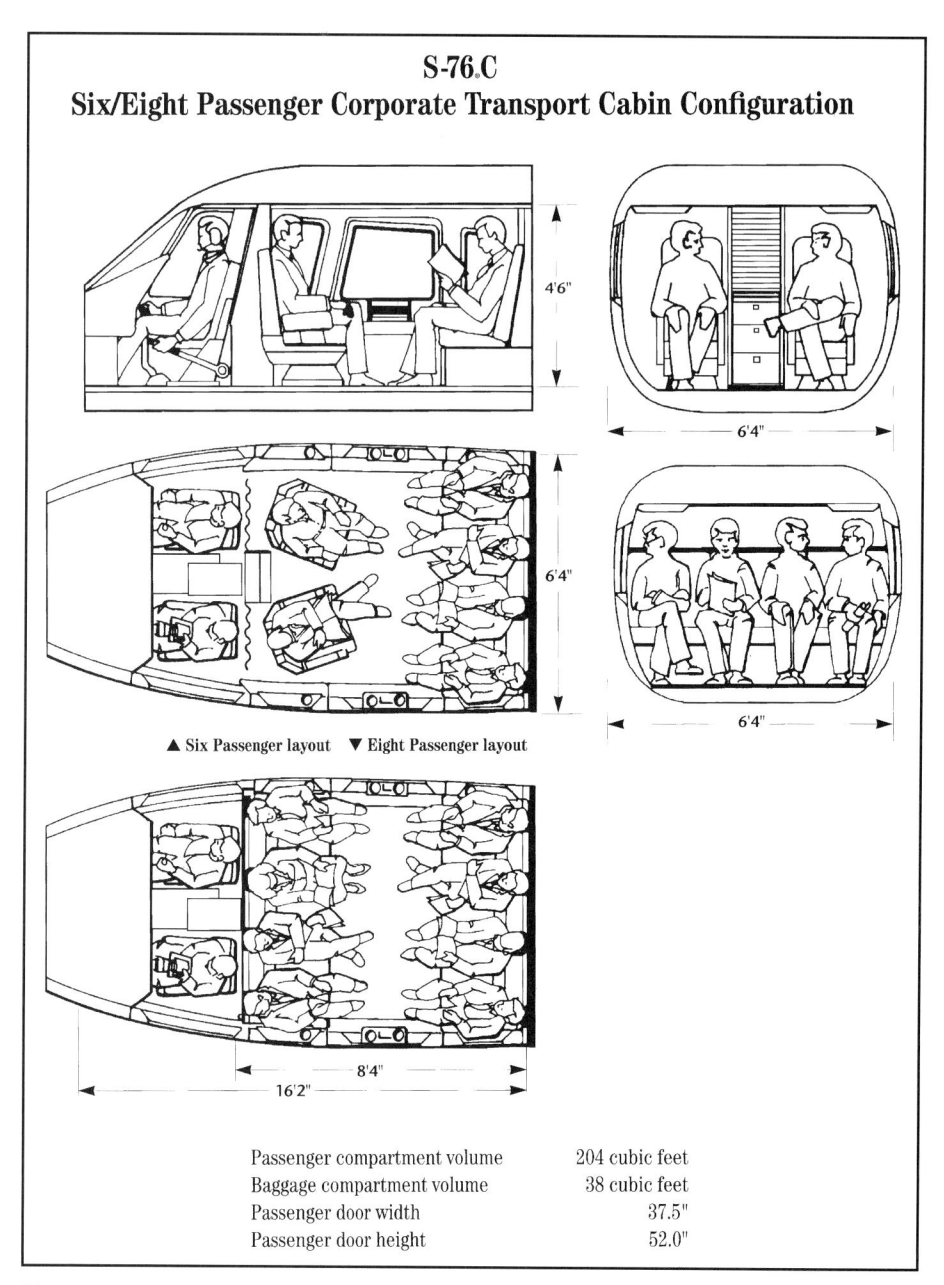

Figure 5-72. S-76C Six/Eight Passenger Corporate Transport Cabin Configuration. Courtesy of United Technologies Sikorsky Aircraft.

S-76C
Cost of Operation - Corporate Transport Service

FIXED COSTS

Crew costs	Sikorsky Estimate	Operator Estimate
Salary ($/year)*	62,000	
Salary with benefits (x 1.3)	80,600	
Pilot hours per year	800	
Crew cost per hour	100.75	
Total crew cost for two pilots ($/hour)	**201.50**	

Annual costs	Sikorsky Estimate	Operator Estimate
Assumed price ($)		
Hull insurance, assumed rate (%)	1.33	
Hull insurance, annual cost ($)		
Depreciation, assumed rate per year (%)	7.5	
Depreciation, annual cost ($)		
Total annual cost ($)		
Assumed flight hours per year (hours)	500	
Annual costs ($/hour)		

TOTAL OPERATING COST SUMMARY

	Sikorsky Estimate	Operator Estimate
Total direct cost ($/hour)	651.08	

Figure 5-73. S-76C Cost of Operation—Corporate Transport Service, Fixed Costs. Courtesy of United Technologies Sikorsky Aircraft.

S-76C
Cost of Operation - Corporate Transport Service

The following information is supplied to aid in the preparation of cost estimates for the operation of the S-76C helicopter in corporate transport service. Costs have been calculated in general accordance with the practices described in *Guide for Presentation of Helicopter Operating Cost Estimates*, published by the Committee on Helicopter Operations Costs. This estimate presumes a mature operation in which there has been opportunity for costs to stabilize and assumes no benefit for warranties.

Direct operating costs are calculated for an S-76C flying 500 hours per year for 6 years using 1994 component prices, average overhaul costs, and 1994 power-by-the-hour rates.

VARIABLE COSTS

Fuel and lubricants	Sikorsky Estimate	Operator Estimate
Average fuel consumption (gallons/hour)	84	
Fuel cost per gallon ($)	1.95	
Cost for fuel ($/hour)	163.80	
Cost for lubricants (3% of fuel)	4.91	
Total cost for fuel and lubricants ($/hour)	**168.71**	
Labor		
Salary ($/year)*	43,400	
Salary with benefits (x 1.3)	56,420	
Labor rate ($/hour)	28.21	
Direct maintenance (MH/FH)	2	
Indirect maintenance (MH/FH)	2	
Total maintenance (MH/FH)	4	
Total labor cost ($/FH)	**112.84**	

Figure 5-74. S-76C Cost of Operation—Corporate Transport Service, Variable Costs. Courtesy of United Technologies Sikorsky Aircraft.

Activities

1. List the steps taken to choose a suitable aircraft for a company.
2. Prepare a list of fixed and variable costs for an aircraft operation and have this list reviewed by aviation experts to determine its validity.
3. Contact an aircraft broker to discuss the various types and costs of aircraft available to a company.
4. Contact an aircraft manufacturer to obtain information about a specific type aircraft.
5. Communicate with company managers using timesharing programs to learn more about this method of operation.
6. Review current literature on aircraft sales in order to compare aircraft prices and characteristics.
7. Obtain the current *Planning & Purchasing Handbook* published by *Business and Commercial Aviation* to review aircraft types and specifics.

Chapter Questions

1. Evaluate the advantages and disadvantages of buying, leasing, and chartering corporate aircraft.
2. Prepare a list of aircraft fixed and variable costs for a specific aircraft model and make.
3. Why is the team concept of project management desirable for aircraft selection?
4. Explain what is meant by aircraft fractional ownership and give an example.
5. Discuss when an individual should select a business aircraft without help from others?
6. Assess the use of general aviation aircraft for corporate and business aviation. Explain the types and numbers of aircraft used.
7. Discuss the number and type of public airports available to the business aircraft operator and explain how this can be advantageous.
8. Either individually or as a team member, develop an aircraft selection matrix for a specified company mission.
9. Discuss aircraft costs of acquisition and possession.
10. Refering to Table 5-1, explain why corporate managers continue to use business aircraft after low utilization rates have been reported.
11. Select three different type aircraft that fall within the same performance criteria and make a selection comparison.
12. Discuss why the number of pilots, especially the more qualified certificated pilots, is a major concern for corporate managers.

CHAPTER 6

Aviation Department Management

Overview

After deciding to use business aircraft, managers have to develop an aviation department or use one of various alternative aviation management methods. In this chapter, the use of a fixed-base operator, an aviation management firm, and company aviation management is presented. Many, but definitely not all, managerial concerns of managing an aviation operation are reviewed, including organizational development, position descriptions, scheduling and dispatching, budgeting, people management, administration, maintenance, operations manual, and planning. The use of good management practices applicable to strategic planning, leadership, and finance are reviewed. There cannot be enough planning in an aviation operation.

Goals

- Understand aviation management alternatives.
- Discuss human resource management.
- Develop an operations manual.
- Understand organizational concepts for corporate flight departments.
- Discuss planning in aviation management.
- Discuss documentation requirements.
- Understand good management practices.

Introduction

After a major managerial decision has been made, that is, the use of business aircraft or corporate aviation, the next major decision is to determine whether to develop a flight department or use one of a number of other managerial options. For a businessperson or company officials who use a business aircraft less than 200 hours a year, using an aviation management firm or contracting a fixed-base operator (FBO) may be more advantageous than establishing a company aviation department. For the single business flyer, the use of an FBO is essential, especially for fueling, hangaring, maintenance, and other services.

Certain chief executives and top managers will not consider establishing an aircraft department because they consider the costs too high or simply do not want to be bothered. Others worry about what the stockholders will say or that they cannot defend the high costs involved. Remember that corporate aviation is a cash-negative activity unless flight operations under FAR Part 135 are approved. It is necessary to determine if a business aircraft is advantageous for a firm and to determine the advantages and disadvantages of managing a business aircraft. Many aviation consultants, aircraft management firms, and even a firm's own employees can analyze the pros and cons of establishing an in-house aviation department or contracting an outside agency. The main purpose of this chapter is to present what it takes to establish and manage an in-house aviation flight department after a brief review of other flight management options.

Fixed-Base Operators

Fixed-base operator (FBO) refers to any general aviation business at an airport. In other words, the commercial airlines are normally large enough to own and operate their aircraft servicing, maintenance, fueling, catering, and other aircraft support activities, but general aviation users are not. Some airlines do contract various servicing needs; however, for the general aviation aircraft operator, having the resources to service an aircraft is not usually cost effective and most operators contract their aircraft support services.

The services of an FBO, then, become essential to the business aircraft operator. One source to review to determine which FBOs are available in the United States and Canada is AC-U-KWIK's *The Corporate Pilot's Airport/FBO Directory*. Contact a local FBO or call AC-U-KWIK at 1-800-400-5945 or 913-967-1719 (www.ACUKWIK.com). Detailed information about FBOs, maps, and additional information for corporate operators is available in the directory. For example, aircraft operators going through North Carolina might use Asheville Jet Center.

After communicating with Asheville's operations liaison, the following information and pictures were provided (see Figures 6-1 through 6-3). Asheville Jet Center is located at the Asheville Regional Airport, Fletcher, North Carolina. It is outside the city of Asheville, in the scenic hills of the Blue Ridge and Smoky Mountains. The aircraft ramp is large; 24-hour fuel service is provided; and aircraft parking, servicing, and maintenance are available. The center's facilities include communications, flight planning, satellite weather feed and DUAT, briefing rooms, conference rooms, pilot's lounge and rest area, and transportation services (G. Kovach, personal communication, February 13, 1997).

Fixed-base operators usually can provide as much service as desired—for a price. Fueling is a main function; however, aircraft maintenance is also a major requirement for small and large aircraft operators alike and is provided by most FBOs.

Asheville Regional Airport

ASHEVILLE JET CENTER

(AVL) VAR 4.0°W
N35-26.2°-W082-32.5°
Elev. 2165 Ft.

ARINC 129.82
UNICOM 122.95

ATIS............120.20
GROUND........121.90
TOWER..........121.10
DEP/APPROACH
    EAST..........125.8
    WEST.........124.65

ASHEVILLE JET CENTER

(704) 684-6832
1-800-833-3490
FAX (704) 687-0135

Asheville Jet Center
Asheville Regional Airport
Fletcher, NC 28732

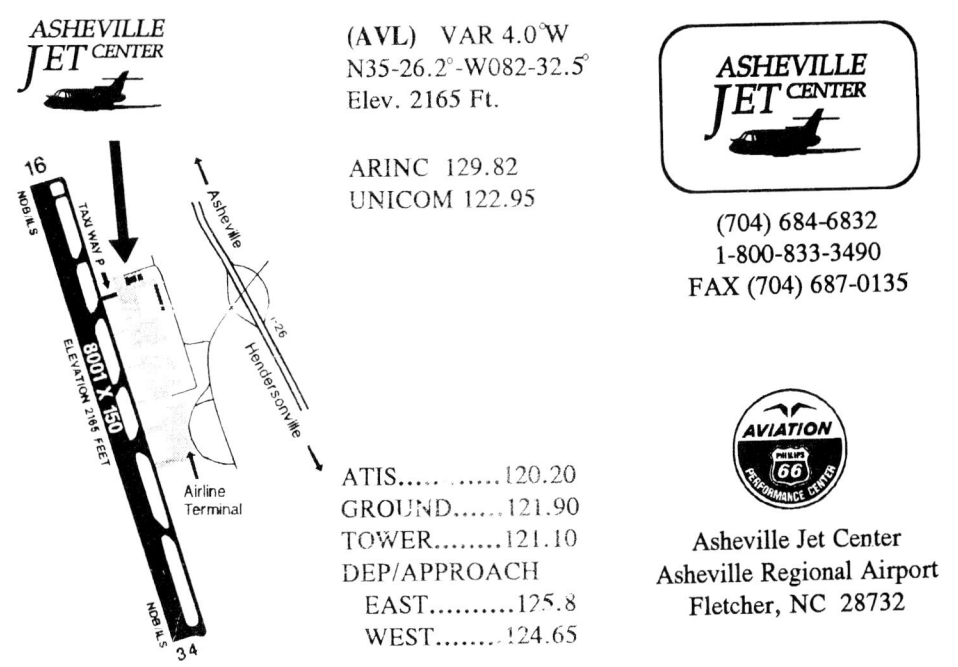

Figure 6-1. Asheville Regional Airport.

Figure 6-2. Asheville Regional Airport. Photo courtesy of G. Kovach and American Aerial Photography.

6: Aviation Department Management 163

Figure 6-3. Asheville Jet Center. Photo courtesy of G. Kovach and American Aerial Photography.

Another FBO, but one located outside the United States, is MAGEC Aviation Limited. Named by *Professional Pilot* as the best European FBO year after year, MAGEC provides outstanding aircraft and management services from its Luton, England, based operations. Luton is approximately 30 miles north of central London. Trevor King, Commercial Director, oversees a 24-hour operation for the business flyers and aircrew. Aircraft handling, crew planning, fueling, catering, maintenance and engineering, and ground transportation are but a few of MAGEC's many services (see Figures 6-4 through 6-10). For a listing of the European FBOs, MAGEC publishes *FBOs & Aircraft Handling Agencies in Europe, The Middle East & North Africa* (tel +44 1582 724182, fax +44 1582 455453).

Figure 6-4. Trevor King, Commercial Director at MAGEC Aviation Ltd, Luton, United Kingdom.

Figure 6-5. Receptionists at MAGEC Aviation Ltd, Luton, UK. Maureen Durrant, Barbara Manner, Carol Meads.

Figure 6-6. Reception Area at MAGEC Aviation Ltd, Luton, UK. Henry Hewitt and Sally Kovach.

Figure 6-7. MAGEC Aviation Ltd Hangar. Clean Spacious Inside Aircraft Parking Area for Customers.

Figure 6-8. MAGEC Aviation Aircraft Parking Area, Luton Airport, United Kingdom. Courtesy of Steven Gannon and Russell Search.

Figure 6-9. MAGEC Aviation Aircrew Ground Transportation, Luton Airport, United Kingdom.

Aircraft Management Companies

While an FBO provides a business aircraft operator with services on demand, an *aircraft management firm* will provide contractual services for aircraft operators. These services may include every aspect of administering, operating, and controlling an aircraft, even providing the aircrew. A business manager who purchases an aircraft can simply contract his/her entire aircraft operation to the aircraft management firm. In turn, all administration, budgeting, aircraft servicing, maintenance, fueling, reporting, and other activities will be accomplished by the aircraft management firm.

For example, MAGEC Aviation Limited provides full aircraft operations and control for aircraft buyers who request such management. Another firm is Jet Aviation, West Palm Beach, Florida, which offers complete aircraft management. Budgets, reports, operation, maintenance, fueling, aircraft outfitting and refurbishment, avionics support, ground handling, and other services are completely provided by Jet Aviation. After obtaining an aircraft and contracting with Jet Aviation, the turn-key management is accomplished (see Figure 6-11).

Aviation Department

When top management determines that it is more advantageous or wants to operate its own flight department, various organizational and managerial concerns must be faced. Full control and flexibility result when company officials establish an aviation department. Many of the issues that should be addressed in establishing a flight department are presented in this chapter; however, it is suggested that contact with the NBAA or other business aircraft associations will be most helpful. Additionally, aviation training programs and seminars have been developed and are offered by professional associations like the NBAA and the University Aviation Association (tel. 334-844-2434/fax 334-844-2432).

The *purpose* of an aviation flight department is the effective use of business aircraft to achieve the air transportation objectives of the company through safe, professional, and quality services. The key ingredients for operating the aviation department are human resources. First, an aviation department manager should be selected.

Aviation Department Manager

Should the aviation department manager (ADM) be a pilot? Debate is ongoing in answering this question, but it is generally accepted that a person's skills at managing resources is most important. Having aviation knowledge is vital, and while it is deemed important to be a pilot to understand piloting, this requirement is not mandatory. What resources are to be managed? People must be managed, because management is basically defined as getting things done through people. Equipment, facilities, and monies must also be managed as they relate to a firm's mission. The corporate aviation manager, then, must clearly understand the firm's

Figure 6-10. Dave Quinn, MAGEC Aviation Chef/Manager of Catering Activities.

mission and then use his/her skills, knowledge, understanding, abilities, and leadership to meet this mission using the resources provided. Vitally important to managing an aviation department is the manager's ability to provide strategic planning and direction.

While management means getting things done through people by directing and controlling, *leadership* involves motivating and influencing people to follow. The best outcome is having a good manager who also is a good leader.

Olcott (1996) stated in an editorial in *Business Aviation Management* that *leadership* is an "atmosphere that surrounds a person who motivates employees" (p. 1). He emphasized that leaders

      a. possess vision,
      b. understand the mission,
      c. establish values,
      d. are creative,
      e. are courageous,
      f. possess and practice a sense of community, and
      g. possess and extend trust.

These talents, according to Olcott, are most important in being a good leader.

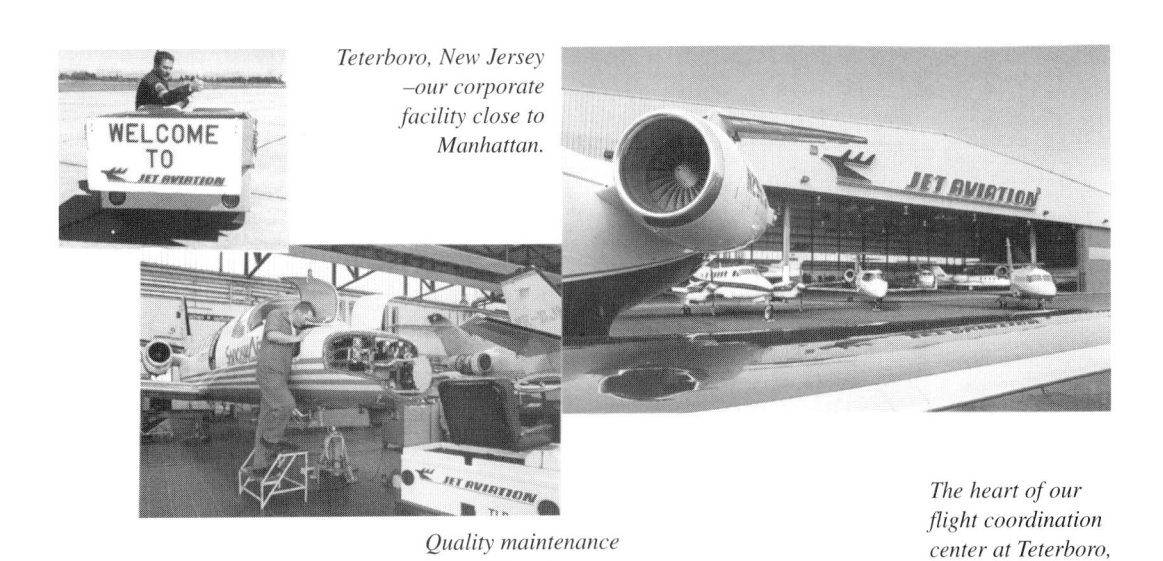

Teterboro, New Jersey – our corporate facility close to Manhattan.

Quality maintenance for safety and reliability.

The heart of our flight coordination center at Teterboro, NJ.

Whatever the engine – we have the skills for the job.

A courteous, personal service from first to last.

Maintenance times that meet deadlines.

Figure 6-11. Jet Aviation. Captions and pictures provided by Jet Aviation.

Leadership can be learned through personal understanding and thoughtful interrelationships. For example, an aviation manager who studies situational awareness after first learning his/her personal leadership styles may react more effectively with flight department personnel. Management styles can be categorized as participative and directive, as well as supportive and task oriented. Knowing one's primary leadership style and then using value analysis to determine the maturity and competence of the other person(s) involved can result in better leadership. A good manager/leader will know and understand how others are affected by his/her style. There are many management and leadership short courses, informative texts, and practice exercises for managers and leaders to make self-improvements. People can make changes to their behaviors and become more effective through better interpersonal relationships and the use of power.

Managers accomplish their tasks through the standard functions of management; that is, planning, organizing, staffing, directing, and controlling. An aviation department manager accomplishes these functions while meeting administrative, operational, maintenance, and miscellaneous duties. A chief pilot at Ford Air Transport in England stated years ago that his main concern was budgeting. Flying was understandably easy for him, but meeting administrative tasks took time and effort (B. Barrow, personal communication, August 1, 1991).

Managing a flight department requires excellent management and human relations skills. An ADM must market his/her department against competitors—the FBOs, aviation management firms, and others. The key to this service function is building an awareness of the aviation department's value-added services. In other words, the ADM must develop a sense of worth for the aviation department. How many aviation managers provide outstanding services and then are faced with justifying the use of the flight department? Knowing your customers and having them state the value of the aviation services will enable managers to justify the operations. Usually, everyone in a firm knows that the flight department exists; the question being asked most often is if the function is worth the money. Have the users answer this question through surveys, interviews, trip reports, savings identified, business obtained as a result of the use of the company's aircraft, and other benefits realized and talked about but not recorded at the time. ADMs should value every trip and keep accurate records. Occasional reports to top management of the benefits and value reported by the aircraft users may prevent the need for a defense against corporate aviation critics. Everyone then *will know* that the aviation service is vital, absolutely needed, and should be maintained. Building awareness is only one major function of the ADM; identifying the value/benefit of aviation is another.

Between 1994 and 1997, the question of whether an ADM should be a pilot was asked of aviation students (79 people), and three categories of responses were identified. Forty-seven percent favored the ADM being a pilot, 16 percent reported distinctly no, and the remaining 37 percent responded that it depends (percentages are rounded). Extracts of various responses for these three groups are provided in Table 6-1.

From Table 6-1, it can be seen that flying expertise and know-how were considered most beneficial in managing an aviation department, while those who pointedly reported no were apparently concerned with managerial skills, not flying skills. A large number of people

Table 6-1. Responses to the Question, Should an ADM Be a Pilot?

For:	Be a pilot with considerable experience.
	Only way to go. Pilots know the aircraft.
	Makes them more well-rounded.
	Should hold at least a private pilot's license. Is involved in decisions on weather, aircraft performance, etc.
	More effective in making changes and working problems.
	Brings operational know-how to the job.
	Has technical knowledge and understanding.
	Runs the flight department; needs the qualification.
Against:	Not essential; others run companies without technical experience.
	Too many areas to work besides piloting; e.g., accounting, maintenance, administrative matters.
	Primary job is managing all aspects of the business.
Depends:	If small operation, yes; if large, just strong management skills.
	Depends on nature and size of flight department. Different skills.
	Understanding is all that is required; depends on what can be afforded.

Note: Verbatim extracts from aviation student responses to the question in informal surveys in corporate aviation operations courses from 1994 to 1997.

reported that the size of the aviation department made a difference. If the department was small (1–3 aircraft), then the manager may also be the chief pilot. As the aviation department gets larger, more emphasis falls on business management skills; in other words, meeting the many administrative requirements like budgeting, reporting, and planning.

From this brief survey, it can be induced that company managers who want to begin an aviation department should hire a pilot with specific qualifications. The person fulfilling the duties and roles of the ADM must make many immediate and strategic decisions. Experience in flying, coupled with a strong managerial background and good academic schooling, should enable the manager to establish an aviation department to meet the needs of the top executives and company financial and personnel limitations. ADMs who are not pilot certificated are recommended to become familiar with flying through private flight schools or introductory courses, even taking company flights would be useful. Most important to these individuals would be the hiring of a chief pilot who has the expertise and the communication skills to work with the ADM and air crew. Pilots familiar with international flight operations may be invaluable for flight planning and operations.

What the aviation department manager brings to a company is the direction and control of an important company tool, the business aircraft. For smaller flight operations, the manager should be familiar with flying, but for larger operations, having skilled managerial expertise, combined with the technical expertise from a good chief pilot, is considered more advantageous. Some top executives may be least concerned about flying skills because they expect pilots to be qualified; but they may be very concerned about good scheduling, high-quality passenger handling, low maintenance downtime, excellent ground transportation support, good catering, and other activities. Their displeasure of the support services could result in the demise of the department.

Good communication is a skill needed by every aviation manager, not only to communicate with aviation department personnel, but with top executives, community leaders, and local and governmental authorities and agencies. A firm's image and reputation may depend on an ADM who represents the company at civic functions. Piloting at this point is immaterial; promoting company community contributions may be crucial. The ADM may even represent the firm in important legal cases or in potential company contracts. ADMs will accomplish many of the normal human resource management and bureaucratic tasks others face; for example, hiring and firing, salary negotiations, personnel counseling, motivating employees, and record keeping.

So, what should be considered? First, the CEO or board of directors needs to state any major desires or managerial qualifications needed by the aviation department manager. Then, the personnel manager must review any company hiring rules and ascertain compliance with any governmental or relevant hiring practices. Recruitment follows to obtain the right person to fit the profile established. All applications for this position should first be screened by the personnel manager or a selection committee for acceptable qualifications, then a short list of qualified people should be sent to the firm's executive officer or board. Besides the personnel officer, key executives who will be using the firm's aircraft and responsible for the flight department should be on the committee. With the right job description and good recruiting practices, several qualified applicants can be easily obtained. The NBAA, aviation agencies, and even other firms having a flight department can be helpful in recommending required qualifications. For example, Embry-Riddle Aeronautical University trains pilots in its Master of Aeronautical Science program. Managers of flight departments should possess higher education qualifications to use effective writing, speaking, and research skills. Aviation managers should continue to develop their personal skills through available formal education or seminar programs.

Once chosen, the aviation department manager begins a major task. The ADM has to plan the aviation department and work to select suitable aircraft, develop job descriptions, formulate operational policies, and accomplish extensive planning. Quilty (1996a) reported that the flight department manager "must be able to pick good people for your department and manage and motivate them properly. And you need enough of the technical aspects of this profession to do an effective job, and understand enough accounting information and principles . . ." (p. 26).

In NBAA and University Aviation Association surveys of flight managers, chief pilots, and company officials, Quilty (1996a) reported that "the job of chief pilot or flight department manager calls for more than superior flight skills" (p. 28). Two skills needed were identified in the survey results—job knowledge and interpersonal skills. Aviation people surveyed listed

Table 6-2. Skill Areas for Aviation Managers

Aviation	Non-Aviation
Aviation regulatory compliance	Interpersonal communications
Flight operations	Business and professional ethics
Operating procedures	Management principles
Aviation safety/program management	Personnel management
Vendor selection	Small group communication
Aircraft selection, purchase, and outfitting	Basic business operations
	Conflict management

Source: Quilty, S. M. (1996b). What does it take to be chief? *Business Aviation Management, 3,* 26–29. Used with permission, John Olcott.

Table 6-3. Corporate Executive Expectations for Flight Managers

Educational	Skill Areas
Business and professional ethics	Leadership
Interpersonal communications	Decision-making
Governmental regulation	Ethical conduct
Basic business operations	Planning
Personnel management	Organizing capability
Management principles	Team building
Organizational communication	Interpersonal communication
Occupational safety	Listening
Small-group communication	Time management
Conflict management	Budgeting/accounting

Source: Quilty, S. M. (1996a). Meeting company expectations. *Business Aviation Management, 3,* 30–33. Used with permission, John Olcott.

the topics in Table 6-2 as significant. Top management wants aviation managers who have human relations skills and managerial abilities gained at a young age.

For corporate executives, the topics listed in Table 6-3 were considered important for flight department managers. It was noted that company and aviation managers expected aviation managers to be aviation experts. Common themes that Quilty (1996a) reported were leadership, communication, aviation management, and business administration.

Organizational Structure

Typical aviation department organizational structures are shown in Figures 6-12 and 6-13. Three main activities exist in any aviation organization: 1. Operations, 2. Maintenance, and 3. Administration. A fourth activity can be labelled *miscellaneous*.

Operations involves managing the flying activities, schedules, and training. Although FAR Part 91 does not require an operations manual, many managers decide to develop and use one to regulate and follow prescribed company procedures. How to develop an operations manual will be presented later in this chapter, but the responsibility of this task falls on the ADM. Within flight operations, air crew training is a major activity, and a separate chapter will be devoted to this function.

Other operational concerns include flight-, duty-, and rest-time considerations; cockpit resource management; air crew fitness and health; air crew procedures; passenger and cargo handling; flight operating procedures; fueling, accident reporting, and many other matters affecting flight performance and mission accomplishment. Preflight and postflight activities also require planning and policies. The many FARs and other rules that govern flight operations must be met by pilots, notwithstanding the many rules that apply to international flight operations.

Each of the above activities and many more mean that aviation managers and pilots devote themselves to knowing their jobs, ones which corporate managers expect air crews to do without compromise. Flight operations requires continual review for regulatory compliance and flight safety.

Maintenance operations are no less important, and regulatory compliance is critical to the flying mission. The business aircraft must be in an airworthy and dispatchable condition. Within the maintenance activity, equipment, facilities, and personnel have to be managed for the pilots to be able to use the aircraft when needed. Aircraft maintenance training requires expert knowledge and understanding and falls under strict federal aviation rules. Not even pilots can fix an aircraft without proper training and certification. For ADMs who contract their aircraft maintenance, maintenance coordinators are necessary, and a few aircraft mechanics may be required for routine operator's maintenance.

Administratively, aviation managers develop the guidelines for personnel to follow. ADMs are required to know, understand, and make contractual obligations, be familiar with purchasing, company reporting, legal matters, and other important documentation requirements. Procedures are to be followed in preparing department budgets, complying with company personnel rules, managing offices and supplies, and following company personnel rules. Administering the aviation department means that details in reporting and communicating be clear, and that facilities and equipment be effectively used.

Top management will use the company aircraft more than others in the firm, and oftentimes there is a conflict in flight requirements for senior managers. It is not the aviation manager's job to decide which manager has priority over another. The ADM provides the air services required that top management believes will be needed most.

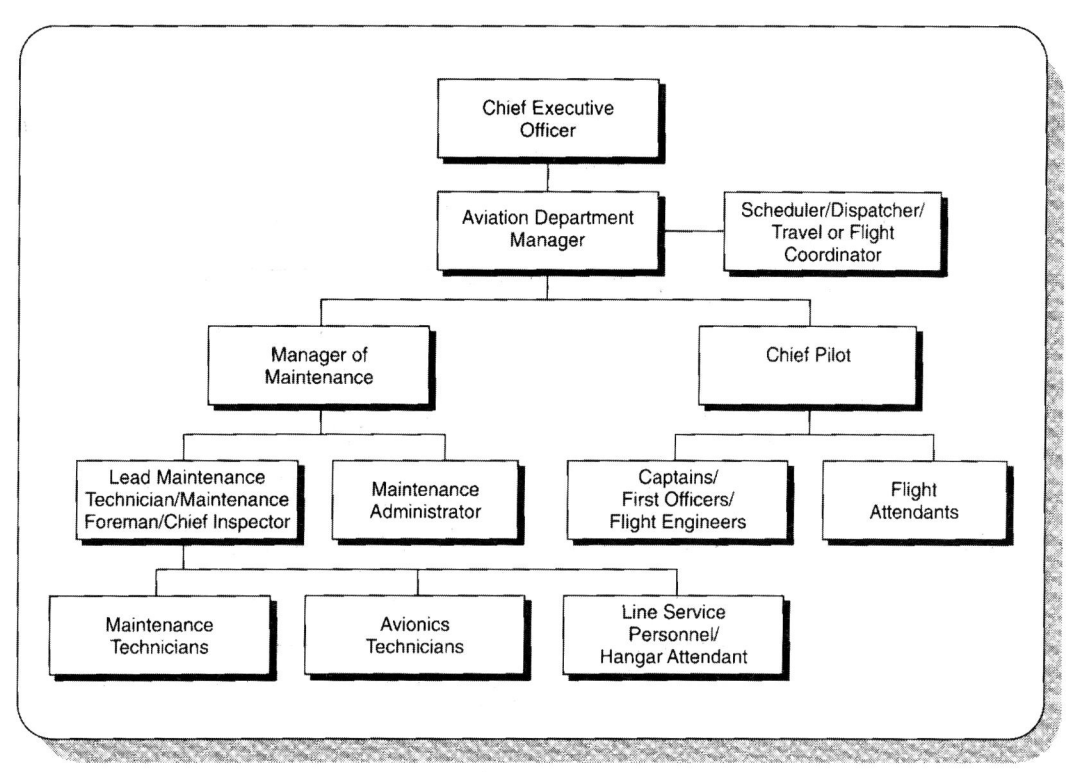

Figure 6-12. Sample Corporate Flight Department Reporting Chart—Larger Flight Department. Provided by NBAA.

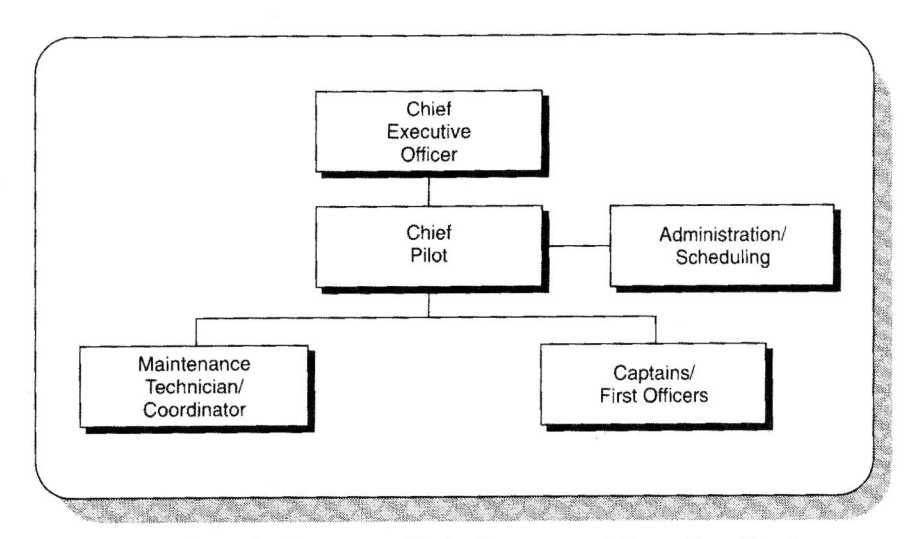

Figure 6-13. Sample Corporate Flight Department Reporting Chart—Smaller Flight Department. Provided by NBAA.

6: Aviation Department Management **175**

Table 6-4. Domestic/International Differences for Consideration by Aviation Managers

Factor	U.S.	International
Currency	Common	Mixed
Politics	Stable	Mixed, volatile and uncertain
Language	English	Local
Labor	Skilled	Mixed
Transportation systems	Excellent	Mixed
Government regulations	Open	Mixed
Culture	Homogeneous	Heterogeneous
Communications	Excellent	Varied

Miscellaneous ADM duties and responsibilities may involve serving as a liaison to the community, participation on local and governmental committees, serving as the company representative in important events, and other activities. Firms that are members of the NBAA or other aviation associations often allow employees to do various studies and to participate in standing committees to promote business aviation. Developing new guidelines, presenting corporate issues to government lawmakers, and speaking to concerned people about noise and air pollution may be miscellaneous tasks not thought about when establishing an aviation department.

Within each of these activities, an aviation manager must learn and apply good management practices. Sheehan (1996) reported that aviation managers need to "talk the talk" and "walk the talk" (p. 3). By this he meant that every aviation manager must learn good management, know the fundamentals, keep current in management techniques, and practice management through effective interactions. This requires personal dedication and effort in a dynamic aviation industry environment. Additionally, aviation managers must thoroughly understand the different managerial requirements for managing a department responsible for providing international, as well as domestic, flight services. People and national rules differ from country to country. While a firm's strategic and tactical plans require company managers to conduct business in a global marketplace, an aviation manager may have to similarly review and assess operational differences. For example, in Table 6-4, differences are identified for various factors that an aviation manager must consider. A factor that is reported to be *Mixed* means that each national interest has its own rules, which may be different from other authorities. Flight control and operations procedures may vary significantly from place to place, and the range of variance internationally is significantly higher than in the U.S. for the factor. ICAO and IBAC attempt to standardize flight procedures as much as possible; however, sovereigns remain individualistic and have unique requirements.

Placement

The larger the aviation department and the greater the number and type of aircraft to be operated, the more complicated and detailed management planning becomes. Placement of the aviation department within the firm depends on top management, but it is wise to have the aviation department manager report as high in the organizational hierarchy as practical. If a CEO shows no interest or does not have the time to oversee the aviation department, then someone below this level should become responsible. Placement of the aviation department too low in the organizational hierarchy may cause operational problems, personality battles, or other undesirable events/actions.

An early NBAA study (NBAA Recommended Standards, 1982) provided interesting information about a survey taken of actual aviation department placements. The aviation department was placed at just about every level imaginable, at the CEO, senior vice-president, director, midlevel manager, executive secretary, son of the president, and so on.

What is vital to the operation of the flight department is direct communication to top decision makers. Placement that results in poor or ineffective decision making only hurts the company; placement at the organizational level where important decisions for the effective use of the firm's aircraft can be made results in productivity and efficiency.

Aviation Personnel

Within each function identified above (administration, operations, maintenance), qualified people are most important. People who can accomplish quality office work, logistics, secretarial work, documentation, and other administrative duties must be hired based upon the size and need of the aviation department and company objectives. Within operations, a chief pilot, captains, first officers, flight attendants, flight engineers, and others are required. Within maintenance, a chief of maintenance, line mechanics, avionics experts, fuel experts, and other line personnel are needed.

The following job descriptions are extracted from an actual operations manual and indicate information about position descriptions required for an aviation operation (all names are withheld upon request). For more information about position descriptions, contact the NBAA or other aviation associations. These position descriptions are only samples and do not include every position or duty requirement that may be found.

Director, Aviation Services

> Responsible for the aviation organization. Develops, implements, and administers operating policies and procedures for executive flight operations. Studies new developments in aviation, including aircraft and equipment. Negotiates the cost of aircraft and equipment, including purchase price, warranties, and maintenance. Has financial responsibility for developing, administering, and monitoring the department operating budget.

Administrative Assistant

Provides administrative support for the director of aviation services. Maintains personnel records, aircraft permits and licenses, contracts, and leases. Conducts a full range of administrative support functions for department managers and associates.

Manager, Safety, Standards, and Training

Responsible for implementing an accident prevention program to ensure safe and standardized flight operations of all corporate aircraft department flight personnel. Plans, monitors, and evaluates flight crew vendor-provided training programs to ensure that a high degree of proficiency is maintained.

Manager, Finance and Operations Analysis

Responsible for the overall planning, measurement, analysis, reporting, and control of financial policies and programs. Manages the corporate aircraft department budget process. Performs administrative tasks and special projects, including audit functions, purchase and sale of aircraft, and operational analysis for senior management.

Operations Assistant

Performs administrative support functions, including audits, file management, local accounting and invoice processing, telecommunications, supply inventory management, budget tracking, aircraft scheduling, and operational support.

Manager, Flight Support

Responsible for the planning and implementation of aircraft scheduling for executive transportation. Monitors daily air operations to ensure safe and efficient use of aircraft. Facilitates the coordination, planning, and scheduling of international flights for senior management.

Supervisor of Flight (SOF)

The SOF functions as the principle contact for the corporate flight department beyond normal business hours and on weekends and holidays. Normally, during the business work week the flight support manager is the SOF after business hours. During weekends and holidays, a rotating list of supervisory captains and flying staff managers function as the SOF. When assigned SOF duty, the SOF is available and on call at all times.

Flight Operations Coordinator

Schedules company aircraft for executive transportation in accordance with company policies and government regulations. Provides coordination and support services for such trips with all appropriate parties.

Chief Pilot

Manages flight crew personnel. Ensures operational compliance with FAA and company regulations. Participates in the development and implementation of flight policies and guidelines. Maintains budgetary control of flight operations and approves flight-related expenditures. Functions as a captain on company-owned or leased aircraft.

Supervisory Captains

Develop and conduct programs to ensure standardization and maintain pilot-in-command proficiency for their designated aircraft type. Conduct training flights that prepare pilots for qualification-check flights. Assist the chief pilot with supervision and development of pilot personnel. Serve as technical advisors for aircraft maintenance and operational matters on specific types of aircraft.

Pilot, Executive Aircraft

Functions as the pilot-in-command on company-owned or leased aircraft. Has direct responsibility for the safety of passengers and crew and the comfortable, timely operation of the aircraft. Supervises the flight crew. Ensures that all flight and ground operations comply with federal regulations.

Copilot, Executive Aircraft

Acts as second-in-command on specified aircraft during air transportation of company personnel and their business associates.

Flight Attendant Supervisor

Supervises the activities of flight attendant personnel. Functions in a dedicated flight service position and is professionally trained in cabin safety and evacuation. Provides for the care, comfort, and well-being of passengers on assigned flights. Ensures that commissary supplies and food are properly stocked before every flight.

Manager, Maintenance and Technical Services

Responsible for overall aircraft maintenance, repair, modification, inspection, and ground service activities. Ensures compliance with FAA regulations and manufacturer service bulletins. Prepares specifications for bids on outside repairs, maintenance, and inspections. Reviews maintenance contracts and makes recommendations to the director of aviation services. Functions as the senior environmental health and safety coordinator for the aviation department. Provides direction and leadership to associated maintenance personnel.

Maintenance Planner

Responsible for planning, developing, maintaining and administering aircraft maintenance programs on assigned aircraft. Keeps abreast of improvements and progress in the aviation industry with particular reference to maintenance and repair methods and procedures on particular aircraft. Supervises and provides technical direction to vendor-supplied aircraft technical personnel as well as performing hands-on repair of aircraft.

Technician, Aircraft Maintenance

Performs maintenance, repair, and complete ramp activities on company aircraft in accordance with company and FAA maintenance policies and procedures.

Materials Controller

Provides administrative support to the maintenance department, including vendor-supplied procurement of aircraft and non-aircraft parts and supplies. Performance of this function has a direct effect on the aircraft maintenance budget and parts availability.

Recruitment Sources

The total number of pilots in the United States forecasted by the FAA was around 713,900 by 2007, a 0.7 percent annual growth rate (FAA, 1996). At the end of 1995, the FAA identified 123,877 airline transport and 133,980 commercial pilots in the United States (Figure 5-10). Air transport and commercial pilots are expected to increase by a rate of 1.5 and 0.4 percent annually, respectively. Many of these pilots will come from aviation flight schools that are training students interested in aviation. Available pilots from the military, however, are decreasing because of military downsizing. To counter the decrease in available ex-military pilots, some airlines are establishing ab initio flight training programs. Students in these programs progress from having no flying experience to obtaining commercial airline licenses. Some pilots have even been trained in simulators and have earned their commercial ratings without ever doing any actual flying. Baty (1996) reported at the General Aviation Forecast Conference that various actions have been taken to encourage more people to become pilots. The Experimental Aircraft Association has a *Young Eagles* program to attract people ages eight to seventeen to aviation. The Aircraft Owners and Pilots Association's *Project Pilot* encourages new aviators, as well as the *Learn to Fly* program, which GAMA promotes. Embry-Riddle Aeronautical University encourages new student pilots in its *Sun Flight* program at Daytona Beach, Florida—having students solo in just two weeks. Women in Aviation, International, is promoting more women to fly as women make up 50 percent of the population and 70 percent of the workforce (Baty). Additionally, *General Aviation Team 2000,* an independent organization sponsored by 49 founding companies, is trying to interest young people in flying. Its goal is to boost new-student enrollments in flight schools to 100,000 by the year 2000 (Higdon, 1996). All aviation-oriented activities are being encouraged to promote this goal and motivate young flyers in order to establish a pool of qualified pilots in the millennium.

As more general aviation aircraft are sold because of the liability rule changes, as aircraft are used more as a business tool, and as other benefits are realized by businesspersons, the demand for qualified pilots will similarly increase. President Clinton's 1993 Commission on Airline Competition reported that there would be a need for 11,851 cockpit jobs in 1997. Of the total, it was estimated that 7,604 pilots were required for corporate, business, air taxi, law enforcement, and other non-airline positions (Higdon, 1996). NBAA reported also that Aviation Data Services estimated an increase in aviation department jobs after aircraft operators went from 6,859 companies in 1990 to 7,322 companies in 1996. With this increase, the number of aircraft being flown rose from 9,642 jets and turboprops to 10,592 aircraft. Despite these reported increases, the number of qualified commercial- and airline transport-rated pilots has fallen; for example, from 710,000 in 1986 to 639,000 at the beginning of 1996 (Higdon).

In summary, pilots come from the airlines, the military, students with appropriate flight hours, foreign airlines, new flight programs, and other flight operators. Similarly, maintenance personnel come from the military, training schools, other company operators, and foreign sources. Over 150 institutions offer FAA-certificated maintenance training in the United States. Recruitment agencies such as Air Incorporated (formerly Future Aviation Professional Association) provide names of qualified pilots and maintenance personnel to requesters. Whether by word of mouth, through academic institutions, or through other aviation operators, personnel trained in aviation specialties are available for the corporate aviation operator, albeit not in the numbers desired.

Selection Process

Prior to recruitment, company managers should begin their search for employees (hiring model is provided in Figure 6-14) by accomplishing a needs assessment to determine what type of people and how many are required for the aviation department. Hiring the aviation department manager first and giving this person a major role in this task is most advisable.

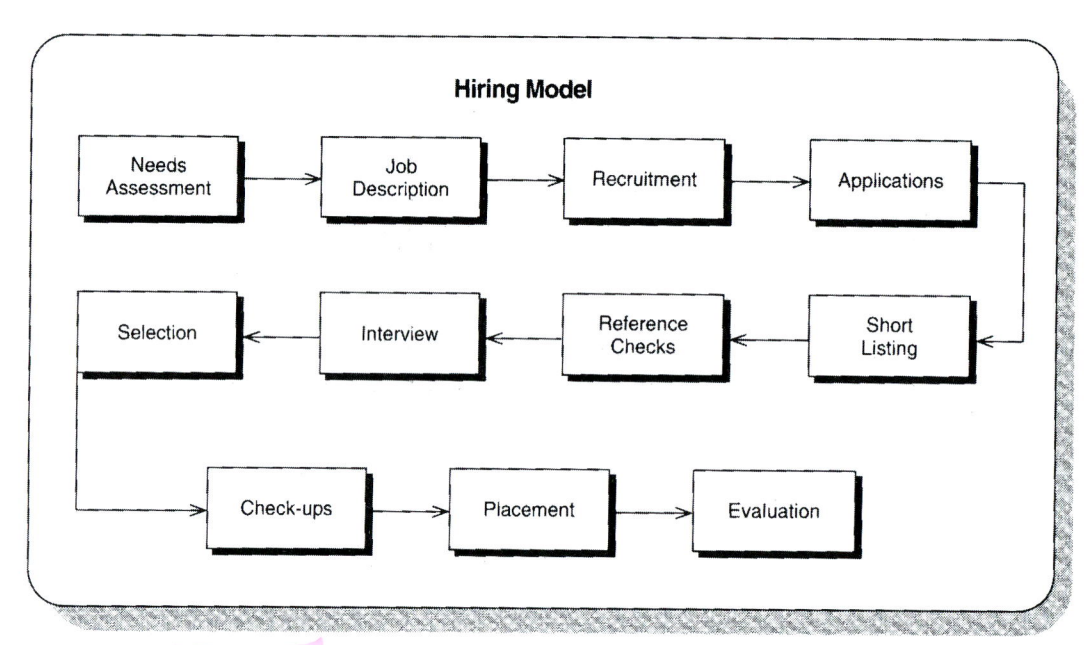

Figure 6-14. Hiring Model.

Between the ADM and the Personnel officer and with the advice of the NBAA or other aviation agencies, various job descriptions can be developed. Advertisements can then be placed, and company communications with interested agencies and others can be made to obtain a pool of likely candidates.

All applications should be received by the firm's personnel officer for initial review of completeness and acceptability. The ADM should be involved in short listing the applicants to a more manageable pool of suitable candidates. After short listing, reference checks and initial contact with the applicants need to be made. A quick check with the Federal Aviation Administration's Operational Systems Branch, AVN-120, P.O. Box 25082, Oklahoma City, Oklahoma 73125, to request an airman's accident and enforcement history may be useful for employees short listed. This may save untold time and effort in the long run.

The time for a selection committee to interview the final group of short-listed candidates comes either before or after personal testing; the latter may involve physical and/or mental aptitude tests. Who should be on the selection committee? The personnel officer and ADM have already been active and should continue to provide excellent recommendations; however, other committee members may include a few of the top executives who will use the services being planned or who will have a major role in its budgeting or operation.

By the time the final interviews are conducted, the technical qualifications of the applicants will have been satisfactorily reviewed. During the final selection, hiring may depend on personal bearing, attitude, intelligence, appearance, career plans, or other individual merits. After selection, the process continues with final checks, job placement, and probationary evaluations. The overall process is not difficult, but may be very time consuming and requires close attention to details. An employee who will fly the firm's top executives or will provide critical support services for the company must be just as satisfied with the company doing the hiring as the company will be with him or her.

Number of Pilots and Mechanics

Identifying the number of pilots and mechanics needed to support a corporate flight department is also not an easy task, mainly because of budgetary constraints and the concerns about the availability of qualified people. The type and number of aircraft have a major effect; however, other factors such as aircraft use, company developments, and managerial preferences also must be considered. The NBAA (National Business Aircraft Association [NBAA], 1997), identified eight factors to consider in determining the number of pilots. These included:

1. Number of aircraft,
2. Company flight time and rest considerations,
3. Availability of the ADM or chief pilot to fly,
4. Number of concurrent trips,
5. Number of consecutive flights,
6. Number of night flights,
7. Vacation policy, and
8. Training policy.

The basic computation in determining the number of pilots involves taking the total number of days or hours a pilot is available for flying per year and dividing it into the total time required. The number of days required for time off, vacation, holiday, sick time, and training/certification requirements is estimated.

For example, normally pilots are allowed two days off per week, two weeks of vacation per year, one to two weeks for sick time, 11 holidays, and two weeks (10 days) of training time. This equates to approximately 140 total days of non-availability (using 14 days for vacation, 5 days for sickness) and 225 days of availability. Next, a manager determines the total number of flight days needed per year for the company aircraft. For a six-day-a-week flight operation using a two-crew aircraft, 624 flight days need to be supported, or 312 times the two-crew requirement. This requirement divided by 225 gives 2.77 or three pilots required for a one-aircraft operation. For a seven-day-a-week operation, 3.23 air crew members are needed.

If a two-crew aircraft is planned to be available six days per week, 52 weeks per year, sixteen hours per day, then the total flight hours estimated for manning purposes are 9,984 hours. For more aircraft with similar flight requirements or different duty hours for the pilots, different piloting hours may be required. But staying with 9,984 hours means that 5.55 pilots are required, if each pilot is available 1,800 hours. The difference is in the number of hours of pilot availability per day. If instead of eight hours per day, a sixteen-hour pilot availability is used, then the estimate is back to 2.77 pilots.

If 5.55 was rounded down, five pilots are needed for one aircraft with a two-person crew. Perhaps the ADM can pick up some flight time, or the pilots will be used for more than eight hours per day. Whatever the situation for a company, flight duty hours or flight days need to be determined as a basis for hiring pilots. Added to this is the concern for hiring captains or first officers. Hiring all captains means a huge salary budget; hiring only two captains doesn't give flexibility for illness, special programs, and other activities. Perhaps three captains are needed. A *simple method* of determining the number of pilots is to use a factor of 1.5 crews, or three pilots, for a two-crew aircraft.

Determining the number of maintenance personnel follows a similar pattern. The NBAA recommends consideration of six factors in determining the number of maintenance personnel required:

1. Type and number of aircraft,
2. Home-base location,
3. Flight-route structure,
4. Utilization rate by hours flown,
5. Proximity to overhaul and repair facilities, and
6. Supply points for spare parts (NBAA, 1997).

First determine the total number of maintenance hours needed, then deduct non-availability time. Normally, it takes *four hours* of maintenance for every flight hour. If an aircraft is used 500 hours per year, then 2,000 hours divided by the total availability of the maintenance person equates to the number of workers required. The NBAA estimates a maintenance worker is available 1,577 hours per year. This gives 1.27 or two maintenance employees. In this case, more than one maintenance expert is recommended to cover for time off and work involving more than one person. Tailoring the number of employees to the company is the task of the ADM.

Salaries

Important not only to each employee but also to the company is the amount earned per flight department employee. The obvious answer is to pay people adequately for what they are worth, but what is adequate? *Business & Commercial Aviation* and *Professional Pilot* publish average salaries of surveyed corporate aviation employees. Additionally, the NBAA provides guidance upon request. These results will provide reviewers with the salary range for types of aviation positions; however, it is apparent that several major factors have to be considered. The type of aircraft being operated, the size of the corporation, geographic location of the company, aviation department size, individual experience and expertise, and, as reported by aviators, even the relationship of the ADM to the CEO. If the flight department is considered vital by company managers, then salaries will be considerable; if the department is not overly promoted, then salaries may be less than desirable.

Several managerial concerns revolve around the salary issue. During past and present times, corporate flight departments have been used by pilots to gain flight experience before working for the commercial airlines. Pilots accepted lower pay to gain flight experience and to wait for an opening with the airlines. Corporate managers were satisfied to keep labor costs down, but what was not considered was the cost of recruiting and retraining new air crew. Approaching the new millennium, pilot pay is increasing, along with improved work rules and company benefits. This has encouraged many pilots to remain in corporate aviation, even though a slightly higher wage could be earned with the airlines. Longevity with a company has become an issue versus working for an airline, which may not continue to operate or may drastically change its operation. On the other hand, smaller corporate aircraft operations may quickly disappear due to budget cuts or changing management with different operating philosophies. Some pilots have been suddenly informed that there was no job. One aviation manager reported that the more volatile the company operation, the better the pay. Conservative managers were seen to provide lower pay, but with more certainty of operation (E. White, personal communication, October 8, 1995).

During personal surveys of corporate aviators, a wide range of salaries was reported based on a variety of aircraft types and company sizes. Overall, the more complex or expensive the aircraft, the more the pilots and other aviation personnel earned. Aviation department managers controlling a large fleet realistically earned more. Some aviators reported that the salaries reported in the professional journals were on the low side, but statistics, whenever reported, have to be reviewed in context. Averages do not identify the outliers—those extreme high or low amounts. The median is also not a very good identifier for those interested in reviewing salaries. What is needed are the mode, range, and averages identified, along with particulars of the sample or surveyed group.

Surveyed air crew members were well aware of salary ranges for their specialties, and usually reported that they earned slightly less than the median. For example, two copilots reported a range of $19,000 ($35,000 minus $16,000) with a median of $25,500, but stated that they earned $24,000 for flying a light jet (< 20,000lbs MTOW). Company managers and/or ADMs can accomplish personal surveys to determine salary ranges for air crew members; however, contacting the experts at NBAA or at other aviation agencies is highly recommended to establish air crew pay. Many factors have to be considered; for example, the size of the company's flight department, type of aircraft being flown, state of the economy, geographic location of the firm, time in service, company pay policies, and others.

Other Department Personnel

Various employee job descriptions have been presented earlier; however, while pilots and mechanics have federal certifications to acquire, not every employee in a flight department is required to be certified by the federal government. Academic experience and good communication skills are most important. Academic programs help people improve writing, speaking, thinking, and research skills. Other special skills learned in schools include good office procedures, computer operations, time management, and group interaction skills. Good communication abilities enable workers to get the job done together effectively and efficiently.

Flight dispatchers, flight attendants, schedulers, general office workers, flight controllers, cleaners, caterers, receptionists, and others are needed for corporate flight operations. When an ADM in a smaller flight department can have employees accomplish a variety of jobs, then economies of scale result. Additionally, some aviation department functions may be accomplished within an already-established company operation. Air transportation planning and scheduling may be accomplished by a company's transportation office so that any requests for the use of a firm's aircraft would go to the transportation office, which, in turn, would provide the aviation department its air transport requirements. Other managers might have all matters relating to air transportation routed directly to the aviation department; however, executives use aircraft for only part of their travels. Ground transportation, hotels, and other matters may be handled by another office. If an aviation manager has to arrange all necessary functions for an executive's travel plan, then appropriate personnel have to be provided.

Budgeting

A major managerial activity is budgeting. While accountants probably enjoy their work, other managers dread the financial computations. This is one area where an aviation department manager should excel. Knowing the firm's mission and strategic plans is a first step in managing; the next step is to plan for people, facilities, and equipment. These assets cost money, and budgeting is when a good aviation manager proves his or her value. The firm's top management expect to hire an aviation department manager who can establish and coordinate what is needed to organize and operate the aviation department.

How much money is needed and how should the aviation department be costed? The ADM can identify the resources needed and determine costs of acquisition and possession for the equipment necessary. Vendors, other operators, publications, catalogs, and personal experience are used in budgeting. Facility planning and costing requires additional help from appropriate experts. Advice and direction from the firm's financial officer are mandatory. Also, outside agencies (NBAA and others) can provide ideas and suggestions for the equipment necessary depending on the type and size of aircraft operation. In the end, the ADM prepares a listing of people, equipment, and facilities needed to operate in support of the firm's mission and identifies the costs involved in these assets.

Top management usually assigns a budget level based on the identified needs provided by the ADM and recommendations of the financial officer. To assure good management, it is highly recommended that the ADM identify a budget based on the flying mission; the problem is that this mission can change at a moment's notice and cannot be controlled by the ADM. Therefore, it is advisable to identify a *flexible budget.* This means that the ADM should set a budget to operate an aircraft operation at so many flight hours at a specific level.

If the flying hours increase or decrease, then a sliding budget scale should be used. Stories have been heard of financial officers and top executives criticizing ADMs for poor budgeting skills when, in fact, the ADM operated the firms' aircraft as requested. For example, with a 400-hour flying requirement, the ADM should set a budget; for 600 hours another budget, and so forth. Then the ADM should clearly state that if top management uses its aircraft at so many hours, a set amount of money is needed. The budget is then managed based on the company's need, and the major effort of the ADM is *accurate costing.*

Keeping accurate records of expenses is necessary, and planning ahead for new equipment, personnel, and facilities takes careful consideration. ADMs who try to keep costs down by using old equipment or not purchasing new items on a scheduled basis use false economies. Unknown or unplanned expenses dictated by top management, the economy, or regulatory officials may also cause budget changes.

Fixed expenses, including salaries, hangar fees, maintenance, training, employee benefits, administration, and many others, can be costed. *Variable expenses* based on flying hours and aircraft use have to be planned as best as possible. These expenses include fuel, oil, landing fees, travel costs, special equipment, maintenance, and other critical items.

Allocation of the expenses of the flight department must also be determined. Some top managers simply charge the entire operation to the company; others charge the expenses to various company departments. Still other costing methods are used. A percentage of the flight operation may be charged to the company, and then passengers may be charged on a trip basis, mileage basis, hourly basis, or a combination of these. Some managers cost their flight operations like a commercial airline and charge customers based on the trip fare. All costs are ultimately borne by the firm; it may just be a managerial matter to identify who is using the firm's aircraft so that company budget planning is specific. Charging costs to specific departments helps to identify the departments and the extent of use; however, if air travel is changed because of the record of use, then a question of need exists. The bottom line is that managers should plan to use the firm's aircraft if essential to the firm's mission, is cost effective, or is efficient.

The NBAA's Travel$ense® program previously identified can help managers make a cost comparison of airline, charter, and corporate flights. Declaring and managing an aircraft as a business tool means that costs are kept to an optimum level. What optimum means depends on specific aircraft users and management.

Budgeting as an element of planning and control is a skill that can be learned. Any manager who has a sense of dread in budgeting should make this task his or her top priority because the aviation activity depends on this being effectively completed. Understanding income statements, balance sheets, financial ratios, zero-based budgeting and other methods and how they are used is the key to successful budgeting. A company's financial officer may take pride in helping an aviation manager to learn this process, because in the end, the company's mission is at stake. Using weekly, monthly, quarterly, and other periodic financial reports enables a manager to control the operations and to plan for future operations. Combining the skills of being able to plan and control a budget with management and leadership skills in operations will prove extremely successful.

Operations Manual

FAR Part 91, General Operating and Flight Rules, does not require an operations manual for a private aircraft operator, but if a firm's management decides to operate under FAR Part 135, Air Taxi Operators and Commercial Operators, an operations manual is mandatory. Why the difference? Certificate holders who serve the public must maintain certain minimum safety and operating standards recognized by the FAA. These standards are set within the FARs. Company managers who decide to use aircraft for private use are not required to meet public operating standards, although safety is important for both operators. Certain corporate managers have stricter flight operations rules than the airlines.

It is considered practical and highly recommended to develop and use a corporate operating manual for the aviation department as its use greatly outweighs its non-use; an NBAA member must use an operations manual. Company aviation policies and procedures can be clearly established and followed. Also, company manuals have the legal effect of binding employees to a contract. While the FAA reviews FAR Part 135 operators' operations manuals and may levy fines for non-compliance, the FAA inspectors may consider a violation of a FAR Part 91 company operations manual as a company matter only. An excellent policy to follow is that if a policy is unwarranted or not followed, it should be reviewed for deletion, change, or enforcement. During special situations, altering a company policy based on a critical mission requirement does not warrant a review for non-compliance. The key point here is that the *critical situations* do not become the norm, because the main purpose of the operations manual is to establish standard procedures. Deviations from recognized and accepted procedures mean that the procedures need to be reviewed for possible change or the deviations be stopped through careful planning and attention by management.

What should be contained in the company's operations manual when operating under FAR Part 91? The answer is that each manual should be tailored to the specific company. Some manuals have been seen with 80 sections and others with only 4 sections. Extracts of operation manuals for a FAR Part 91 and 135 operator and only a FAR Part 91 operator are provided in Appendix H; the sections and topics are excellent examples of what is found in good manuals. In Appendix H, eight sections are identified as follows:

1. Duties and responsibilities,
2. Personnel policies,
3. General operating procedures,
4. Standard operating procedures,
5. Standard callouts,
6. Emergency procedures,
7. International procedures, and
8. Administrative.

Also note in Appendix H, that at the bottom of the page of the FAR 91 and 135 operator, FAA acceptance is noted, and the date of revision is recorded. All names have been excluded upon request.

Reviewing the FARs provides guidance for those developing an operations manual; also talking to others who use manuals, to the NBAA and other aviation experts, and to company managers will be most helpful. Some basic headings that should be considered are *Policies and Organization, Flight Operations, Maintenance, Emergency Procedures,* and *Record of Revisions.*

The contents of any operations manual should supplement the FARs, but may, in some instances, be duplicated as a matter of emphasis. The bottom line is that any manual should be written clearly and concisely for the specific company involved. Contents should be kept to an absolute minimum, but everyone in the company should be aware of its contents. A copy should be given to each pilot, one placed in each company aircraft, and one copy distributed to each company department and other offices as desired. Below are sectional topics identified by Whempner (1982, pp. 137–141) that remain pertinent for consideration.

1. Policies and Organization
    a. Goals, objectives, and company management philosophy.
    b. Organizational chart.
    c. Personnel duties and responsibilities.
    d. Scheduling authority.
    e. Scheduling priority.
    f. Authorized travel groupings.
    g. Statement of authority.
    h. Passenger conduct.
    i. Route changes.
    j. Flight crew duty limitations.
    k. Personnel procedures.
2. Flight Operations
    a. Dispatching procedures.
    b. Aircraft operating limitations.
    c. Weather minimums.
    d. Cockpit management.
    e. Pilot training and proficiency standards.
    f. Flight crew physicals and health limitations.
    g. Passenger briefing and cabin duties.
    h. Aircraft refueling procedures.
    i. Handling and securing aircraft.
    j. Flying equipment.
    k. Customs procedures.
    l. Aircraft stores (supplies).

m. Communications with home base.
n. Over-water operations.
o. Fuel conservation procedures.

Maintenance and Emergency procedures generally focus on a firm's capabilities, facilities, and actions for emergencies. Aircraft inspection procedures, maintenance procedures away from home base, maintenance training, equipment use, and related activities involving maintenance should be presented. Emergency procedures relating to aircraft operations, security, safety, accident reporting, and communications should be clear to all users.

Standardization of activities involving the aviation department is a main reason for developing and using an operations manual. The procedures to be followed in the manual should support management's substantial investment in the flight department, its people, equipment. aircraft, and other vital resources. Some managers keep the manual very brief in order to make it more effective; others provide more details in order to make actions more definite. Every air crew member should be familiar with the FARs and other regulatory procedures, but company procedures and policies may not be so well known. Some of the many items that an aviation department manager and company management should consider in its manual are identified in Table 6-5. This is not an all-inclusive list, but it does identify pertinent topics for management.

Table 6-5. Flight Management Items

Smoking policy	Crew briefings
Departure runway requirements	Use of drugs
Departure airport weather rules	Use of intoxicants
Carriage of weapons	Hazardous materials handling
Aircraft weight/balance	Enroute requirements
Crew rest away from home	Crew expenses
Destination airport requirements	Ground deicing
Reporting for duty	Aircraft fueling
Alternate airport requirements	Emergency procedures
Passenger handling procedures	Preflight operations
Using the MEL	Cockpit management
Severe weather operations	Postflight procedures
Extended over-water flights	Use of autopilot
Crew readiness/alertness	Trip expenses
Scheduling/dispatching	Crew billeting

Air Crew Expenses

During the early days of corporate aviation, air crews took wads of dollar bills, credit cards, and blank checks. Today, air crew need to plan ahead by checking how services will be paid, what credit is available, and plan for those unexpected bills. Fuel, cleaning, catering, hotels, parking, rental equipment, and other items may need to be paid for before the aircrew is allowed to depart a destination, especially in a foreign country. The aviation department manager should ensure that proper procedures have been planned to meet these costs when realized.

Air crew travel costs and reimbursements should be known in advance. Expenses beyond the norm should be justified by the air crew, and if possible, large expenses should be approved prior to being made. A policy to check with the ADM or representative can be easily followed, but at remote locations, good judgment is all that is needed. An opportunity for advance travel payments should be provided to all air crew members.

Air Crew Certificates and Ratings

Can a private pilot fly a corporate aircraft? Which type of air crew rating is required? The NBAA stated that "Business aircraft operators should require that the pilot-in-command hold an airline transport pilot (ATP) certificate" and that other pilots "hold at least a commercial certificate with an instrument rating" (1997, p.12). Besides this, the NBAA also recommended that each captain or PIC "have logged a minimum of 3,000 flight hours with 500 flight hours in the type of aircraft" and that each first officer or second-in-command "have logged a minimum of 1,500 flight hours." FAR Part 61, Sections 61.5 and 61.55, should be reviewed for certificates and ratings and second-in-command qualifications, respectively.

A private pilot may fly him- or herself in the furtherance of business. FAR Part 61.118 provides that a private pilot may, for compensation or hire, act as pilot-in-command of an aircraft in connection with any business or employment if the flight is only incidental to that business or employment and the aircraft does not carry passengers or property for compensation or hire. FAR Part 61.139 provides that the holder of a commercial pilot certificate may:

1. Act as pilot-in-command of an aircraft carrying persons or property for compensation or hire.
2. Act as pilot-in-command of an aircraft for compensation or hire (FAA, 1997).

Subpart F, FAR Part 61, provides detailed information concerning the eligibility and ratings for airline transport pilots. NBAA's recommendations for airline transport and commercial certifications for corporate pilots are strongly favored because corporate aviation operations require highly qualified and experienced pilots. Review FAR Part 65 for certification of airmen other than flight crew members.

Other Flight Management Issues

To meet the mission of the aviation department, an ADM has many administrative and operational matters to work. Keeping costs down while meeting the aviation department's mission is a major task, and ADMs have to use good human resource management skills, apply knowledge and understanding to general and technical activities, and provide expert direction and control. The use of business aircraft has proved to be significantly beneficial in company developments. Better management of this activity requires demanding effort and personal skills. Most demanding is the continued justification of a function that clearly adds to the cost of doing business, but does not clearly show the tangible value. Yet, corporate operators who use business aircraft attest to the fact that it is a value-added activity. From November 1996 through May 1997, a group of 28 aviation-oriented students and aviation pilots and managers were asked to identify the top 10 concerns for corporate aviation operations. In addition, these respondents were asked to prioritize their top two items. In Table 6-6, their top two aviation management priority issues are identified.

Table 6-6. Aviation Management Issues

Top Priority	Second Priority
Safety (7)	Fraudulent parts
Experience (2)	Time and money (2)
Flight department operations manual (4)	Air traffic congestion
Planning	Flight training (2)
CFIT prevention training	Maintenance
Crew briefings	CRM training
Economy (3)	Cockpit management
Human resources (3)	Planning
Budget	Crew duty times (3)
Hub and spoke changes	Security
Aircrew certificates	Scheduling/dispatching
Crew rest	Budgeting (4)
Emergency procedures (2)	Emergency operations
	Regulatory compliance
	Qualified ADM
	Standardization of operations
	Operations manual (2)
	Comfort
	Tax issues
	Equipment

Note: Items count once unless higher frequency is indicated in parenthesis. Total responses = 28 in each grouping. CFIT = controlled flight into terrain; CRM = cockpit resource management.

From the items identified in Table 6-6, it can be seen that safety was the top priority and having an operations manual was number two. Human resource management, budgeting, and training appeared to be major concerns. Most of the people who responded were pilots and aviation department staff members, with only one ADM responding. The issues that the aviation personnel considered important varied by job position and experience. It was significant to find that the focus was on personal jobs, not on overall accomplishment of the flight department mission. Perhaps surveying aviation personnel by position will result in identifying specific needs and concerns that can then be used to develop an awareness program for the aviation department managers.

Flight operations is the heart of the aviation department, but an ADM must develop a sense of personal worth and belonging to the entire operation in order to successfully meet the demanding mission. Corporate managers should provide the opportunity for not only the ADM, but for every flight department employee to improve his or her communication and job-related skills. Strategic planning will lead to long-range improvements and a stable and effective flight operation.

Summary

Various alternatives exist when deciding to use business aircraft for company operations. Top management may use the services of a fixed-base operator, contract with an aviation management firm, or develop the firm's own flight department. Fixed-base operators provide the full range of services, including providing aircraft, flight crew members, aircraft servicing, fueling, catering, and other functions—for a price. Fueling is a main function of a FBO. Examples of FBOs in the U.S. and Europe are provided.

If management purchases an aircraft but desires to have an aircraft management firm operate and control the aircraft, air transportation services may be supported but not necessarily with the firm's aircraft. To optimize the use of aircraft managed, management firms operate aircraft on an as-needed basis and use aircraft most suitable from their inventory.

The purpose of an aviation flight department is the effective use of business aircraft to achieve the air transportation objectives of the company through safe, professional quality services. Establishing an aviation department requires the hiring of an aviation department manager or chief pilot, depending on the size of the operation. Whether the ADM should be a pilot depends on many factors, most notably the size of the operation and the quality of the person. For smaller flight operations, the manager may be the chief pilot and fly on an as-needed basis, especially to maintain flight currency. For larger operations, the ADM may be too involved in the day-to-day and long-range planning to be able to spend time flying on company trips. The ADM has many administrative, operational, maintenance, and miscellaneous duties to perform. Management skills, especially human resource management skills, are critical to the success of an ADM. Getting things done through people is management, while being able to influence others is leadership. For the ADM and other flight department personnel, effective communication is a primary requirement.

In order to develop a flight department, the aviation department manager should be hired first. The ADM can then help write the job descriptions and be active in the hiring process of the other department personnel. If an ADM is not a pilot, it is critical to hire a chief pilot with the necessary technical and communication skills. Various skill areas and expectations are identified in the chapter tables. Placement of the flight department is also critical to the success of the activity. Having the ADM report as high up in the organizational hierarchy as possible, yet maintaining effective communication and direction, is recommended. Descriptions of the duties and responsibilities of various aviation department personnel are provided from a firm's actual operations manual and give an indication of personnel requirements. Recruiting and selecting qualified personnel are reviewed, along with a hiring model. Important also is the determination of the number of pilots and mechanics necessary to have an effective and efficient operation. Having reasonable salaries for the people is extremely important, while determination of the value of the department may be based more on intangible factors. A critical duty for the ADM is to justify the existence and operation of the flight department, and this is best accomplished by presenting the value of the operation through reports and records of service.

Budgeting plans can also be developed from trip reports. Important to the entire flight department operation is the use of an operations manual. Various topics were presented that can be considered for development, and an extract of a firm's operations manual is provided in Appendix H. Finally, many flight management topics are presented, including personal appearance and conduct, travel expenses, and air crew certificates and ratings. A personal survey resulted in identifying safety as the major issue in the operation of the flight department.

Activities

1. Review the literature or contact a fixed-base operator to determine the range and type of services provided to aircraft owners and operators.
2. Contact an aircraft management firm to determine the type and extent of contractual services available.
3. Visit a firm that has its own flight department and talk to the managers and employees about its operations and requirements.
4. Review a firm's operations manual for duties and responsibilities designated.
5. Prepare a comparison chart of corporate aviation salaries relative to various major job positions within a large corporate flight department.
6. Review the requirements in FAR Parts 91 and 135 to differentiate between the rules to follow.
7. Talk to corporate aviation personnel about their job responsibilities and determine the job qualifications appropriate to each activity.

Chapter Questions

1. Justify a policy for air crew travel expense.
2. What is the purpose of the aviation flight department?
3. Why should an operations manual be developed for a corporate flight department when FAR Part 91 does not require one?
4. Why should staffing the flight department be a joint effort? Who should be involved in staffing?
5. What are the advantages for placing the aviation department as high in the organization as practical?
6. Identify major sections for an operations manual and explain their importance.
7. Develop an organizational chart for a large flight department. Explain the organizational levels.
8. Explain when it is best to use a fixed-base operator or aviation management firm to control a firm's aircraft.
9. Should the ADM be a pilot? Explain.
10. How does leadership differ from management?
11. Identify and explain five major skills that an aviation department manager should possess.
12. Where should flight department personnel be obtained?
13. Explain an effective method to use in determining the number of pilots and mechanics for a flight department?
14. Explain the significance of being able to effectively budget for a flight department.

CHAPTER 7

Flight Operations

Overview

After establishing an aviation department, managers have many important functions and activities to manage. Flight operations involves more than just flying aircraft; it involves many support activities leading up to and beyond the actual flight. Scheduling, dispatching, catering, administration, maintenance, and planning may begin and continue long after a flight ends.

Several of the preflight, inflight, and postflight activities are reviewed in this chapter, with emphasis on scheduling/dispatching, fueling, and international flight operations. Flight planning software are emphasized.

Goals

- ✈ Understand domestic and international flight operations activities.
- ✈ Discuss aircraft scheduling and dispatching.
- ✈ Discuss aircraft self-fueling and contract operations.
- ✈ Understand flight planning requirements.
- ✈ Discuss international flight activities and concerns.
- ✈ Understand methods and procedures for controlling flight operations.
- ✈ Discuss documentation requirements.
- ✈ Explain the meaning of SARPS and PANS and other international rules.
- ✈ Understand the role and activities of international flight handlers.
- ✈ Discuss the activities of IBAC.
- ✈ Explain the significance of MNPS/RVSMs.
- ✈ Discuss flight planning programs.

Introduction

To begin flight operations, a request for service has to be received and then flight planning accomplished. Corporate aircraft operators usually identify an established transportation department to coordinate air services or have the aviation department manager handle requests through a scheduling office. Preflight activities for the air crew begin when they are notified of a scheduled flight. For international flights, planning involves careful attention to many dissimilar flight activities that air crews do not have to plan for in U.S. domestic operations. Documentation becomes more of a major concern for these flights. The air crew must maintain flight currency in their aircraft, but must also ensure that additional personal documentation, that is, visas, passports, and others, are ready.

Flight operations require compliance with many federal, state, and local regulations and company policies and rules. Applicable operations manuals must be followed to provide quality air transportation service; additionally, ICAO and other international rules also apply when managers use their aircraft for traveling to foreign destinations. Corporate managers who have invested substantially in their aircraft, personnel, equipment/facilities, and other resources desire to meet their transportation needs with safety first and foremost in mind, but also with efficiency of operations. Some of the key topics for flight operations are listed in Table 7-1. To prepare air crew and other flight department personnel in these areas, the NBAA, flight training schools, and other helpful sources are available and should be used in flight department training. A firm's operations manual should also be used to standardize the desired flight items.

In a brief, informal April 1997 survey of 15 air crew members, it was interesting to note that only language/cultural differences repeated as the top item of concern for international flight operations. Different items were identified by these air crew members as topical flight

Table 7-1. Flight Operations Items

Diverts	Crew briefings	Use of drugs
Departure runway requirements	Use of intoxicants	Hazardous materials
Airport weather rules	Carriage of weapons	Crew expenses
Reporting for duty	Oxygen requirements	Crew alertness
Using an MEL	Use of autopilot	Emergency operations
Ground deicing	Preflight	Postflight
Severe weather operations	Fueling	Maintenance
Passenger briefing	Crew rest	Passenger handling
Smoking policy	Communications	Security
CRM	Documentation	Flight speeds

Table 7-2. Flight Issues Identified in Air Crew Survey	
Top Priority Issues:	On-time travel, financial considerations, personal documentation, training of air crew, air crew currency, aircraft documentation, maintenance facilities, airport security, safety, planning, flight planning, navigation and communication equipment, language/cultural differences (2), MNPS certification
Other Issues:	Aircraft reliability, operations manual, flexible scheduling, access to management, funding of flight, regulatory compliance, airfield compatibility, flyaway kit, passports, visa, spare key, GPS navigation aids, IBAC ID cards, health check, airman certification, proof of citizenship, airworthiness certification, aircraft flight manual, flight authorization letter, credit cards, Plan B, extended range fuel tank, flight planning, accommodations, handling agents (4), customs (6), cash (9), aircraft documentation (7), survival equipment (6), aircraft tiedown/locks (2), maintenance spares (3), overwater equipment (3), aircraft maintenance (5), security/anti terrorism measures (5), communications (5), ground transportation (2), airport restrictions, weather briefing (3), fuel, ICAO rules, ATC, MNPS.

Note: April 1997 survey of 15 air crew members taking corporate aviation operations course, Embry-Riddle Aeronautical University. Frequency in parentheses; otherwise item reported only once.

operations issues to be addressed for international flight (Table 7-2). From the survey results, it appears that air crews expect their flight qualifications to be met when considering flight planning, but concern themselves with many of the issues that affect documentation and ease of movements. Several of these issues will be addressed later in this chapter.

General Flight Topics

In Table 6-4 and in Appendix H, various flight management topics were identified. Aviation associations like the NBAA, GAMA, University Aviation Association, and others provide training and guidance in many of these areas. Additionally, generic issues (e.g., management, accounting, etc.) can be studied and learned through academic institutions and training programs. Managing and working within an aviation department requires general and specialized skills for effective and efficient operations. For flight operations, however, different issues must be considered and managed.

Transportation Requests

When a firm's aircraft exists solely for the top executive (s), scheduling may be easy to do. At the other side of the spectrum, when company aircraft can be used by any company employee, then scheduling may be a real headache. Most managers of large aviation departments operate somewhere in between; in other words, company aircraft are scheduled on a routine basis the greater percentage of the time and provide ad hoc or as needed air transportation the remaining time. In England, for example, the previous aviation department manager of Ford Air Transport set a regular flight schedule 80 percent of the time and provided on-demand flight service only 20 percent of the time. This enabled Ford's many employees to know and depend on an excellent air transportation service.

When top management declares its aircraft to be available for everyone in the company, then it is recommended that a transportation request form be used. Various company personnel require air travel services based on sometimes different company missions, but may require it at the same time. Who gets the support when limited aircraft are available? The ADM should not decide this, but the firm's top management should through setting air travel priorities in an established company policy. The critical decision of who should get served depends on the *company needs* at the time. Having a transportation request form (NBAA-recommended example in Figure 7-1) enables the appropriate authority to assign the aircraft in the company's best interests. Perhaps a senior manager may be denied use of the firm's aircraft when a maintenance expert has to be flown to fix a vital piece of company equipment at a distant location.

The ADM's job is to manage the firm's aviation assets as needed, not to decide the mission priority of company employees. Having the aviation department high up in the organizational structure helps in this situation by allowing senior management to establish the importance of the flight. Senior managers know better than others what the mission importance of the travelers will be. Additionally, budget concerns may be a major consideration when deciding to use company aircraft.

Scheduling/Dispatching

Corporate flight scheduling/dispatching appears to be one of the most difficult and seemingly complex requirements in some flight departments. In others, it is no problem at all (Figures 7-2 and 7-3). Why? Basically, it depends on a company's aviation-use policy, the aircraft available, the passengers/cargo requirements, air crew, aircraft maintenance, logistics support, regulatory issues, airport authorities, customs, and others. FAR Part 65, Subpart C, identifies certification requirements for aircraft dispatchers. Importantly, an applicant for an aircraft dispatcher certificate must have at least two years experience within the three years prior to certification in one of several work areas. A written examination on dispatch regulations, weather, and meteorology must also be passed. Skills must be shown through a practical test on aircraft weight and balance, using cruise control

SAMPLE

Company Aircraft Travel Request Form & Passenger Manifest

Aircraft:		Requested by:		Dept:	Date:	
Flight Requested	Date	Time	Flight Confirmed		Date	Time
From						
To						
From						
To						
From						
To						
From						
To						

Passenger Name, Affiliation, and Address	Department to be Charged	Employee #	Authorizing Name & Title	Aviation Dept. Confirmation
1.				
2.				
3.				
4.				
5.				
6.				

Figure 7-1. Company Aircraft Travel Request Form and Passenger Manifest. Extract from *NBAA Management Guide,* 1994, p. 20. Used with permission.

charts, fuel and oil computations, using an operations manual, and on many aspects of air routes and airport requirements (FAA, 1997).

Gormley (1990) identified several items reported by flight department managers as causing scheduling concerns. These included weather difficulties, unrealistic passenger requests, executive privileges, air crew manning, aircraft availability, and others. U.S. domestic operations differ significantly from international aviation operations, and aviation managers providing company air transportation services to foreign countries realize that many scheduling and dispatching issues have to be planned as far ahead as possible to result in the service desired. The NBAA lists nine factors to consider for scheduling guidelines. These focus mainly on determining who has authority to request and approve the use of company aircraft, passenger handling, time requirements, and aircraft use.

It has been known that some executive wives can make various air transportation demands on company flight managers; likewise, executives themselves can ignore scheduled takeoff

Figure 7-2. MAGEC Aviation Flight Dispatch. Trevor King (left) and Steve Gannon (on the phone) helping Don Walbrecht in flight planning.

Figure 7-3. MAGEC Aviation Flight Dispatch. Trevor King (left), Steve Farmerns and Katy Preece showing Sally Kovach flight operations.

times or cause unexpected flight delays. For international flights with scheduled aircraft slot times, this may mean a delayed aircraft until new times can be arranged. The passengers who cause the delays may be unaware of the complexity of the problem and may even criticize the aviation managers for poor service.

Formal procedures to schedule company aircraft are necessary in many flight operations while other flight managers follow general use policies. When aircraft and air crew members are limited in a company and travel demand exceeds supply, then the aviation manager must resolve the problem by applying an established company policy or by seeking direction from top management. The aviation manager is not the person to decide who gets to use the company aircraft; company officials who understand the business at hand and its urgency and priority should be responsible.

The use of a company travel request form (Figure 7-1) can be extremely helpful in this procedure. First, information required for the flight scheduler can be obtained from the form. Second, the approval for the transportation can be routed to a senior executive responsible for prioritizing conflicting requests. Information needed for air services should include as a minimum (a) the name(s) of the travelers, (b) origin and destinations, (c) times desired, (d) cargo/mail/baggage requirements, (e) funding office, (f) special requirements, and (g) the reason for the request. A comment section allows the requester to provide important information justifying the air transportation service.

The sophistication of the scheduling process generally varies with the size of the aviation department. For small operations or companies having only one or two aircraft, it can be a simple matter of checking the availability of the aircraft and the calendar. The use of a schedule board allows a scheduler a quick overview of the big picture of committed air services. Software is available for more complex scheduling operations and these can be obtained by contacting the NBAA or reviewing Avcomps programs. *Business & Commercial Aviation International* publishes a directory of Avcomps programs usually in the fall of each

Table 7-3. Scheduling Software Vendors

Name	Address	Software
Air Support	P.O. Box 24 DK-7190 Billund Denmark	Pre-Flight Planning Systems-PPS
CAMP Systems, Inc.	999 Marconi Avenue Ronkonkoma, NY 11770 516-588-3200	Andromeda FS for Windows 95
Professional Flight Management	555 E. City Avenue Suite 530 Bala Cynwyd, PA 19004 610-668-2001	PFM Windows
SeaGil Software Company	3187 Corsair Drive Suite 250 DeKalb-Peachtree Airport Atlanta, GA 30341 404-455-3006	Bart and Bart 4 Windows
Universal Weather & Aviation	8787 Tallyho Houston, Texas 77061	FlightPak Pilot's Choice

year. Check the latest issue. Some of the more effective software programs are identified in Table 7-3 and at the end of this chapter.

Important to all systems, manual or computerized, is keeping accurate records. Besides air transportation services, a scheduler may have to arrange ground transportation services, hotels, and other support items. Companies that operate a transportation department may have air transportation requests routed to it, and the aviation manager receives only an approved air service order. Coordination is made by the transportation office with the flight department for aircraft availability. For corporate aircraft schedulers, estimated at around 600 in 1996, formal training programs to further their knowledge, skills, and contributions to their aviation departments and companies are recommended.

Crew Duty Limitations

Fatigue can be a serious issue for air crews who do not have established flight crew duty limitations. FAR Part 91 does not set duty or flight time limits, while FAR Part 135, Subpart F, has limits for scheduled and unscheduled flight duties. In the past, some corporate air crew personnel reported extremely long duty days involving long hours of flight, waiting delays, and then additional next-day flight commitments. Safety of flight and crew must be a top priority for aviation managers and corporate managers. The ADM should coordinate with the air crew and recommend crew duty limitations that can then be established in the company operations manual. Only in very extreme situations should any deviation from these limits be made. An example of established flight- and duty-time limitations for a company can be found in Appendix H, Section 2.03.

The FAA defined *flight time* as the "time from the moment the aircraft first moves under its own power for the purpose of flight until the moment it comes to rest at the next point of landing" (Jackson & Brennan, 1995, p. 1–8). This is normally called block-to-block time. Cannon (1988) defined *duty time* as the "total time involved in the performance of flight" and *rest time as* the "time free from the duties involved in the planning, coordination, or execution of a flight" (p. 68).

Should a company manager establish duty and flight time limits? Concern for safety dictates yes to this question. Some people report that air crews might use the time limits as a tool or excuse against management or even the passengers; however, proper crew rest is necessary for the safe operation of corporate aircraft, or any aircraft for that matter. The concern for everyone's safety overrides any manager's wish to go the extra mile for the sake of the company or the executive. Good personal communications throughout the aviation department and company should leave no doubt in anyone's mind that an air crew operates on behalf of everyone. Employees who limit the boundaries of mission accomplishment and have poor personal interrelationships will soon be looking for other work; however, managers who extend the boundaries of unreasonableness jeopardize the mission and people. Besides safety, morale and well-being of flight personnel must be considered to maintain quality staff.

Two areas to consider in setting time limits include aircraft trips and remain overnights (RONs). Good judgment should be used to establish each time limit. Crews need adequate rest, and this involves consideration for individual privacy. Setting duty-, flight-, and rest-time limits should be a joint effort of the ADM, chief pilot, other air crew members, and top management.

Pilots can also prepare to prevent and manage fatigue, the most common cause being the lack of sleep. Getting plenty of sleep prior to the day of flight enables a pilot to start without sleep debt. Eating well and exercising regularly also helps in conditioning. Avoiding alcohol is a major step in preventing fatigue. During a flight, drinking a variety of liquids, using lumbar supports, snacking frequently, and keeping communications awareness helps. The latter involves checking flight and other data and air crew talk. Some pilots have planned their time prior to crossing several time zones by changing their sleep patterns at home so that physically and mentally they are prepared. Aviation managers assist their air crew by giving them plenty of advance notice, if possible, and by providing facilities, food, and equipment for air crew members to condition themselves.

Figure 7-4. MAGEC Aviation's Chief Pilot, Captain John Robinson. Photo courtesy of Raytheon Corporate Jets.

Personal Appearance and Conduct

When transporting company personnel, crew members are expected to be neat in appearance and well groomed at all times (Figures 7-4 and 7-5). Their conduct should reflect favorably upon themselves and their companies. Loud, abusive, offensive or other negative language or actions are never to be permitted. Company uniforms, although not required by the FAA, serve well to provide a good company image and instill good crew morale and behavior. In international flight operations, the wearing of uniforms by the air crew may be a major advantage in conducting flight operations in foreign countries.

Passenger conduct also must be of the highest order. Intoxicated passengers may become a serious flight safety matter; unacceptable personal conduct during flight may cause a pilot to end a flight before the final destination, and attention to this not only will hurt the people involved, but the company. Air crews should never be placed in the situation where flight safety is jeopardized by passengers or air crew personnel hesitate to properly act because of a concern for continued employment with a company. The senior passenger on board should

Figure 7-5. United Technologies Hawker 1000 and Captain David Isaacson after successful cross-country flight (Hartford to San Diego). Slight thunderstorm deviation over Kansas, but still great flight for United Technologies.

be held responsible for the conduct of all passengers; this policy can be presented in the company's operations manual. A brief but important notice prior to flight by the captain that a person is designated as the senior passenger on board and any deviations or passenger requests will go through him or her should suffice. Obviously, if the senior executive is on board, everyone will know—only when it may not be clear who is senior should notice be given.

Use of Intoxicants or Drugs

FAA regulations stipulate alcohol and drug limitations; for example, FARs 61.15, 91.17, 135.121, 135.249, and others. Every aviation manager should know and understand these regulations. In addition, company policies may be very strict in this regard and be emphasized in the company's operations manual. These issues are equally applicable to the air crew as well as the passengers. Passengers who bring unknown material aboard a company's aircraft may be asked to open the package or describe its content by a pilot who has the ultimate responsibility for the flight and all passengers on board the aircraft.

It is great to win the big sales contract or bring in a multi-million dollar deal, but a few celebratory drinks may result in aircraft damage or a serious incident or accident. Officials of aircraft management firms have stopped certain celebrities and executives from flying on aircraft because of improper flight conduct. The gain of income from flight services does not make up for the possible loss of life, damage to equipment, or poor conduct to those providing the service.

Carrying Drugs and Weapons

Qualified medical personnel, passengers carrying prescriptions, authorized armed guards, and other personnel known to the pilot-in-command (PIC) may need to be allowed aboard corporate aircraft. Again, the FAA rules need to be known and followed. Company policies about the carriage of weapons or drugs aboard company aircraft may be strict. An embarrassing situation can be avoided through the communication of company policy to passengers before any flight. The PIC of any aircraft must never be hesitant to communicate the rules and policies when suspicious or when a concern exists. Unforeseen factors may present themselves to the PIC at flight time or even during a flight; the PIC must be able to make the right decisions in these moments. In addition, the carriage of other items such as illegal funds, pornography, animals, plants, gifts, jewelry, tobacco, and art may be prohibited or restricted. It is the responsibility of the PIC to be in charge; therefore, it behooves the PIC to be cognizant of the applicable rules and to deal correctly and firmly with any situation. Suspicious baggage or packages may be checked, as well as checks made for the proper permits for weapons or other items, especially when traveling to foreign countries. Schedulers/dispatchers can use preflight time to notify passengers of important matters involving such transportation rules. An informative flyer or notice sent to scheduled passengers can be a normal procedure, and may prevent a serious incident.

Air Crew Expenses

During the early days of corporate aviation, air crews took wads of dollar bills. Today, air crews need to plan ahead by checking how services will be paid, what credit is available, and plan for those unexpected bills. Fuel, cleaning, catering, hotels, parking, rental equipment, and other items may need to be paid for before the air crew is allowed to depart a destination, especially in a foreign country. The aviation department manager should ensure that proper procedures have been planned to meet these costs when realized.

Air crew travel costs and reimbursements should be known in advance. Expenses beyond the norm should be justified by the air crew, and if possible, large expenses should be approved prior to being made. A policy to check with the ADM or representative can be easily followed, but at remote locations, good judgment is needed. An opportunity for advance travel payments should be provided to all air crew members.

Air Crew Certificates and Ratings

Can a private pilot fly a business aircraft? Which type of air crew rating is required? The NBAA (1994) stated that "business aircraft operators should require that the pilot-in-command hold an airline transport pilot (ATP) certificate and that other pilots hold at least a commercial certificate with an instrument rating (Section. 2.2)." Besides this, the NBAA recommends that each captain or PIC "have logged a minimum of 3,000 flight hours with 500 flight hours in the type of aircraft" and that each first officer or second-in-command "have logged a minimum of 1,500 flight hours" (Section 1.5). FAR Part 61, Sections 61.5 and 61.55, should be reviewed for certificates and ratings, and second-in-command qualifications, respectively.

A private pilot may fly him- or herself in the furtherance of business. FAR Part 61.118 states that a private pilot may, for compensation or hire, act as pilot in command of an aircraft in connection with any business or employment if the flight is only incidental to that business or employment and the aircraft does not carry passengers or property for compensation or hire. FAR Part 61.139 also states that the holder of a commercial pilot certificate may:

1. act as pilot in command of an aircraft carrying persons or property for compensation or hire, and
2. act as pilot in command of an aircraft for compensation or hire (Office of Federal Register, 1994, FAR Part 61).

Subpart F, FAR Part 61, provides detailed information concerning the eligibility and ratings for airline transport pilots. The NBAA Recommendations for corporate pilots are strongly favored. Corporate aviation operations require highly qualified and experienced pilots. Review FAR Part 65 for certification of airmen other than flight crew members.

Aircraft Fueling

One of the major expenses besides aircraft and crew members for an aviation manager is aircraft fuel. Fuel handling and fuel use must be efficiently managed by the aviation department manager as an important operating expense (these can average about 35 percent of the

Figure 7-6. MAGEC Aviation Aircraft Fueling Operations, Luton Airport, United Kingdom.

aircraft operating budget for some corporate operators). Some managers use self-fueling because they have no alternative, may desire to have full control, provide more timely service, have better security, and save money through bulk buying. Fixed base operators normally provide fuel service when self-fueling is not used (Figure 7-6).

Disadvantages of self-fueling arise from having to purchase a fuel truck, having an underground storage tank (UST) or other storage method, maintaining equipment, handling hazardous materials, and hiring and training qualified personnel. Compliance with strict regulatory requirements of environmental, airport, equipment, and personnel rules requires serious managerial attention and adds to the costs of self-fueling. Because airport FBOs make their living mainly with fuel services, having a positive and friendly working relationship with an FBO may be difficult for a corporate aircraft operator that has its own fuel operation. At some airports, state, airport, and other local regulations prohibit corporate self-fueling operations. Additionally, besides the equipment and training necessary for self-fueling, corporate operators may need to lease or build facilities on airport property. This may be a major expense in itself. Although there are many disadvantages for self-fueling, companies do own and effectively control their aircraft fuel operations.

When an FBO is not available or fuel is too expensive or cannot be obtained, fuel tankering is another option for aircrew. This is simply carrying extra fuel in the aircraft tanks. Flight planning charts can be used to calculate trip fuel values for the minimum fuel landing weight with and without tankered fuel. The additional trip fuel required for tankering and the net tankered fuel then can be computed to find the break-even cost for fuel at the destination. When fuel at the destination costs more than the costs of tankering, the latter should be done. Aircraft manufacturers provide fuel tankering charts to help managers determine the break-even point for costs.

A detailed analysis of the disadvantages and advantages, both tangible and intangible, has to be made before a final decision to self-fuel is made by top management. To lower fuel costs, the NBAA (1994) recommended several control actions. These included:

1. keeping the aircraft clean and light,
2. planning the flight carefully to fly at the most efficient speed,
3. minimizing engine operations on the ground,
4. using reduced power on takeoffs,
5. climbing cleanly and quickly to optimum altitude,
6. maximizing specific range,
7. optimizing enroute descents,
8. reducing aircraft weight,
9. maximizing the use of simulators, and
10. auditing flights for performance data.

Flight Preparations

Time limits for reporting to duty for air crew personnel may be set by a company or be left to the PIC for specific flights. A pilot's reading file is useful to keep abreast of changing company policies and other important items and should be checked at least monthly and *before* each flight. The PIC should ensure that the aircraft assigned is appropriate to the mission, has no outstanding maintenance discrepancies, and has the required books and documents on board. These include the aircraft flight manual and checklists, airworthiness certificate, registration, FCC radio station license, operations manual, appropriate weight and balance data, performance charts, and current navigation charts. For guidance in flight planning, review Sections 2 and 3 of the NBAA Management Guide, Flight Operations and International Operations, respectively (NBAA, 1994). Most importantly, flight operations rules in FAR Part 91 (FAR Part 135 also when applicable) should be closely reviewed and followed.

Preflight Planning

Following are items of consideration for preflight actions/planning for air crews and aviation managers. Corporate international flight planning is reviewed since this is complex and takes longer than domestic flight planning. The success of every flight depends greatly on the effectiveness of preflight planning.

Air crew and aircraft documentation (as noted above), airport entry requirements, and identification of international flight handlers need to be reviewed for international flights.

Some companies leave it to the PIC to ensure that air crew have personal documentation, such as passports, visas, and others available; other firms have a travel department or office to ensure personal documentation is available. Documentation is especially critical for international journeys; air crew members and aircraft without proper documentation can be stopped.

Air crew personnel must have appropriate visas for foreign countries. Working permits may be required in some instances. Personal documentation requirements for international flights are found in the International Flight Information Manual (IFIM), which specifies passport, immunization, visa, and other requirements by country. IBAC and NBAA also are helpful in providing planning information for international flights.

Minimum Navigation Performance Specifications

Flights within the boundaries of the North Atlantic minimum navigation performance specifications (MNPS) require a certificate approval by the FAA (FAR Part 91.705). Basically, an aircraft operator must have approved navigation capability. Appendix C of FAR Part 91 identifies the North Atlantic airspace specifications. This region basically is the area outlined by a cone shape beginning at the North Pole, proceeding south along 0° longitude to 27° N latitude, then west to 60° W and back to the North Pole again between FL275 and FL400. Some aircraft operators can avoid MNPS airspace because of their high performance aircraft; however, any flight changes or emergencies may cause certain legal or other problems. A gross navigation error (GNE) is reported when an aircraft is 25 nautical miles or more off course. ICAO reporting rules are followed when this occurs.

Reduced Vertical Separation Minimum

Early in 1997, *reduced vertical separation minimum* (RVSM) standards were implemented. The simple concept is to reduce vertical separation from 2,000 feet to 1,000 feet in the airspace over the North Atlantic between FL290 and FL410. This means that 13 flight levels will exist instead of seven levels. A phase-in plan began between FL330 and FL390.

The concern for corporate aircraft operators is the equipping and certification of their business aircraft. Bradley (1995) reported that "two independent altitude-measuring systems, at least one secondary surveillance radar altitude reporting transponder, an altitude alerting system and an autopilot" (p. 86) are needed. Certification requires altimetry error to be less than 200 feet and the autopilot to hold altitude within 65 feet. ICAO is overseeing the RVSM standards since the airspace is international. Corporate operators argue that substantial monies are being paid to gain little from meeting these rules. The FAA estimates that general aviation accounts for one-third of the aircraft but three to five percent of the traffic frequency of the North Atlantic. Cost estimates are about $200,000 per aircraft to meet RSVM requirements. Corporate pilots will not likely fly above the RVSM tracks because of the temperature, even for aircraft with 45,000-plus feet service ceilings. For the international corporate mission, RVSM has a significant effect, one which must be considered in flight planning. FlightSafety International and SimuFlite Training International offer training courses on flying in MNPS airspace.

Documentation

Besides the aircraft documentation already identified, still other documents may be necessary. Insurance certificates, engine logbooks, diplomatic clearances, landing permits, and import and export papers may be needed. Obtaining landing and overflight permits should be the first planning actions for international flights. Personal documentation includes required immunization records. Personal documentation of family matters (power of attorney, will, etc.) are not necessary but should be checked.

Communications

Proper communication is vital to every flight, but for flights over international waters, more than 30 minutes flying time, or 100 nautical miles from the nearest coastline, aircraft are required to have communication equipment outlined in FAR Part 91.511. Radio equipment includes transmitters, microphones, headsets, and independent receivers. Navigation equipment includes independent electronic navigation units.

Single-sideband HF radios are used for most oceanic communications because of line-of-sight limitations on very high frequency (VHF) radios. Pilots maintain constant listening unless the radio is equipped with the selective calling (SELCAL) feature. SELCAL is an aircraft-specific, coded alerting system whereby the annoying HF radio station can be turned down until an alert message is received (Wrobel, 1995). Automatic dependent surveillance (ADS) systems will track aircraft conducting transoceanic flights. Real-time aircraft position reports to ground stations via datalink and satellites using the global positioning system (GPS) can be made.

Flight Crew Health and Fitness

Each aviation manager should ensure employees are healthy and fit, but air crew health during flight operations is most important. Diet, rest, proper exercise, medicines needed, and other health matters cannot be ignored. Illness during a flight may be critical to an operation. Health of the air crew is a matter for everyone involved, but air crew members must hold themselves as primarily responsible for their health.

Flight Handlers

While fixed-base operators and airport service organizations (ASOs) help air crews in the U.S. and other locations, flying in foreign countries not having many general aviation operations may require the services of an international flight handler. Assistance from a flight handler may range from obtaining overflight and landing permits, security information, flight plans, international notices to airmen (NOTAMs), and weather forecasts to handling passenger and aircraft requirements. An example of a recognized handler is Houston-based Universal Weather and Aviation, Incorporated. Others include Air Routing International,

Jeppesen Dataplan, BaseOps International Incorporated, AMR International Aviation Services, and Spectrum Air Services.

The handlers only charge for services provided, except for Universal, which requires a monthly retainer for each airplane to be handled and a deposit to draw against (Parke, 1990). Universal uses its own high-frequency facility for communications, and Air Routing, Jeppesen, and the others use Arinc for HF communications. Handlers operate 24 hours-a-day, every day of the week. For corporate aircraft operators who have not made international flights, using a flight handler until air crews can plan flights is recommended. Even then, using a handler may provide a trouble-free flight, which may cost slightly more, but be less stressful and encounter no problems. Handlers in foreign countries may save a company's trip mission, not because they have better people, but because their people can handle the many details required in a specific country.

ETP Computations

Planning ahead for international flights over water requires pilots to determine the equal-time point (ETP). While most new corporate jets can be flown on long-range flights (check *Business & Commercial Aviations's Planning & Purchasing Handbook),* computations should be reviewed for overwater flights. The ETP is a geographic position along the route where the time required to return to the departure point is equal to the time required to proceed to the destination, assuming the departure and arrival airports are the closest airports. Problems before reaching the ETP mean that a pilot should return to the airport of origin; problems after the ETP mean continuing on to the destination. Aircraft not having the range may have a *wet footprint.*

Calculating the ETP in nautical miles is done by multiplying the total flight distance (D) by the ground speed to return (GSr) and dividing this product by the sum of the ground speed to continue (GSc) and the ground speed to return. The formula is:

$$\text{ETP} = \frac{D \times \text{GSr}}{\text{GSr} + \text{GSc}}$$

The key to the ETP computation is the wind factor. The greater the wind, the farther the ETP moves from the geographic midpoint. Three common reasons for determining ETPs are medical emergencies, engine failures, and pressurization problems. Review Aarons (1988) for an example of a medical emergency description of ETP computations.

For a medical emergency, the question is whether to return to the origin or proceed to the destination. Flying eastbound over the North Atlantic on a 1,900-nm leg with a true airspeed (TAS) of 424 knots at FL430, burning fuel at 990 pounds an hour, with an average wind factor of plus 50 knots (using a Learjet 35A) results in the following computation.

$$\text{ETP} = \frac{1{,}900 \text{ nm} \times 374 \text{ kts}}{374 \text{ kts} + 474 \text{ kts}} = 838 \text{ nm}$$

In other words, if you passed 838 nautical miles (963 statute miles) from your origin, proceed to your destination. At the ETP, it is your choice. An engine failure changes the calculation. When you lose an engine, usually you decrease altitude and speed. The aircraft's performance manual may indicate that the aircraft speed will decrease to 300 knots at FL250, burning 970 pounds of fuel per hour. The forecast winds at FL250 could be plus 25 knots. Now the ETP is recomputed to 870 nm. The key is the wind factor. What about the wet footprint or having to ditch in the water because you cannot make a landing? Losing an engine results in reduced speed, but fuel burn remains about the same. At FL430 and TAS of 424 knots, 990 pounds of fuel are burned per hour, or 0.428 nm per pound of fuel. Drifting down to FL250 with one engine results in a range of 300 nm for 970 pounds of fuel, or 0.309 nm per pound of fuel. Figuring out how much fuel is needed at the new airspeed to reach the destination will let you know if you will end up in the water or will have enough fuel to reach your destination.

Using Boeing's *Winds on World Air Routes* for a given route will help you plan your flight. Different percentages of reliability for winds are given by Boeing. Aircraft performance today is better than ever, but air crews still need to be knowledgeable and use good flight planning.

Miscellaneous Topics

Planning for a corporate flight, especially for an international trip, takes time and effort. The more advanced notice an aviation manager can provide the air crew, the more time there will be for flight and personal preparation. Air crews may be technically qualified to fly aircraft assigned, but each PIC will have sole responsibility once airborne for not only the aircraft but the air crew and passengers. When unexpected changes happen, advanced planning may save the day. Reviewing weather and divert airfields; checking hotel reservations, ground transportation, fueling and maintenance contracts, diplomatic clearances, aircraft stores and necessary spare parts, language books, and other trip items; and using a self-developed or company flight/trip checklist are highly recommended actions. FAR Part 91.103 must be complied with for preflight actions. The old adage that it is better to be safe than sorry fits here.

Enroute Operations

Equally, if not more important, is enroute operations. This is when things can go wrong and affect the trip or flight. Air crews must remain alert and exercise caution to prevent mistakes. Good cockpit resource management is required and effective communications with all concerned—the passengers, home-base operators, air traffic control, and many others. Planning for MNPS and RVSM flight becomes a reality. FAR Part 91, Subpart B, outlines many flight duties and responsibilities for the PIC and air crew. These rules, along with the other FAR requirements, should be known and followed when applicable. Who should brief the passengers? When can flight crew leave their duty stations? What about shoulder harnesses in an aircraft? How many flight attendants are necessary? FAR Part 91 covers the answers to these questions and many others.

Monitoring aircraft navigation systems, enroute weather, fuel consumption, aircraft positions, and aircraft systems is an ongoing process. Checking flight publications, diplomatic clearances, enroute emergency airfields, and passengers is required. Reviewing aircraft and personal documentation, aircraft support contracts, and the other numerous documentation items can be accomplished. Perhaps long flights or international flights are common to the air crew, but those who do not take time to check or follow good practices may not have a good excuse when things just do not happen the way they were planned. Even a military C-141 pilot who landed in Scotland was amazed that his aircraft was impounded when correct aircraft and cargo documentation was missing; he just could not understand why it happened to him. Planning and checking go hand-in-hand.

Cockpit Resource Management

During the 1970s, the National Aeronautics and Space Administration developed a training program called cockpit resource management, or CRM. Its purpose was to reduce the high rate of human error in flying. Diehl (1992) reported that accident rates reduced between 28 and 81 percent for organizations trained in CRM. The NBAA highly recommends application of CRM concepts for corporate aviators involved in aircraft operations. Even single-crew member aircraft operations involves the pilot closely coordinating with mechanics, airport ground personnel, air traffic control personnel, and passengers.

Usually corporate aviation operations result in air crews flying with each other more than in a typical large airline operation. Flight crews tend to develop their own unofficial set of policies based on personal limits. CRM allows flight crews to address these deviations and assess the effects of compliance/non-compliance. Aviation managers need to review safety records of their air crew to help in implementing CRM training when needed. One step to take in deciding to implement CRM is to ask the air crew. Van Cleave (1996) developed a table of intangible and tangible benefits for cockpit resource management (Table 7-4). It is up to the ADM and air crew to convince a firm's management that the training provides value to the corporate aircraft function.

Once it is decided that CRM training would be useful, air crew personnel could be sent to professional programs or a consultant can be contracted to begin the training in the company. Five fundamental training skill areas are reviewed in CRM training: (a) attention management, (b) crew management, (c) stress management, (d) attitude management, and (e) risk management. A firm's flight safety and training official should be responsible for conducting the CRM program. Check with other corporate pilots and organizations for current training programs and benefits being realized through CRM applications.

Table 7-4. Tangible and Intangible Benefits of Cockpit Resource Management

Tangible	Intangible
Reduced flight mishaps	Concern for fellow pilots
Reduced serious incidents	Teamwork/team spirit
Increased operating efficiencies (*on-time takeoffs, 100% qualified pilots, etc.*)	Improved morale (*due to feeling management cares about their well-being*)
Decreased air traffic control violations	"Close call would-have-beens"
Decreased standard operating procedures	Savings on life insurance deviations payouts (*because no one knows if death would have occurred; over long period can be tangible*)

Source: Van Cleave, F. (1996, March 22). The development of a cockpit resource management (CRM) program in a corporate or business aviation flight department. Unpublished graduate report. Daytona Beach, FL: Embry-Riddle Aeronautical University.

Extended Range Operations

The commercial airlines adhere to extended range operations (ETOPS), and corporate aviation activities may be affected by possible Joint Aviation Authority (JAA) rulings in Europe. The concern arises from interpreting whether regulators attempt to control the activity of flying or the equipment, the aircraft. Attention to rule changes for flying is an ongoing process for air crews, and ETOPS is only one item of many to keep reviewing.

Postflight Actions

For corporate pilots who make successful flights to their interim destinations, the work may have really only just begun. Passengers have to reach their destinations, have appropriate ground transportation, be properly billeted, and have their cargo or baggage handled. The air crew may be the company representatives to ensure these actions are taken. For example, if an air crew member or a passenger is carrying over $10,000 or more in cash, a declaration to customs is required. Ensuring customs requirements are met, securing the aircraft, making sure proper fueling occurs, and planning for the return trips begin.

Customs for the corporate aircraft operators are serious activities to address. Aircraft and passengers are subject to inspections and possible fines (minimum fine $5,000). Customs rules for business aircraft operators fall under private flyers, which has been defined by

Customs as aircraft not being used to transport persons or property for money. The U.S. Government Printing Office (Appendix B) provides a copy of the *Guide for Private Flyers* produced by the U.S. Customs Service. Obtaining a copy of the *International Flight Information Manual* (IFIM), which covers customs as well as international flying, is also recommended. Air crews and passengers need knowledge of permitted items, limitations, and prohibited goods. Again, the PIC becomes responsible for his/her aircraft and all on board, and an aircraft must be properly registered and appropriate documents available. Flight crews must notify U.S. Customs of the aircraft's type and registration, departure point, PIC, border-crossing time, arrival time, and citizenship status of persons on board. Customs clearance procedures may be easier at home base, but at a foreign airport, air crew personnel must be aware of all correct procedures to follow.

Planning to get to the interim destinations is replaced with planning to return to home base. Communication becomes a major activity because not only does the air crew want to know what happens next, the flight department manager wants to know what is happening and how well everything is going. If a flight handler is not contracted, then maintenance and other aircraft services may have to be personally handled by the air crew.

When finally back at the home base, the air crew's work is not completed until other actions are completed. Trip reports, travel forms, aircraft checks, maintenance debriefings, aircraft association reports (IBAC, NBAA International Feedback Form, etc.), and other necessary documents and activities need to be properly and accurately completed. If any incident or accident occurred, then additional paperwork and discussions are required of the air crew.

Other Flight Operations Topics

Flight operations require planning, planning, and more planning, especially for international flight. The NBAA and IBAC provide helpful programs and information to assist corporate aviators for these operations. Figure 7-7 is a checklist provided by the NBAA, but again, this is not all-inclusive. Each ADM must review specific flying requirements and plan for their successful completion through good policies and preparation.

It is important for the air crew to maintain a good sense of humor. People, no matter where they are located, will work more easily with others who can communicate in a friendly and gracious manner. Being aggressive will surely result in a poor reputation and perhaps less-than-satisfactory working relationships with those who are involved in the corporate flight.

Some top executives have their favorite pilots, and if flying skills and expertise are satisfactory, then a flight planner may have no choice assigning the desired air crew. Using an air crew roster may be a concern when executives self-select air crew members. Overall, a flight planner should select the best crew for the specific flight being planned.

Most countries are members of the International Civil Aviation Organization (ICAO). ICAO publishes mandatory rules of the air. These standards and recommended practices, or SARPS, require international aviation operators to comply with jointly developed rules for using international airspace. Every country is free to file exceptions to these rules, and most do.

EXHIBIT 8
NBAA International Operations Checklist

Documentation
Flightcrew
- Current and Qualified in International Operations
- Rested
- Trip Itinerary
- Passenger Manifest (full name, citizenship and date of birth)
- FAA Airman's Certificates
- Passports
- Visas
- Tourist Cards
- Proof of Citizenship (Not driver's license)
- Immunization Records
- Customs' Forms
  PAIRs CF 178
  General Declarations
- IBAC Flightcrew Cards
- IBAC Luggage Tags

Aircraft
- Airworthiness Certificate
- Registration (not "Pink Slip")
- Radio licenses
- Aircraft Operations Manual with Weight & Balance
- Minimum Equipment List (MEL)
- MNPS Authorization
- Metric Conversion Tables (with preconverted aircraft size and weights)
- Airframe & Engine Logs
- Certificate of Insurance (U.S., military, & foreign, as applicable)
- Import papers for aircraft of foreign manufacture
- Single Long Range Navigation System Operations Manual
- IBAC Decal

Passengers
- Trip Itinerary
- Passenger Manifest (full name, citizenship, and date of birth)
- Passports
- Visas
- Tourist Cards
- Proof of Citizenship (Not driver's license)
- Immunization Records
- Traveler's Checks
- Credit Cards
- Cash

Operations
Permits
- Overflight & Landing
- Export Licenses
- Diplomatic Licenses
- MNPS Airspace Authorization
- Military "Civil Aircraft Hold Harmless Agreement"

Services
- Inspections
  Customs
  Immigration
  Agricultural
- Ground
  Security
  Catering
  Handling Agents
  ASOs & FBOs
  Fuel (credit cards, carnets, & contracts)
  Prist
  Methanol
  Anti or De-ice
- Maintenance
  Technician
  Flyaway kit (spares & tools)
  Fuel contamination check kit
  List of FAA Foreign Repair Stations

Financial
  Credit cards, Carnets, & Contracts
  Letters of credit
  Banks
  Servicing air carriers
  Handlers
  Fuelers
  Travelers checks
  Cash

Communications
Equipment
- VHF
- HF (3A3J)
- Headphones
- Microphones
- SELCAL
- Portables (ELT's etc.)
- SATCOM

Agreements
- ARINC
- SITA
- BERNA
- Stockholm
- British Telcom International
- INMARSAT (COMSAT)

Navigation
Equipment
- VOR
- DME
- ADF
- Inertial
- VLF/OMEGA
- LORAN
- GPS (Satellite)

Publications
Flight Deck
- En Route charts (VRF, IFR)
- Plotting charts
- Approach charts
- NAT track message (current)
- Flight Management System (Current)

Other
- Company Operations Manual
- IFIM
- INOTAMS
- Manufacturer's maintenance manual
- World Handling Agents Manual

Survival Equipment
- Area survival kit with text
- Medical kit with text
- Emergency Locator Transmitter
- Life preservers
- Life rafts

Facilitation Aids
- U.S. Department of State (Office of Aviation & U.S. Embassies)
- U.S. Customs Service
- FAA Office of International Aviation
- FAA Office of Security
- FAA International Representatives

Other Considerations
- Professional Planner
- Aircraft locks
- Spare keys
- Commissary supplies
- Ground transportation
- Hotel reservations
- Camera (use with discretion)
- International Feedback cards

Figure 7-7. NBAA International Operations Checklist. Extract from *NBAA Management Guide,* 1994, page 73. Used with permission.

These exceptions must be reviewed and complied with by corporate operators. As planning for an international flight may start months ahead of a flight, a last minute review of the SARPS and procedures for air navigation (PANS) is recommended. PANS are the advisories for international flight operations.

The *International Flight Information Manual* (IFIM) is another publication to be checked since not every airport is an approved port of entry for foreign travelers. Other useful items to use when planning flight operations are the National Geographic globe for a quick route

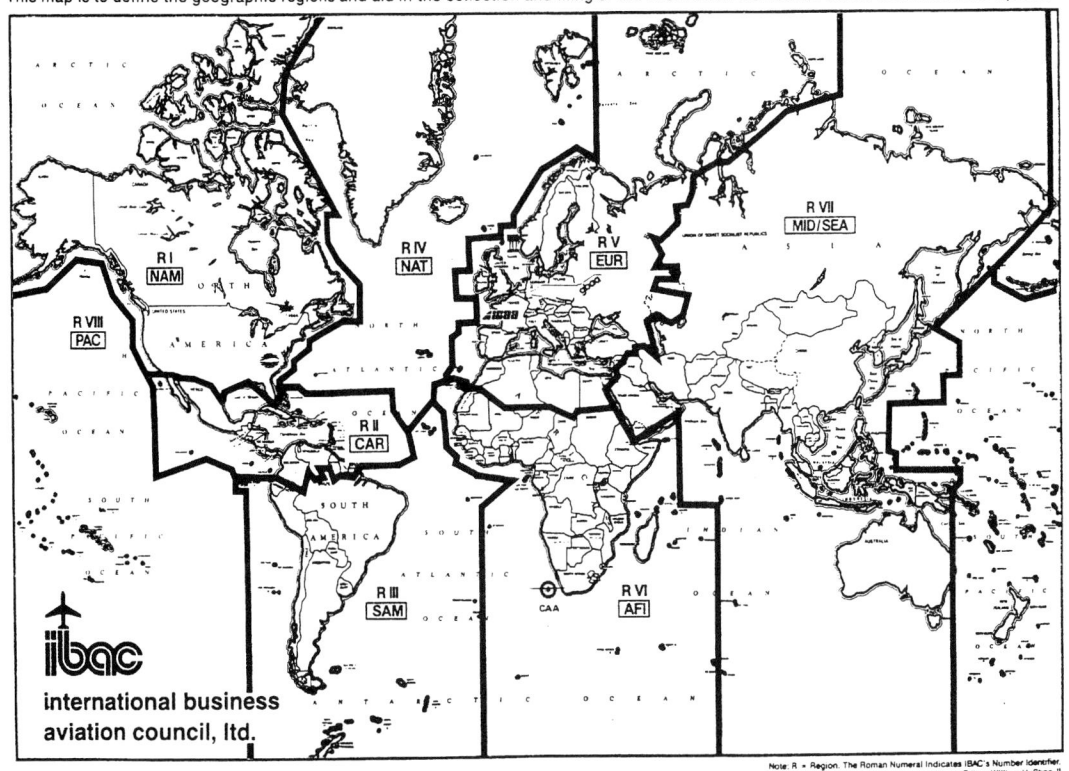

Figure 7-8. IBAC Geographic Regions.

review and IBAC's *International IBAC Update*. The latter provides air crew information pertaining to flight operations in IBAC's regions (see Figure 7-8).

Finally, flight planning software programs are readily available, and corporate operators should review their benefits. An example of one of these excellent programs was provided by Brian Leutschaft (personal communication, February 19, 1996) of Mentor Plus Software, Incorporated. Mentor's FliteStar corporate program is sophisticated, and besides the program features identified in Figure 7-9, additional features were later added to help corporate aircraft operators. These included:

1. Digitized approach plates on CD-ROM, which enables automatic updating;
2. User-editable flight log enabling last minute changes before takeoff;
3. Comprehensive AOPA airport information library, which also provides hotel, rental car, FBO, and customs information;
4. Advanced aircraft modeling for improving fuel burn and time predictions; and
5. 24-hour technical and customer support.

7: Flight Operations

JEPPESEN
A Times Mirror Company

Jeppesen Mentor
22781 Airport Rd. N.E.
Aurora, Oregon 97002
Tel: (503) 678-1431
Fax: (503) 678-1480

Dear Corporate Operator,

Thank you for requesting information about our FliteStar Corporate. FliteStar Corporate is sophisticated, yet easy to use and contains features desired by Corporate and Part 135 Air Taxi operations. FliteStar Corporate does what the expensive on-line services do without the on-line charges.

In addition to the features listed in the attached FliteStar Corporate brochure, the following features have been added to make your flight planning tasks faster and simpler than ever before.

- Optional Digitized Approach Plates on CD-ROM will eliminate the need to manually update your paper chart subscription. Just slip the new CD-ROM into your computer and the "updating" is done automatically!

- A user editable flight log, which will allow you to make those last minute changes before take-off.

- A comprehensive Jeppesen Airport information library is at your finger tips that provides you hotel, rental car, FBO and customs information.

- Our new advanced aircraft modeling, will provide you fuel burn and time predictions to within 0.5% accuracy.

- 24-hour technical and customer support. So you won't have to worry about questions concerning installation and operation.

In summary, FliteStar Corporate is a proven asset for corporate managers interested in reducing time, cost and effort. For further information, please feel free to contact me personally at (800) 990-9263 Ph, (503) 678-2918 Fax, or bplmentr@teleport.com email.

Sincerely,

Brian Leutschaft

Brian Leutschaft
Jeppesen Marketing

Figure 7-9. Mentor Plus FliteStar Corporate Flight Planning Software. Courtesy of Brian Leutschaft, Jeppesen Mentor.

Features

- Windows & Macintosh
- Graphic user interface
- Scrollable Graphic Charts
- Jeppesen NavData
- NOS Approach Plates
- Profile View
- Graphic W&B
- Navigation Log
- Low Alt. Airway depiction
- Quick Planner
- Fuel stop planner
- Imports DUATS Winds
- Color weather charts
- Automatic DUATS access
- DUATS flight plan filing
- Auto-route point to point
- Auto-route IFR RNAV
- Auto-route Great Circle
- Auto-route Low Airways
- Auto-route High Airways
- Preferred Routes
- Intersections
- Alternate Airports
- FAA & ICAO Flight Plan
- Costing Report
- SIDs and STARs
- AOPA Aviation USA
- Jet Aircraft Library
- Variable Moment Arm
- Pilot/crew Log Book
- Quick Quoter
- W & B Manifest
- Custom Aircraft Performance Model
- Standard 28 day Subscription Updates

Data Coverage

North America

Covers the USA, Mexico, Canada, Alaska, Hawaii, Bahamas, and Central America North of the equator. (Requires about 12 Mb hard disk space).

Europe

Covers all of Europe, parts of Russia and North Africa. (Requires about 10 Mb hard disk space).

World Wide

Covers the entire world in one continuous database. (Requires about 20 Mb of hard disk space).

Technical Support

Your time is valuable. That's why we've made FliteStar Corporate exceptionally easy to learn and to use. And we back it with the best support in the industry — from the comprehensive manual and Quick-Start tutorial to 1 year of unlimited toll-free technical support. Whatever your operation or kind of flying, MentorPlus will provide a high level of service, tailored to your needs.

Demo and Tutorial Video

If you'd like to see FliteStar Corporate in action, get the demo/tutorial video for just a $7.50 shipping and handling charge. This 25 minute video takes you through a typical flight plan including getting a DUATS weather briefing and filing a flight plan.

Computer Requirements

FliteStar runs on true IBM compatible and Macintosh computers.

For the Windows version you need:
486 or Pentium processor, Windows version 3.1 or higher, at least 8 Mb RAM and 20 Mb hard disk space.
The Windows95 version provides enhanced 32-bit speed and performance. For best performance, 486 or better processor, 16 Mb of RAM and 20 Mb of hard drive space are recommended.

For the Macintosh version you need:
A MacOS computer with at least 8 Mb RAM, System 7.0 or later with a hard disk.

For More Information on FliteStar Corporate or Other MentorPlus Products or To Order for Same Day Shipment Call 24 Hours
1-800-990-9263
or see us at our web site www.mentorplus.com

MentorPlus Software Inc., 22781 Airport Rd. N.E. Aurora, Oregon 97002-0356
Ph: (503) 678-1431   FAX: (503) 678-1480   www.mentorplus.com

© copyright 1990-1997 MentorPlus Software. All rights reserved. Printed in the U.S.A.
FliteStar and FliteStar Corporate are registered trademarks of MentorPlus.

specifications subject to change without notice.

Figure 7-9. Continued

The Best Routing Available

Automatic Routing

FliteStar automatically plans your IFR flight, computing optimum altitude and route considering Great Circle distance, winds aloft, aircraft performance and your flight-specific mission requirements.

It's as easy as clicking on or typing in your destination (airport, VOR, NDB, intersection or user-defined waypoint) and letting FliteStar do the rest. Route direct, point to point, on low or high altitude airways or via the Great Circle with RNAV or GPS/LORAN waypoints. FliteStar displays your routes with on-screen Jeppesen NavData, or lets you print and take them into the cockpit. FliteStar also prints a convenient knee board-sized flight log showing your climb, cruise and descent legs as well as time, fuel and distance.

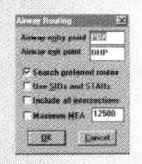

Flexible Routing Capability

Great Circle

FliteStar's Great Circle algorithm quickly finds the shortest route between any two points on the planet and generates the waypoints you need to navigate that route. Choose RNAV waypoints to get waypoints defined by distance and radials from VORs or GPS/LORAN to get equally spaced waypoints defined by latitude and longitude.

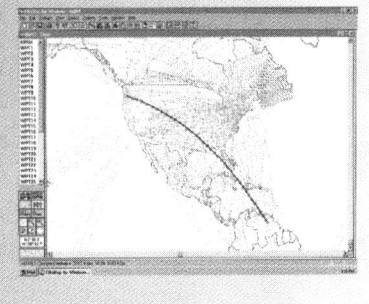

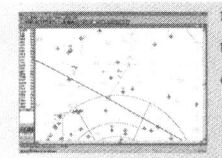

Low altitude airways

FliteStar's 'traveling salesman' algorithm reliably finds the shortest path in the complex network of airways. It automatically routes you to the nearest airway entry point and keeps you on the airway system to an exit point as close as possible to your destination. If necessary, you can set entry and exit points that expedite your departure and arrival and FliteStar will remember them the next time you plan that route. You may also use Preferred routes, if available, or SIDs and STARs. You can even tell FliteStar to try to find a route at a specific minimum en-route altitude (MEA).

High altitude airways

If your aircraft goes into the Flight Levels, FliteStar offers precision High Altitude Airway routing. As with Low Altitude Airway routing, FliteStar's 'traveling salesman' algorithm quickly finds the shortest route and lets you choose Preferred Routes or SIDs and STARs.

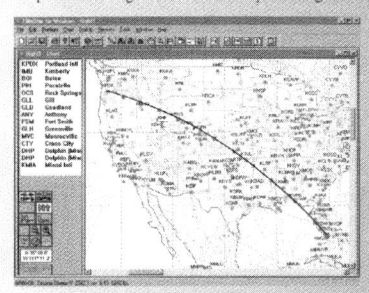

Figure 7-9. Continued

Figure 7-10. Air Support of Billund. Per Jensen (left) and Jens Pisarski demonstrate PPS at the European Business Air Show, May 1997.

Air Support of Billund

Jens Pisarski and Per Jensen (Figure 7-10) demonstrated their Air Support Pre-Flight Planning System, PPS, at the 1997 European Business Air Show. PPS is a PC-based system that can be used to plan for aircraft up to Boeing 747 type. Routes, fuel, weight and balance, winds, and other variables are provided real-time through the use of Swissair Navigation Services. While services are provided for worldwide coverage, Pisarski stated that flying in Europe requires extra attention to complex operations not realized in the United States. Figure 7-11 provides Air Support's data in detail.

Universal Weather & Aviation, Incorporated

Another outstanding corporate software provider is Universal Weather & Aviation, Incorporated. Details of Universal Weather's FlightPak and Pilot's Choice programs are illustrated in Figures 7-12 and 7-13, while additional details of its flight operations, ground handling, weather services, and fuel programs are included in Figures 7-14 and 7-15. For current details of these and other programs, contact the appropriate service provider. Using one of these systems should provide an operator with more information and improved flight planning and support.

AIR SUPPORT

P.O. BOX 24
DK-7190 BILLUND
DENMARK

Billund, the 1st January 1997.

Information concerning PPS Flight Planning software for Airline operators.

THE PPS - PREFLIGHT PLANNING SYSTEM Main Module:
PPS is an advanced PC-based flight planning software solution for professional users. PPS can handle performance data of aircrafts ranging from Citation Jets, Falcons to Fokker 50´s, Dash 8´s and all the way up to Boeing 727, 737, 747 and 777 as well as the Airbus series. PPS Version 5.08 has been designed to be able to handle complex aircraft performance and a complex ATC environment by performing all necessary calculations under actual conditions.

AIR SUPPORT today offers a unique integrated flight planning solution which offers manual selective as well as full automatic Airway Auto Routing capability (and/or charter price calculations based on actual ATC routing distances). Our update service is based on Airport, Navigation and Airway updates from SWISSAIR Navigation Services as AIR SUPPORT is a licensed SWISSAIR Navigation Data supplier. PPS is fully network compatible on all PC network types.

Actual Forecasted grid winds and temperatures will automatically be downloaded by your network modem from the AIR SUPPORT BBS server station at each wind release and will automatically be used for all your actual flight planning calculations, including optimum flight level considerations like best economical flight level, minimum time flight level and minimum fuel flight level based on the actual weight. Furthermore, TAFs, METARs, SIGMETs etc. can be downloaded via modem. Tailored NOTAM download will be available from fall 1997.

The PPS program can calculate a full flight log within the limitations of the aircraft given under actual conditions with navigation data, fuel data, weight & balance data, balanced field length requirements all presented on a flexible flight plan lay-out designed according to your requirements.

PPS has been sold in 12 countries today and is one of the strongest professional software products in the marketplace which can be implemented at realistic cost levels.

When implementing PPS you will experience that AIR SUPPORT takes care of the other "half" of your software purchase namely all the installation requirements where we can mention: entire network set-up, set-up of stand-alone computers, printer installations, aircraft performance data with required operational performance profile installations, design of the flexible company flight log lay-outs/aircraft specific flight log lay-outs as well as introductory training of your operations staff and key-personnel.

AIR SUPPORT guarantees full on-site implementation service & a unique aftersales support. It is our business to secure that your requirements are covered by the delivered software solution as

Figure 7-11. Air Support of Billund. Provided by Jens Pisarski and Per Jensen, May 1997.

well as it is our standard to cover later adjustments without extra costs, which in some cases cannot be identified by either party before the system has been fully implemented and your operations dept. as well as your pilots have been working with the system for a period of time. It is our company goal to fulfil your expectations a little better than you dare expect.

The PPS Flight Planning Main Module is the basic tool in a complete PPS solution with its full automatic Routing capability, Departure- and Arrival point capability, Navigation data and Weight & Balance data. Below we have listed the typical modules that most scheduled carriers prefer in an integrated PPS software solution in order to cope with either a scheduled traffic program or even a mix of scheduled- and charter traffic programs + mixing of any ad hoc flights:

The accessibility of the entire PPS software system is secured via an easy-to-use pass-word system which secures that only authorized personnel can enter into the program's databases.

SCHEDULED FLIGHT PLANNING Module:
This module is a Flight Plan Management and Dispatch system for scheduled and charter airline operators. The Scheduled Flight Planning module features a database of pre-made routes (your traffic time table) ready to be used in the processing and dispatch of ATC flight plans. The ATC flight plans can be sent to a printer, via modem to any fax number or via SITA directly into IFPS/CFMU in Bruxelles in the correct correctly addressed and in correct format. Operations personnel can trigger the dispatch of prepared ATC flight plans by sending for example a full day's stored plans in one working order, in selected groups or as single ATC flight plans at any convenient time.

The Operations personnel also have access to a function called DAILY FLIGHT PLANNING where they can finish pax load, luggage load and fuel uplift for a specific flight, and print out a finished flight plan log, all based on actual forecasted Grid winds and Grid temperatures en-route.

It is also an ideal tool for dividing work loads between personnel groups, for instance if there is authorized navigation personnel (who are in charge of the routings) and dispatch personnel (who monitor or launch the flights in the operations department).

PILOT FLIGHT PLANNING Module (Multi base flight planning):
The PILOT FLIGHT PLANNING module is a sub-module linked to the SCHEDULED FLIGHT PLANNING module in the PPS Main System. One or several satellite PILOT FLIGHT PLANNING modules can be used in connection with the PPS Main System. For instance in the pilots' flight planning room at the home airport, at remote bases and/or at your Handling agents' locations. The PILOT FLIGHT PLANNING module can be installed at each base you consider relevant (even at a handling agent's office) and it is connected to the PPS Main system either as part of your internal network, as an ISDN on-line connection or via an automatic activity triggered telephone modem connection. Most customers opt for the modem solution which is

Figure 7-11. Continued

extremely effective and reliable with today's modem technology and which carry extremely low implementation and running costs. When a pilot triggers the PPS PILOT icone on the computer screen, PPS PILOT will immediately show him the day's flight program as has been prepared by your Operations Dept. He can then select his own flight(s) and finish fuel uplift based on booked pax load and & luggage load where the whole flight log is then calculated on the basis of actual forecasted en-route Grid winds and temperatures. If the pilot wants to monitor the flight log before print-out he can do so. If he wants to see the traffic program for the following day or other days he can call up this information on the screen. The pilot has the possibility to change the stored Alternate airport if for instance sudden changes in weather happens or he wants to flight-plan with 2 Alternates, on the basis of POR or POD. In such case the program will show the 19 nearest Alternates to the Destination for easy selection with indication of Distance, Magnetic track from Destination as well as Available Runway length for each listed Alternate.

The pilot can normally only choose between finished ATC routings which have been prepared by Operations Dept. The Operations Dept. has the capability to update changes for flight plans or other data changes in the PPS Main System at any given time. Whenever a change is completed by Operations Dept. the changed data are instantly available at all the satellite station modules. If it is the case that an operator wants to enable their pilots to make a full flight plan within PPS from the ground up, the PPS PILOT FLIGHT PLANNING module can be configured so.

Pilots have no direct access to the databases in the PPS Main System via the PILOT FLIGHT PLANNING module. They **cannot** alter any stored data or any weight & balance data, or store any data within the entire PPS Main System. They can only key in the crew names, pax load and luggage load, monitor the loaded Grid Winds and Temperatures (or input manual weather data), decide their fuel policy for the trip (max fuel, min req. fuel, manually selected fuel figure, specific landing fuel load at destination, or fuelling in order to reach a specific landing weight at destination) and then print-out the calculated flight log.

A full PPS network solution with 3 remotely placed modem coupled PPS PILOT FLIGHT PLANNING modules were supplied to AUGSBURG AIRWAYS in Germany in early 1996. RHEINTALFLUG in Austria has implemented a PPS network solution along with 2 PPS PILOT FLIGHT PLANNING modules in Sept. 1996 (also modem coupled). GILL AIRWAYS Ltd. in Newcastle purchased a complete PPS solution in October 1996 and also ordered 1 Pilot Flight Planning module which will be made operational in the pilots' briefing room at the home base in spring 1997.

MNPS Module

AIR SUPPORT is currently developing the MNPS Module for PPS. The MNPS module is designed to display each day's North-Atlantic Tracks (NATRACKS) which will automatically be downloaded via modem together with the Grid winds and temperature forecast from our Server station in Billund. The PPS software automatically implement the downloaded NATRACKS in the navigation database and will display the NATRACKS and their validity in the graphic FAST ROUTING module. MNPS will compare all available tracks (including eventual random routing

Figure 7-11. Continued

> tracks outside MNPS area) and calculate the Minimum Time Track based on actual TOW which is a function of expected payload and the needed fuel **and** in relation to the forecasted winds and temperatures in different flight levels for each of the different available Tracks.
>
> The MNPS module is expected to be available during fall 1997.
>
> **PPS ETOPS FLIGHT PLANNING MODULE**
> AIR SUPPORT is about to offer a complete ETOPS module for operators who operate twin-engined aircraft for trans-oceanic crossings. The PPS ETOPS module will be integrated within the present PPS Flight Planning module. It will include automatic download and storage of each day's available NAT tracks together with actual forecasted Grid winds and Grid temperatures. The module can calculate a full ETOPS flight log with specification of PET between possible en-route Alternates while taking into consideration **loss of 1 engine** as well as **critical fuel scenario** calculations (decompression fault w/descent to 10.000 Feet and contiued flight to correct Alternate under actual forecasted winds and temperatures and based on actual aircraft weight), all in relation to the aircraft's / operator's actual 90, 120, or 180 minutes ETOPS status. The ETOPS module will naturally be linked with the MNPS module so that an ETOPS operator automatically gets an MNPS/Minimum Time Track which fulfils the ETOPS requirements.
>
> The PPS ETOPS module is expected to be available during fall/winter 1997/98.

Figure 7-11. Continued

Summary

Flight operations require significant planning for aviation managers and flight crew. International flight operations require even more planning and preparation than for domestic flights mainly because of additional ICAO international rules and foreign government regulations. A major requirement is the proper documentation for aircraft, airmen, passengers, cargo, and equipment. Pilots expect their technical qualifications to be met, but vary widely in their opinion about flight preparations. Various flight issues for flight planning are presented in the chapter tables, and aviation associations and training agencies help to provide assistance to air crews seeking to improve and/or expand their knowledge of important aviation procedures.

A company's flight begins with an air transportation request and does not end until the mission has been completed and aircraft checked with all required final reports completed. Dispatchers must be certified by the FAA (FAR Part 65) as their importance in flight planning and handling is significant. To help schedulers and dispatchers, various software programs are available from vendors. Some of these are identified in Table 7-3, and others can be found by reviewing the AvComps directory published annually by *Business & Commercial Aviation*.

Although FAR Part 91 does not set flight limits for air crews, aviation managers should establish reasonable duty-and flight-time limits for the primary reason of safety. Additionally, ADMs should ensure good air crew appearance and conduct at all times. Several of the flight operations issues, that is, use of intoxicants or drugs, carrying weapons, and fueling are covered. The importance of having qualified air crew members who know and understand the many flight rules and who can successfully preplan flights cannot be overemphasized by an aviation department manager or company official.

Figure 7-12. FlightPak® Systems. Courtesy of Universal Weather & Aviation Inc. Provided by Marinda Hochadel-Jolly, May 14, 1997.

Figure 7-13. Pilot's Choice^{SM}. Courtesy of Universal Weather & Aviation Inc. Provided by Marinda Hochadel-Jolly, May 14, 1997.

7: Flight Operations 229

Figure 7-14. Flight Operations/Ground Handling. Courtesy of Universal Weather & Aviation Inc. Provided by Marinda Hochadel-Jolly, May 14, 1997.

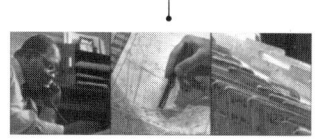

Hand-in-hand with planning a flight goes the actual flight operation itself. Many flight activities might have to be considered, including ETP computations, MNPS/RVSM operations, effective communications, and of course, excellent flying skills. For aviation managers who are not familiar with international flights or desire to use qualified handlers, international flight handlers are available and will provide expert flight and ground handling services.

Cockpit resource management is emphasized to help aviation managers reduce or keep flight problems to the absolute minimum levels possible. To help overall in flight operations, information and systems provided by flight programmers can be obtained from various operators, for example, Mentor Plus Software Incorporated, Air Support of Billund, and Universal Weather & Aviation, Incorporated (sample programs illustrated). The purpose of the flight department is to provide the air transportation needed for the company. Through successful planning, this purpose can be achieved.

Customized forecasting pinpoints conditions that impact the aviation, marine, emergency management, media, sports, and construction industries.

The UVair® Fueling Card is the most recognized card in the industry, accepted at more than 1,350 locations worldwide.

Weather Services

Reliability and expertise. Because weather is dynamic, Universal operates three weather offices in the U.S., and provides a full complement of worldwide forecasting and notification programs.

More than 50 professionally-trained meteorologists access our weather database to prepare and deliver customized briefings. In addition to the latest global satellite imagery, surface weather charts pinpoint storm centers, significant turbulence and jet streams, including upper level wind conditions.

Clients receive text, charts, and customized reports via telephone, facsimile, Telex, modem, E-mail, satellite, SITA, ARINC, AFTN and H-F radio frequencies – 24-hours a day.

Various customized software programs, such as Pilot's Choice,℠ not only allow users to pre-file and update flight plans, but provide on-screen viewing of text and color weather graphics.

NotiFax℠ is our cost-effective 24-hour severe weather warning program. When the National Weather Service issues bulletins that affect your area, they are automatically transmitted to individual fax machines. Users choose only the bulletins they wish to receive, such as tornado, severe thunderstorm, hurricane, marine, or flash flood watches. They also select the time of day and locations to be affected.

With the advent of Internet, registered users can now obtain current radar and satellite images on the World Wide Web.

Learn more about
Universal Weather Services
on the World Wide Web.
http://www.univ-wea.com/univwx/weather.htm

Fuel

Acceptance and savings. Our UVair® Fuel Division provides worldwide fueling support for corporate, commercial, and military aviation through a vast network of suppliers, FBOs, and handling agents.

Thanks to specially-negotiated contract pricing at more than 1,200 airports worldwide, UVair cardholders pay less than the posted airfield price. Should fuel be required in isolated or non-contract locations, UVair can almost always secure credit arrangements.

A monthly contract fuel price list is provided for each aircraft, and cost control is easy to monitor. All charges are detailed as part of Universal's One-Step Custom Billing,℠ and a fuel summary report follows each quarter.

Because there's no membership or monthly retainer, savings accrue from the first purchase.

Learn more about the
UVair® Fuel Program
on the World Wide Web.
http://www.univ-wea.com/univwx/uvair.htm

Figure 7-15. Weather Services/Fuel. Courtesy Universal Weather & Aviation Inc. Provided by Marinda Hochadel-Jolly, May 14, 1997.

Activities

1. Contact a flight handler to determine the extent and costs of services provided.
2. Talk to a pilot who has completed international corporate flights and discuss his or her major concerns for planning the flight (s).
3. Review FAR Part 91 for information about flight operations requirements.
4. Discuss with corporate aviation personnel the value and benefit of using uniforms for corporate aviation operations.
5. Discuss with aviation personnel the three parts of flight, that is, preflight, enroute, and postflight activities.
6. Review the current MNPS/RVSM flight requirements.

7. Obtain a copy of Boeing's *Winds on World Air Routes* to review various statistics.
8. Review the current literature, especially *Professional Pilot* and *Business & Commercial Aviation* for information about flight handlers, aircraft, and flight operations.
9. Use the ETP formula to compute a flight example.
10. Contact FlightSafety and/or SimuFlite to determine current training programs available.
11. Review the NBAA International Operations Checklist and discuss these items with colleagues.

Chapter Questions

1. Develop a list of advantages and disadvantages for corporate self-fueling.
2. Discuss the advantages of using a uniform for corporate aviation employees.
3. Establish and explain a policy for air crew flight-time limits.
4. Discuss what MNPS is and what RVSM has to do with it.
5. Compare the services provided by an FBO and an aircraft management firm.
6. Explain what SARPS and PANS are.
7. Why would it be appropriate to obtain and use flight planning software?
8. Discuss how domestic and international flight planning differ and are similar.
9. Explain the duties and responsibilities of an aircraft scheduler/dispatcher.
10. Documentation requirements for flight operations are extensive. Identify and explain various key documents needed for air crew, aircraft, and passengers.
11. How does IBAC assist international corporate flight operators?
12. Develop corporate flight-crew time limitations and defend the limits established.
13. Explain the difference between flight time, duty time, and rest time for corporate air crews.
14. How should a PIC notify the passengers of the flight rules and handle any disruptive passenger behavior?
15. Who can carry weapons or drugs aboard a corporate aircraft and why?
16. What pilot certificate is necessary to fly a corporate aircraft? Should this be different? Explain.
17. Identify various procedures that a pilot can take to reduce aircraft fuel consumption?
18. Discuss air crew communications requirements.
19. Using the ETP formula, compute the ETP for an eastbound aircraft at FL400, with a TAS of 395 knots and a plus 50-knot wind in the Atlantic area. Trip length is 1,800 nautical miles.
20. Discuss the importance of cockpit resource management and present various tangible and intangible advantages for its application.
21. Describe a flight-planning software program.

CHAPTER 8

Aircraft Maintenance

Overview

Corporate and business aviation operators require the services of professional aircraft mechanics who have been properly certified. Often thought of as holding less glamorous positions, aircraft mechanics provide air crews and company managers with the vital business tool, the firm's aircraft, in an airworthy condition. That is the purpose of the aircraft maintenance department. Various methods may be used to meet this mission: in-house, contract, and a combination of these aircraft maintenance programs. In this chapter, the activities of this function, advantages and disadvantages specific to each aircraft maintenance method, and examples of aircraft maintenance programs are provided. An appreciation of the importance of having qualified maintenance personnel should be gained by reviewing these activities.

Goals

- Specify the purpose of the aviation maintenance department.
- Identify pertinent federal aviation regulations (FARs) applicable to aircraft maintenance.
- Discuss the advantages and disadvantages of using an in-house, contract, or combination aircraft maintenance program.
- Describe the purpose and categories of the Air Transportation Association's 100 Aircraft System Code.
- Discuss the CAMP Systems programs.
- Understand the qualification requirement of an aircraft mechanic.
- Determine the requirements to establish an aviation maintenance department.

Introduction

The *purpose* of an aviation maintenance activity is to provide an aircraft in an airworthy condition. Every maintenance operation should have a clearly defined mission statement in meeting this purpose. The Wright brothers would either have maintained their aircraft or had help from other mechanics. At that time, special certified aircraft mechanics did not exist, but through the years, the aircraft maintenance profession has developed into a significant requirement for all aviation categories. To honor aircraft mechanics, the FAA established the Charles Taylor "Master Mechanic" award to recognize experienced aircraft mechanics. Contact a local Flight Standards District Office (FSDO) for details and requirements of this award.

In the early days of flight, many pilots maintained their aircraft, but as planes became larger and more complex, maintenance specialists were needed. In June 1934, the commercial airlines were greatly affected by the Black-McKellar Act, which required the separation of airline operators from aircraft manufacturers. The simple reason was the concern for safety. Safety remains on the forefront of aircraft manufacturing and aircraft maintenance. Aircraft technological improvements have been significant. New avionics, materials, systems, and many other aviation-related items have resulted in federal certification of aircraft mechanics. The Airframe and Powerplant (A&P) license requirements can be found in FAR Part 65, Subpart D. Numerous other federal, state, and local regulations apply to this important support profession.

PAMA, or the Professional Aviation Maintenance Association, promotes and encourages awareness of aircraft maintenance qualifications and activities. Other associations, like NBAA, GAMA, NATA (National Air Transport Association), and others, similarly support understanding and professionalism in this field.

Of major concern to any aircraft buyer is the maintenance needed for airworthiness certification. "The owner/operator of an aircraft is primarily responsible for maintaining that aircraft in an airworthy condition . . ." (Federal Aviation Administration [FAA], 1997). Other FARs require specific record keeping and meeting various maintenance program requirements. The FAA employs people to specifically review the airworthiness of aircraft. To help meet these rules and regulations, company managers procure aircraft maintenance programs from outside agencies. CAMP Systems, Incorporated and SaSIMS are examples of these agencies and will be presented along with various issues pertaining to this industry.

Inspection Programs

A corporate or business aircraft operator must comply with FAR Part 91.409, which establishes aircraft inspection programs. "No person may operate an aircraft unless, within the preceding 12 calendar months, it has had . . . an annual inspection in accordance with part 43 . . . and airworthiness certificate . . . " (FAA, 1997b). Additionally, under FAR Part 91.409, the operator can not operate an aircraft unless within the preceding 100 hours of time in service the aircraft has received an annual or 100-hour inspection. A progressive inspection program may be approved by the FAA in lieu of the above requirements. Overall, there are three inspection program options for the corporate aircraft operator:

1. Progressive,
2. Annual, and
3. Continuous.

The "FAA Flight Standards District Office having jurisdiction over the area in which the applicant is located" provides a certified mechanic holding an Inspection Authorization, a certified repair station, or the manufacturer of the aircraft to accomplish the *progressive inspection* (Federal Aviation Administration [FAA], 1997a). *Annual inspections* are conducted by a mechanic holding an inspection authorization and who complies with FAR Part 43, Appendix D, or the guidelines of the manufacturer of the aircraft. In an annual inspection program, an inspection is required once every twelve calendar months. An annual inspection may be accepted as a 100-hour inspection. For the *continuous inspection* program, FAR Parts 91.409 (e) and (f) regulate large aircraft, turbojet multi-engine airplanes, and turbine-powered rotorcraft.

The NBAA recommends using the aircraft manufacturer's recommended inspection program to gain experience in this area. It further recommends using the same inspection program for the firm's entire fleet to standardize procedures, records, training, and scheduling (NBAA, 1997). A close review of FAR Parts 43 and 91, other regulations, costs of qualified maintenance technicians, and aircraft mission requirements will be necessary prior to determining the appropriate inspection program to use. Later in this chapter, information about available maintenance activities and programs will be presented and will help in making this important decision.

Aviation Maintenance Department

The primary *purpose* of the aviation maintenance department is to maintain a company's aircraft in a safe and airworthy condition. Another purpose is to provide an aircraft when required but at a reasonable support cost. Companies that own and operate a fleet of business aircraft usually appoint a chief of maintenance to manage the firm's aircraft maintenance program. In Chapter 6, a review of various job descriptions includes aircraft maintenance personnel. Most, if not all, aircraft maintenance supervisors are under constant pressure to have a firm's aircraft ready for flight at all times. If aircraft are unavailable, very good reasons must be given by the supervisors.

The number of aircraft personnel required for a company that has its own maintenance department usually depends on the number and type of aircraft to be maintained and the time flown. The NBAA (1997) indicated six factors to consider when determining how many maintenance personnel to employ: (a) type and number of aircraft, (b) home-base location,

(c) flight-route structure, (d) utilization rate by hours flown, (e) proximity to overhaul and repair facilities, and (f) supply points for spares. An estimated four maintenance hours per flight hour per aircraft was used along with 1,577 available maintenance work hours per employee per year. The latter was based on 2,080 total hours of work (52 weeks × 40 hours per week) available, subtracting 120 hours for vacation (15 days × 8 hours), 88 hours for holidays (11 days × 8 hours), 40 hours sick time (5 days × 8 hours), and 80 hours training (10 days × 8 hours). A total of 175 hours of nonproductive time, or roughly 10 percent of work time, was also subtracted to leave the total of 1,577 work hours. If a corporate jet is flown 600 hours per year, then 2,400 maintenance hours would be necessary, or two mechanics (NBAA,1997). Another example of this method is provided in Chapter 6.

Maintenance Issues

Technicians

Besides maintaining corporate aircraft at the home station, maintenance personnel often are flown to the location where the firm's aircraft has broken down. In some instances, aircraft mechanics are included on the flight as a precautionary measure. Record keeping becomes very important to ADMs and chiefs of maintenance, not only for the work completed on the aircraft but for the technicians' experience, health, training, commendations, criticisms, and other company matters. Scheduling these technicians also becomes a primary concern for supervisors.

Minimum Equipment List (MEL)

Minimum equipment requirements for general aviation operations fall under FAR Part 91 and can be found in Advisory Circular 91-67 (Federal Aviation Administration [FAA], 1991). Certain inoperative instruments and equipment are listed in the circular. The main point is that these items are not essential for safe flight. Related FARs for such flights include FAR Parts 43.9, 43.11, 91.203, 91.205, 91.213, and 91.405. Essentially, all instruments and equipment installed on an aircraft must be operative, except as provided under FAR Part 91.213. The FAA, however, recognized that safe flight could be accomplished with some aircraft items inoperative.

The FAA develops a master minimum equipment list (MMEL) which for a particular aircraft, includes "equipment and other items which the FAA finds may be inoperative and yet maintain an acceptable level of safety" (FAA, 1991, p.1). A FAR Part 91 operator who wishes to operate with an MEL must contact the FSDO that has jurisdiction over the geographic area where the aircraft is based. The FSDO will assign an inspector to advise the applicant about FAR requirements, provide a copy of the MMEL, AC 91-67, and the preamble to the MMEL. When appropriate, the FSDO will issue a LOA (letter of authorization) to the operator. Upon receipt of this LOA, the operator develops a document containing operations and

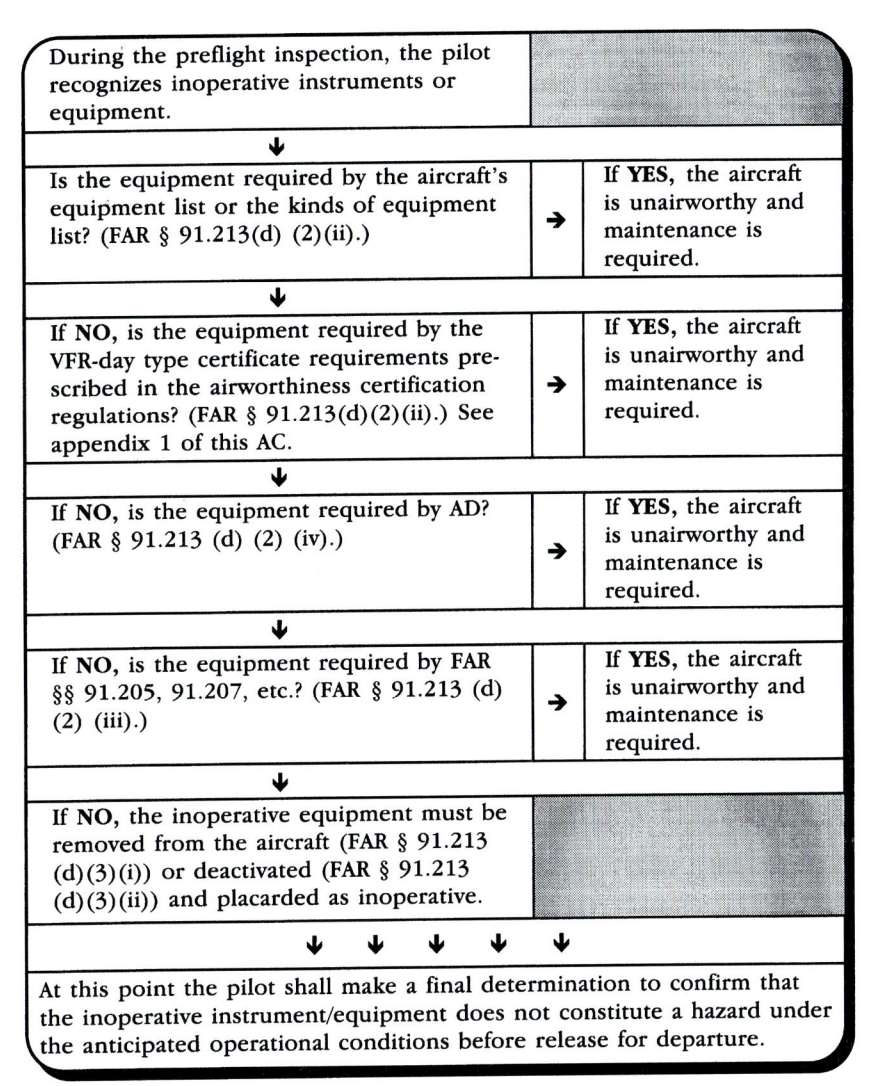

During the preflight inspection, the pilot recognizes inoperative instruments or equipment.	
↓	
Is the equipment required by the aircraft's equipment list or the kinds of equipment list? (FAR § 91.213(d) (2)(ii).) →	If **YES**, the aircraft is unairworthy and maintenance is required.
↓	
If **NO**, is the equipment required by the VFR-day type certificate requirements prescribed in the airworthiness certification regulations? (FAR § 91.213(d)(2)(ii).) See appendix 1 of this AC. →	If **YES**, the aircraft is unairworthy and maintenance is required.
↓	
If **NO**, is the equipment required by AD? (FAR § 91.213 (d) (2) (iv).) →	If **YES**, the aircraft is unairworthy and maintenance is required.
↓	
If **NO**, is the equipment required by FAR §§ 91.205, 91.207, etc.? (FAR § 91.213 (d) (2) (iii).) →	If **YES**, the aircraft is unairworthy and maintenance is required.
↓	
If **NO**, the inoperative equipment must be removed from the aircraft (FAR § 91.213 (d)(3)(i)) or deactivated (FAR § 91.213 (d)(3)(ii)) and placarded as inoperative.	
↓ ↓ ↓ ↓ ↓	
At this point the pilot shall make a final determination to confirm that the inoperative instrument/equipment does not constitute a hazard under the anticipated operational conditions before release for departure.	

Figure 8-1. Pilot Decision Sequence When Operating without a MEL.

maintenance procedures concerning the equipment and instruments under consideration. With this, the operator can begin flight operations, as no further FAA approval is necessary. "The MMEL, preamble, LOA, and the procedures are considered a MEL" (FAA, 1991, p. 11).

The FAA provides decision-tree formats depicting the typical sequence of events a pilot or operator, operating under FAR Part 91.213 (d), should follow when there is inoperative equipment. Figures 8-1 and 8-2 are examples of these decision trees for operations without a MEL and with an MEL, respectively. Additionally, a copy of a firm's operations manual, which relates to MEL procedures, is provided in Appendix H, section 3.67.

8: Aircraft Maintenance 237

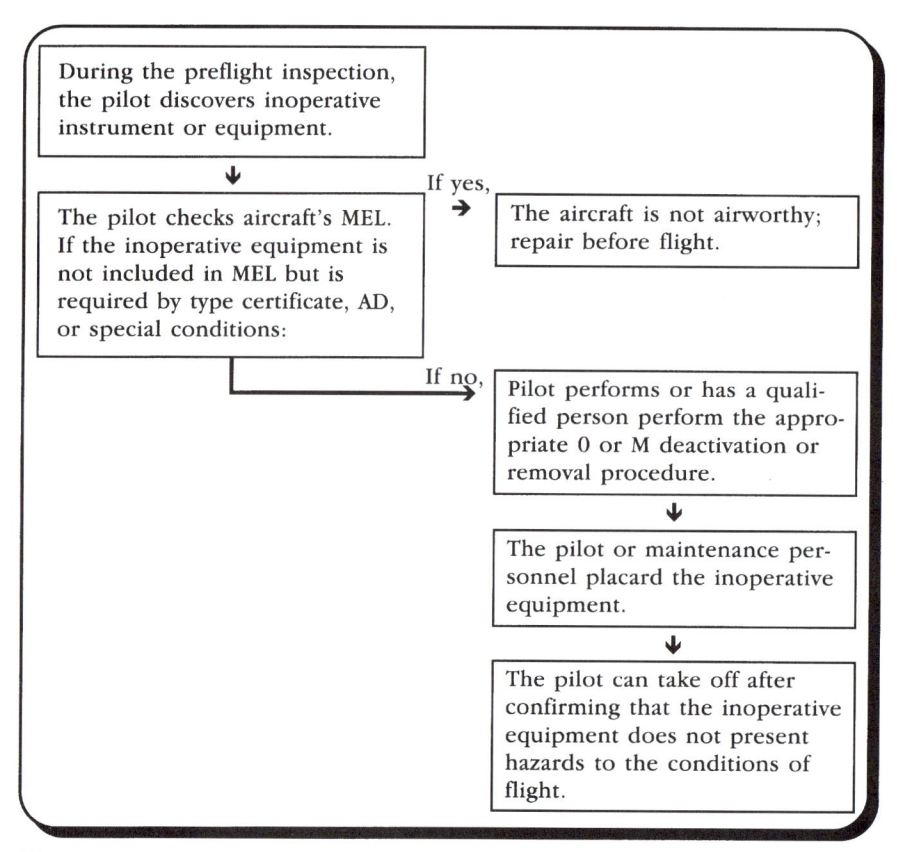

```
During the preflight inspection,
the pilot discovers inoperative
instrument or equipment.
            ↓                      If yes,
The pilot checks aircraft's MEL.    →    The aircraft is not airworthy;
If the inoperative equipment is          repair before flight.
not included in MEL but is
required by type certificate, AD,
or special conditions:
                                   If no,
                                    →    Pilot performs or has a quali-
                                         fied person perform the appro-
                                         priate 0 or M deactivation or
                                         removal procedure.
                                              ↓
                                         The pilot or maintenance per-
                                         sonnel placard the inoperative
                                         equipment.
                                              ↓
                                         The pilot can take off after
                                         confirming that the inoperative
                                         equipment does not present
                                         hazards to the conditions of
                                         flight.
```

Figure 8-2. Pilot Decision Sequence When Operating with a MEL.

ATA System Codes

The Air Transportation Association (ATA) developed a list of codes that identify aircraft systems. The NBAA highly recommends corporate aircraft maintenance managers use the ATA codes when operating under an approved inspection program. ATA's numbering system explanation and an extract of the ATA Specification 100 Manufacturer's Technical Data are provided in Figures 8-3 and 8-4 (C. Atherim, personal communication, undated).

Aircraft Maintenance Methods

In meeting the many FAR requirements and other state and local rules pertaining to aircraft safety and maintenance, business aircraft operators must consider the economics involved. It costs a tremendous amount to hire trained maintenance personnel, obtain equipment, provide appropriate facilities, and ensure other support items are available (administrative offices, supplies, etc.).

ata SPECIFICATION 100
MANUFACTURER'S TECHNICAL DATA

TYPE OF NUMBERING SYSTEM USED

Numbering system

A. General The numbering system is a conventional dash-number breakdown. It provides a means for dividing material into Chapter, Section, Subject and Page. It also identifies the hardware being worked on (Ref. 2-1-4 and 2-13-4). Broad rules for applying the system follow. Specific instructions applying to individual manuals will be found in the detail specification.

(1) Number Composition: The number is composed of three elements which consist of two digits each. For example:

FIRST ELEMENT	SECOND ELEMENT	THIRD ELEMENT	COVERAGE
CHAPTER (SYSTEM)	SECTION (SUBSYSTEM)	SUBJECT (UNIT)	
26 — (SYSTEM) "Fire Protection"	00 —	00	Material which is applicable to the system as a whole.
26 —	20 — (SUBSYSTEM) "Extinguishing"	00	Material which is applicable to the subsystem as a whole.
26 —	22 — (SUB-SUBSYTEM) "Engine Fire Extinguishing"	00	Material which is applicable to the sub-subsytem as a whole. This number (digit) is assigned by the manufacturer.
26 —	22 —	03 (UNIT) "BOTTLES"	Material which is applicable to a specific unit of the sub-subsystem. Both digits are assigned by the

Figure 8-3. ATA Specification 100 Manufacturer's Technical Data-Type of Numbering System Used. Courtesy of ATA.

8: Aircraft Maintenance

ata SPECIFICATION 100
MANUFACTURER'S TECHNICAL DATA

REQUIREMENTS FOR TAB DIVIDERS AND TEXT CONTENT IN STANDARD PUBLICATIONS

	Manual-General	Tab Divider	List of Eff. Pages	List of Eff. T. Rev.	Contents	Part I MM-SDS	Fault Isolation	Part II MM-Prac. and Procedures		Tab Divider	List of Eff. Pages	List of Eff. T. Rev.	Contents	Part I MM-SDS	Fault Isolation	Part II MM-Prac. and Procedures
Title Page	X					X		X	STRUCTURE	X						
Record of Revisions	X					X		X								
List of Effective Pages	X					X		X	51. Std. Pract. - Structures	X	X	X	X	X		X
Table of Contents	X					X		X	52. Doors	X	X	X	X	X	X	X
Record of T.Rs.	X					X		X	53. Fuselage	X	X	X	X	X	X	X
Letter of Transmittal	X					X		X	54. Nacelles/Pylons	X	X	X	X	X	X	X
Service Bulletin List	X					X		X	55. Stabilizers	X	X	X	X	X	X	X
Introduction	X	X				X		X	56. Windows	X	X	X	X	X	X	X
List of Chapters	X					X		X	57. Wings	X	X	X	X	X	X	X
AIRCRAFT GENERAL		X							PROPELLER/ROTOR							
05. Time Limits/Mtce. Chks		X	X	X	X	X			60. Std. Pract. -Prop/Rotor	X	X	X	X			
06. Dimensions & Areas		X	X	X	X	X			61. Propellers	X	X	X	X	X	X	X
07. Lifting & Shoring		X	X	X	X	X		X	62. Rotor(s)	X	X	X	X	X	X	X
08. Levelling & Weighing		X	X	X	X	X		X	63. Rotor Drive(s)	X	X	X	X	X	X	X
09. Towing & Taxiing		X	X	X	X	X		X	64. Tail Rotor	X	X	X	X	X	X	X
10. Parking & Mooring		X	X	X	X	X		X	65. Tail Rotor Drive	X	X	X	X	X	X	X
11. Placards & Markings		X	X	X	X	X			66. Folding Blades/Pylon	X	X	X	X	X	X	X
12. Servicing		X	X	X	X	X		X	67. Rotors Flight Control	X	X	X	X	X	X	X
AIRFRAME SYSTEMS		X							POWER PLANT	X						
20. Std. Pract. -Airframe		X	X	X	X				70. Std. Practices-Engines	X	X	X	X			
21. Air Conditioning		X	X	X	X	X	X	X	71. Power Plant	X	X	X	X	X	X	X
22. Auto Flight		X	X	X	X	X	X	X	72. Engine	X	X	X	X	X	X	X
23. Communications		X	X	X	X	X	X	X	73. Eng. Fuel & Control	X	X	X	X	X	X	X
24. Electrical Power		X	X	X	X	X	X	X	74. Ingnition	X	X	X	X	X	X	X
25. Equip/ Furnishings		X	X	X	X	X	X	X	75. Air	X	X	X	X	X	X	X
26. Fire Protection		X	X	X	X	X	X	X	76. Engine Controls	X	X	X	X	X	X	X
27. Flight Controls		X	X	X	X	X	X	X	77. Engine Indicating	X	X	X	X	X	X	X
28. Fuel		X	X	X	X	X	X	X	78. Exhaust	X	X	X	X	X	X	X
29. Hydraulic Power		X	X	X	X	X	X	X	79. Oil	X	X	X	X	X	X	X
30. Ice & Rain Protection		X	X	X	X	X	X	X	80. Starting	X	X	X	X	X	X	X
31. Ind./Recording Systems		X	X	X	X	X	X	X	81. Turbines	X	X	X	X	X	X	X
32. Landing Gear		X	X	X	X	X	X	X	82. Water Injection	X	X	X	X	X	X	X
33. Lights		X	X	X	X	X	X	X	83. Accessory-Gear Boxes	X	X	X	X	X	X	X
34. Navigation		X	X	X	X	X	X	X	84. Propulsion Augmentation	X	X	X	X	X	X	X
35. Oxygen		X	X	X	X	X	X	X	91. Charts	X	X	X	X			
36. Pneumatic		X	X	X	X	X	X	X								
37. Vacuum		X	X	X	X	X	X	X								
38. Water/Waste		X	X	X	X	X	X	X								
41. Water Ballast		X	X	X	X	X	X	X								
45. Central Maint System		X	X	X	X	X	X	X								
46. Information System		X	X	X	X	X	X	X								
49. Airborne Aux. Power		X	X	X	X	X	X	X								

MAINTENANCE MANUAL

Figure 8-4. ATA Specification 100 Manufacturer's Technical Data-Requirements for Tab Dividers and Text Content in Standard Publications. Courtesy of ATA.

Table 8-1. In-House Aircraft Maintenance

Advantages	Disadvantages
Control	High overhead
Quality work	Equipment costs
Security	Providing facilities
Flexibility	Insurance
Economical for large fleet	Inventories
Economies of scale	Low productivity
Completeness of work	Downtime for training
Pride in work	Significant resources needed
Knowledge and expertise of workers	

Note: Item frequency is not reported. Some items were reported more than once.

Basic daily aircraft maintenance requirements include external and visual inspections of the airframe and powerplant, verification of all fluids (i.e., hydraulic, oil, water, etc.), replacement of needed components such as tires, and other routine maintenance. When major work such as an engine overhaul is required on an aircraft or aircraft system, then specialized equipment, facilities, and skilled mechanics are required. For the small business aircraft operator, it may be uneconomical to hire the people and provide the support equipment and facilities necessary; therefore, contract operations appear appropriate.

In-House Maintenance

What can be accomplished through an in-house aircraft maintenance program depends on management and its willingness to provide the people and items necessary in meeting the company's aviation mission. In Table 8-1, advantages and disadvantages for having an in-house maintenance program were reported by 42 aviation students during an informal survey from 1996 to 1997.

From this brief survey, basic reasons for having an in-house aircraft maintenance program focus on accomplishing necessary maintenance when and how desired and being more economical when a firm owns/operates a large fleet of aircraft. Pilots who know their fellow employees, see how the work is accomplished, and even provide suggestions about items will be more happy and confident flying the aircraft.

This is similar to knowing the pilot who will be flying the aircraft and having the confidence in the pilot's abilities. Disadvantages appear to focus on the high costs involved and the training time for the mechanics.

Whether a crew chief or a job-shop method is used for an in-house maintenance activity is another concern. A crew chief is dedicated to a specific aircraft, while with a job shop

Table 8-2. Contract Aircraft Maintenance

Advantages	Disadvantages
Costs	Quality assurance
Low overhead/labor costs in company	Scheduling
Benefits of experienced workers	Security of aircraft
Familiar with trends	Least reliable
Parts accessibility	Lack of convenience
Frees up company capital	Level of trust
Experience of work on different aircraft	High costs
Pay for only what's needed	Inconvenience

Note: Item frequency is not indicated. Some items were reported more than once.

method, all aircraft are maintained by the maintenance department. Having responsibility for a corporate aircraft results in great personal pride and effort for a crew chief maintaining an aircraft; however, efficiencies of operation are greater using the job-shop method by using the most qualified people for specific maintenance.

Contract Aircraft Maintenance

The services of a fixed-base operator or a firm that provides aircraft maintenance and systems support may be more useful for some corporate operators. The larger FBOs provide a full range of aircraft maintenance, but an aircraft operator may require specialized services for major equipment overhauls. Various listings of FBOs and aircraft handlers are provided in the professional journals and publications (e.g., *Professional Pilot*) or may be obtained by calling the aircraft associations.

In the informal survey referenced above, the advantages and disadvantages for contract aircraft maintenance services were also identified and are presented in Table 8-2. Costs were seen to be an advantage and a disadvantage because a manager can save high training costs of the mechanics yet the costs of having a contract are quite high. The key is to weigh costs against the other advantages and disadvantages. Experience of having experts able to work on various aircraft and to obtain parts more easily is seen to be a major advantage of a contract operation, while not knowing who is actually working on the firm's aircraft is seen as a disadvantage.

Combination Aircraft Maintenance

While there are several advantages and disadvantages for both methods above, many aviation managers decide to use a combination approach to maintaining the firm's aircraft. Many variables have to be reviewed and considered prior to using any maintenance program. The labor force, skill level of the workers, equipment availability, aircraft type and utilization, company's mission, parts supply, specialized services, facilities needed, and others need to be considered. Managers must also complete an environmental impact assessment. Besides the tangible benefits, the intangible benefits weigh into the decision making. Personal pride, having the flexibility to maintain aircraft when desired, tender loving care (TLC), and feeling of confidence from knowing who worked on the aircraft are important. The main advantages for a combination maintenance method approach focus on quality of service, control, and record keeping. This method is recommended by most users because of the experience level of the technicians and the increased attention to detail provided through developed programs.

All aircraft maintenance supervisors and ADMs should review the different methods available to obtain the required maintenance on their aircraft. The team approach in decision making is recommended in choosing a method. Major aircraft maintenance programs like CAMP Systems and SaSIMS provide aircraft users outstanding services—at a price. If the price is right, the service obtained allows aircraft users to have their aircraft ready when needed. A review of CAMP Systems and SaSIMS is provided to introduce readers to the services available.

CAMP Systems, Incorporated

CAMP program features and benefits are listed in Table 8-3, while its locations are identified in Figure 8-5. In Figures 8-6 through 8-8, additional information about CAMP programs, reports, and aircraft are provided.

Table 8-3. Features and Benefits of the CAMP Program

- Facilitates implementation and accurate control of inspection programs under FARs.
- Tailored to meet any maintenance-type operation.
- Provides individual operator with fleet maintenance experience.
- Eliminates nonproductive clerical work.
- Upgrades the quality of maintenance.
- Maintains continuity in the event of changing personnel.
- Recommended by leading aircraft manufacturers.
- Permits responsible person to be in control.
- Provides standardization and operator control of maintenance at FBOs.
- Monitors warranty information.
- Aids in setting annual budget.
- Helps prevent critical inspections being overdue.
- Provides close monitoring of inspection and maintenance needs.
- Increases the resale value of a firm's aircraft.

Note: Extract from CAMP Systems brochure (R. B. Leonard, personal communication, January 9, 1997). Used with permission.

8: Aircraft Maintenance **243**

Figure 8-5. CSI Link Authorized Manufacturers and Service Centers. Courtesy of CSI.

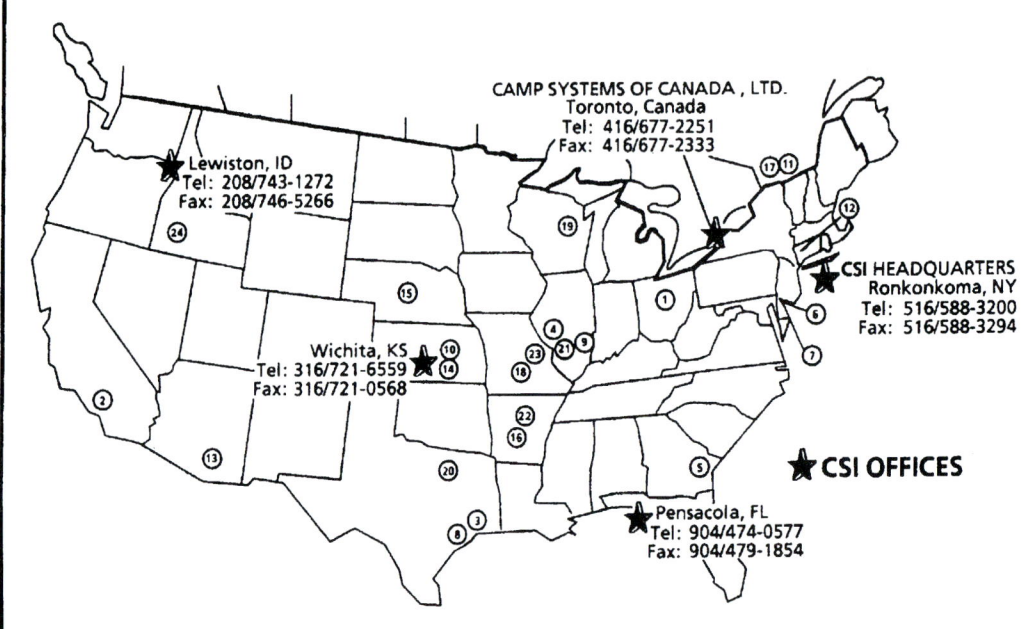

CAMP Computerized Aircraft Maintenance Program

A full-service aviation maintenance management system that uses powerful mainframe computers to support the world's most comprehensive corporate aircraft maintenance database. CAMP is endorsed by many aircraft manufacturers, and is by far the most cost efficient method of assuring tight control over fleet maintenance operations.

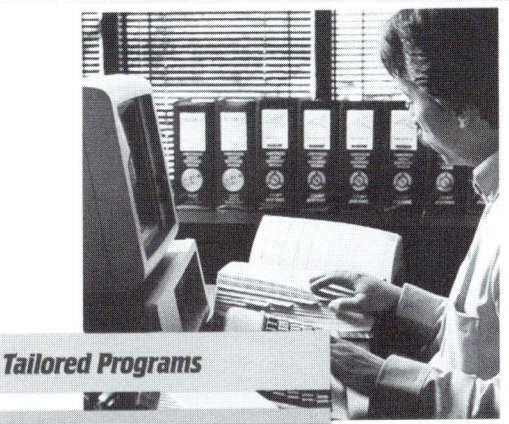

- Tailored Programs
- Maintenance Due Lists
- Procedural Instructions
- Work Compliance Forms
- Cross Reference Feature
- Budget Reports
- Data Quality Assurance
- PC Access to System
- Personal Analyst Support

Tailored Maintenance Programs

When an aircraft is first enrolled in CAMP, the operator and an experienced CSI maintenance analyst establish an inspection and maintenance program especially designed for that aircraft. Included in the program are inspection and maintenance requirements recommended in technical publications issued by the airframe, engine, and component manufacturers. Also included are requirements set forth in Airworthiness Directives, Service Bulletins and mandatory Type Certificate Data Sheets. This data is stored in CSI's central computer and then kept current by an ongoing exchange of maintenance information between the aircraft operator and CSI.

Total Maintenance Control

CSI computers continuously compare an aircraft's current and projected utilization status with the requirements of its stored inspection and maintenance program. From this comparison a Maintenance Due List is generated, prioritizing individual requirements that will need attention. The Due List, which includes support information such as part costs and man-hour estimates, is sent to the operator along with Work Compliance Forms. These forms include step-by-step maintenance procedures and a work compliance sign-off area for each requirement shown on the Due List. After the operator signs off each form, the original is saved to satisfy the FAA-prescribed maintenance records requirement while the carbonless copy is returned to CSI to update the aircraft's inspection and maintenance program. All returned Work Compliance Forms are thoroughly screened and cross-checked by CSI maintenance analysts in conformance with strict quality control and assurance procedures.

Accurate Reports and Forms

Comprehensive yet easy-to-read reports and forms are the heart of the CAMP system. Monthly distribution includes Aircraft Status, Maintenance Due, Aircraft History and Aircraft Update reports. Overview data are provided in a Yearly Budget Report, an Operations

Figure 8-6. CAMP Computerized Aircraft Maintenance Program. Courtesy of CSI.

8: *Aircraft Maintenance* 245

CAMP's comprehensive and easy to use reports and forms have provided unrivaled maintenance control for thousands of aircraft for more than 25 years.

The acclaimed CAMP program continually monitors the full range of maintenance and inspection requirements for an aircraft and brings them to the attention of the operator as they become due. CAMP reports and forms provide operators with the information needed to manage their aircraft with the greatest possible efficiency in all areas of maintenance planning, scheduling and budgeting. Highly flexible, CAMP is ideally suited for use by all types of aircraft.

In addition to the monthly and annual reports and forms, CSI LINK is also available. This unique option allows operators to have on-demand access to their current aircraft maintenance records through the use of a personal computer. CSI LINK gives fleet operators the ability to view and print up-to-the-minute maintenance reports, procedures, and forms for their aircraft when and where they need them.

Work Compliance Form (As Required)

The Work Compliance Form (WCF) is the primary source document for updating the completion of a maintenance task. Prepared by experienced aircraft maintenance analysts, the WCF uses a two-part signoff which conforms to FAR requirements for permanent record retention. The WCF provides the latest specified maintenance procedures which are arranged in a manner that allows a series of related tasks to be performed in the most logical and efficient sequence possible. The amount of detail contained in the WCF is dependent on the complexity of the maintenance procedure. In some instances, a written description of the procedure is sufficient. In others, an illustration is included to clarify the text. The WCF also contains such information as special tools, torque values, types of lubricants, and part numbers of expendable parts required for the job.

Aircraft History Report (Monthly)

The Aircraft History Report is an up-to-date record of all maintenance performed by the operator. It lists all controlled tasks that have been accomplished, and shows why they were performed. In the case of components, the unit time after removal is listed for future reference. The History Report also includes the dollar values of components replaced and the man-hours expended.

Maintenance Due List (Monthly)

The Maintenance Due List encompasses all inspections, services and component replacements that will become due within a two-month period. Derived from the Aircraft Status Report, the Due List is based upon projected aircraft utilization and is divided into sections which show due times in aircraft hours, landings, and/or calendar dates. A summary of component costs and average man-hours required to perform the work is included on the last page of each Due List section.

Aircraft Status Report (Monthly)

The Aircraft Status Report shows operators the present overall status of their aircraft. Included are component part numbers and serial numbers, list prices and average man-hours for component replacement, warranty information, and the unit time for components presently installed in the aircraft. The Status Report also lists the frequencies the operator has selected for replacing components, and performing services and inspections. It also lists the date, aircraft hours and aircraft landings, when the components were installed, when the inspections and services were performed, and when they will next be due.

Figure 8-7. CAMP Systems Reports. Courtesy of CSI.

246 Chapter 8

FAA Approved Inspection Program Manual

An Inspection Program Manual ready for submission to the FAA for approval is provided to those operators who choose to develop and use their own inspection programs as permitted by FAR 91.409 (e), (f) (4). Revisions are provided automatically when the operator makes FAA-approved changes to the manual.

Operations Inspection List (As Due)

The Operations Inspection List contains operator-selected tasks for either progressive or continuous inspection programs. It also contains Work Compliance Form numbers, maintenance manual references, inspection signoff spaces, and man-hour estimates for the individual and total inspection requirements.

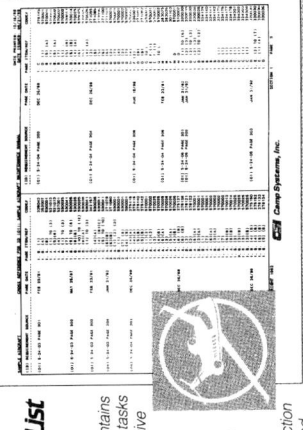

Cross Reference Report (As Required)

The Cross Reference Report is used to locate any requirement stated in the manufacturers' manuals. It is accurate right down to the page revision date of the source document and eliminates any uncertainty about the origin of a requirement.

Aircraft Updates Report (Monthly)

The Aircraft Updates Report reflects those tasks that were updated between the current and prior Aircraft Status Reports. It serves as a checklist for an operator to verify that all tasks submitted for updating were received and processed correctly. The Updates Report lists the tasks by description and includes the compliance information recorded from the sign-off area of the processed Work Compliance Form.

Budget Report (Annual)

The Budget Report lists all components that are scheduled for replacement, and all inspections and services due for compliance within the projected budget period. The last item of the Budget Report is a summary of component costs and average man-hours needed to replace the listed components and to perform the required inspections and services. The Budget Report is divided into sections that reflect the due times according to aircraft hours, aircraft landings, and/or calendar dates.

Performance and Reliability Summary and Trend Report (Semi-Annual)

The Performance and Reliability Summary and Trend Report provides a statistical analysis of all unscheduled component removals. The Report is compiled from the data contained in the Aircraft History Reports of all the aircraft of a specific model enrolled on CAMP. The Report summarizes all unscheduled component removals that occurred over the present and previous 12-month reporting periods, and lists these components according to their rate of removal. The component having the highest removal rate per 1,000 flight hours over the present and previous 12-month reporting period is listed first.

Figure 8-7. Continued

8: Aircraft Maintenance   247

AIRCRAFT MAINTENANCE MANAGEMENT SERVICES AVAILABLE

	CAMP with Text	CAMP Without Text	Beechcraft STARS	Learjet LASER	Andromeda MM
Agusta S.P.A.					
A 109A/II/C	x	x			x
Astra Jet Corporation[1]					
Astra 1125/SP	x	x			x
Westwind 1124/A	x	x			x
Raytheon Aircraft Co. Beechcraft[1]					
Beechjet 400/A		x	x		x
Diamond I & IA		x	x		x
King Air 90 (except F90 series)	x		x		x
King Air 100	x				x
King Air 200/B/C/CT/T	x	x	x		x
King Air 300			x		x
King Air 350			x		x
Starship 2000			x		x
Beechcraft Bonanza			x		x
Beechcraft Baron			x		x
Beech 1900/1900C			x		x
Hawker					
BAe 125-800A/B	x	x			x
Hawker 800 XP	x	x			x
BAe 125-1000A/B	x	x			x
HS 125[1A through 400]	x	x			x
HS 125-600	x	x			x
HS 125-700A/B	x	x			x
HS 125-731 [1A through 600]	x	x			x
Bell Helicopter Textron Inc.					
Bell 206 [all models]		x			x
Bell 222/B		x			x
Canadair					
Challenger [CL600/601/3A/3R]	x	x			x
Cessna Aircraft Co.					
Citation I/II		x			x
Citation SII		x			x

	CAMP with Text	CAMP Without Text	Beechcraft STARS	Learjet LASER	Andromeda MM
Dassault Aviation[1]					
Falcon 10/100	x	x			x
Falcon 20	x	x			x
Falcon 20-5	x	x			x
Falcon 50	x	x			x
Falcon 50 EX	x[2]	x[2]			x[2]
Falcon 200	x	x			x
Falcon 900	x	x			x
Falcon 900 EX	x[2]	x[2]			x[2]
Falcon 2000	x	x			x
deHavilland Inc.					
de Havilland Dash 8	x	x			x
Gulfstream Aerospace Corp.					
Gulfstream I	x	x			x
Learjet Inc.[1]					
Learjet 23				x	x
Learjet 24 BDEF				x	x
Learjet 25/BCDF				x	x
Learjet 31		x		x	x
Learjet 35/A, 36/A	x	x		x	x
Learjet 55/55B/55C		x		x	x
Learjet 60				x	x
Lockheed Aeronautical Systems Co.[1]					
Jetstar	x	x			x
Jetstar II	x	x			x
Jetstar 731	x	x			x
Romaero S.A.					
Rom Bac 1-11		x			x
Saab Aircraft AB					
Saab SF340A		x			x
Sabreliner Corporation[1]					
Sabreliner 40/60/(A,SC)	x	x			x
65/70/80/(A,SC)	x	x			x
Sikorsky Helicopter					
Sikorsky S76A/A+/B	x	x			x

NOTES: 1 Manufacturer endorsed 2. In development

Figure 8-8. CSI Aircraft Maintenance Management Services Available. Courtesy of CSI.

SaSIMS System

SaSIMS stands for Safety by Simplicity Maintenance Management System and was developed by Cassel Aero in Sweden. Because relevant data were needed in a simple and easy-to-use way, Ronny Bengtsson used his 12 years of software expertise and developed SaSIMS. Simplicity was the emphasis. Component total times, inspection lists, service bulletins, and other required items can be easily inputted and accessed. Any number of aircraft can be registered. Many useful advantages exist with the system, for example, managing assemblies, tracing changes to components, receiving warnings of approaching maintenance, using the ATA codes, and others.

SaSIMS operates as a true 32-bit application and runs under Windows 95 and Windows NT 4.0 Workstation. It also runs under Windows NT Server 4.0 and Novell and just needs a

Figure 8-9. SaSiMS. President ULF Cassel (left) and Program Developer Ronny Bengtsson at European Business Aircraft Show, May 1997.

486 processor or above. President Ulf Cassel of Cassel Aero provided extracts of SaSIMS, but for a complete program, fax + 46 660 67016 or call Cassel Aero's Swedish office at + 46 660 67017. Figures 8-9 through 8-13 introduce SaSIMS and the format used and its easy-to-follow form.

Summary

The purpose of the aviation maintenance activity is to provide an aircraft in an airworthy condition for aircraft users when needed. Various federal inspection programs for aircraft are presented, including annual, progressive, and continuous maintenance programs.

Every aircraft owner/operator must consider how the aircraft will be maintained. In setting up an aviation maintenance department, managers must consider the number and type of aircraft used and how the aircraft are utilized. The number of qualified maintenance personnel to employ depends on available work hours and the above factors.

Besides qualified technicians, managers are also concerned with establishing a minimum equipment list or MEL based on the FAA's master minimum equipment list. Using the Air Transport Association's 100 Manufacturer's Technical Data will help managers standardize maintenance activities. Finally, various advantages and disadvantages for in-house, contract, and combined aircraft maintenance methods are presented. The CAMP Systems and SaSIMS programs are examples of several contract aircraft maintenance programs available.

SaSiMS
Safety by Simplicity Maintenance Management System

Staff

The tab **Staff** [Component] [Staff] [Aircraft] contains information about the staff with authority to perform inspections of the company's aircraft.

The [Print] button prints the company's internal certificates (Company Authorisation Certificate)

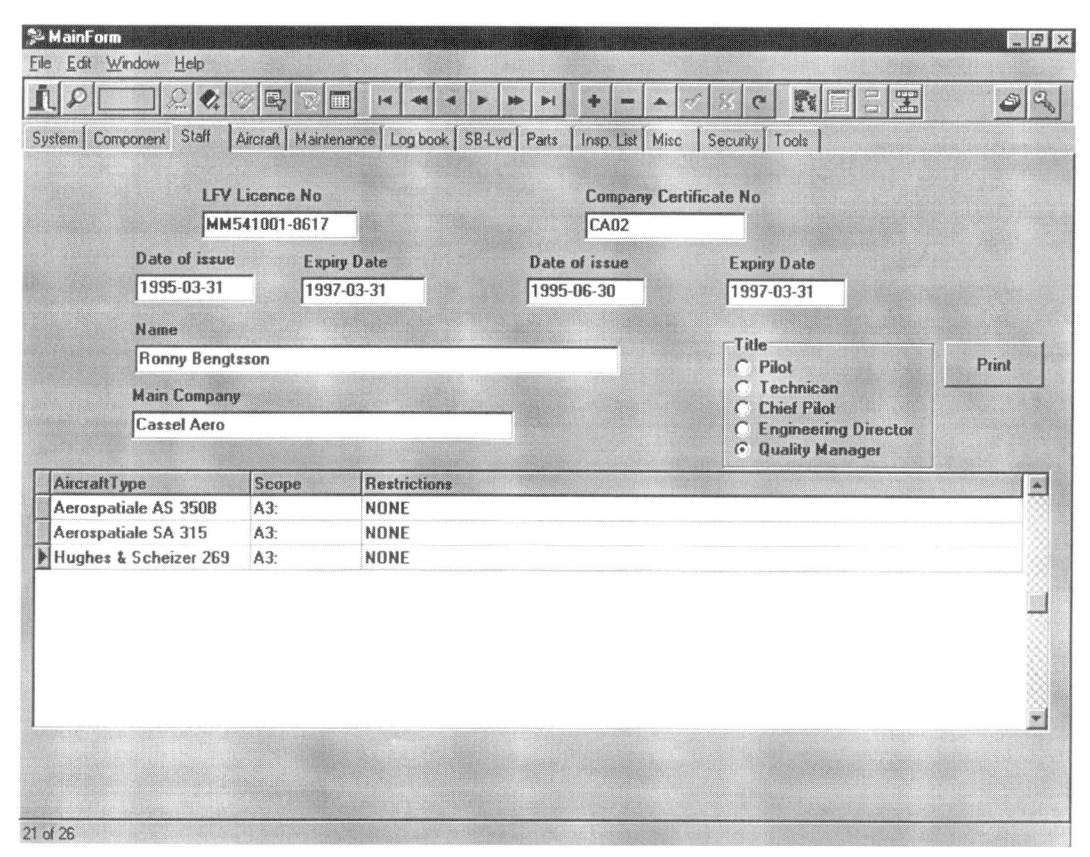

Figure 8-10. SaSiMS Staff. Courtesy of ULF Cassel, President of Cassel Aero.

Basic Aircraft Data

Under the tab **Aircraft** Staff Aircraft Maintenance you insert basic information about the aircraft.
Here you insert all necessary data: Total Time, inspection intervals, inspection lists, references etc.
Total Time, cycles, starts and landings originate from the Flight Log data.

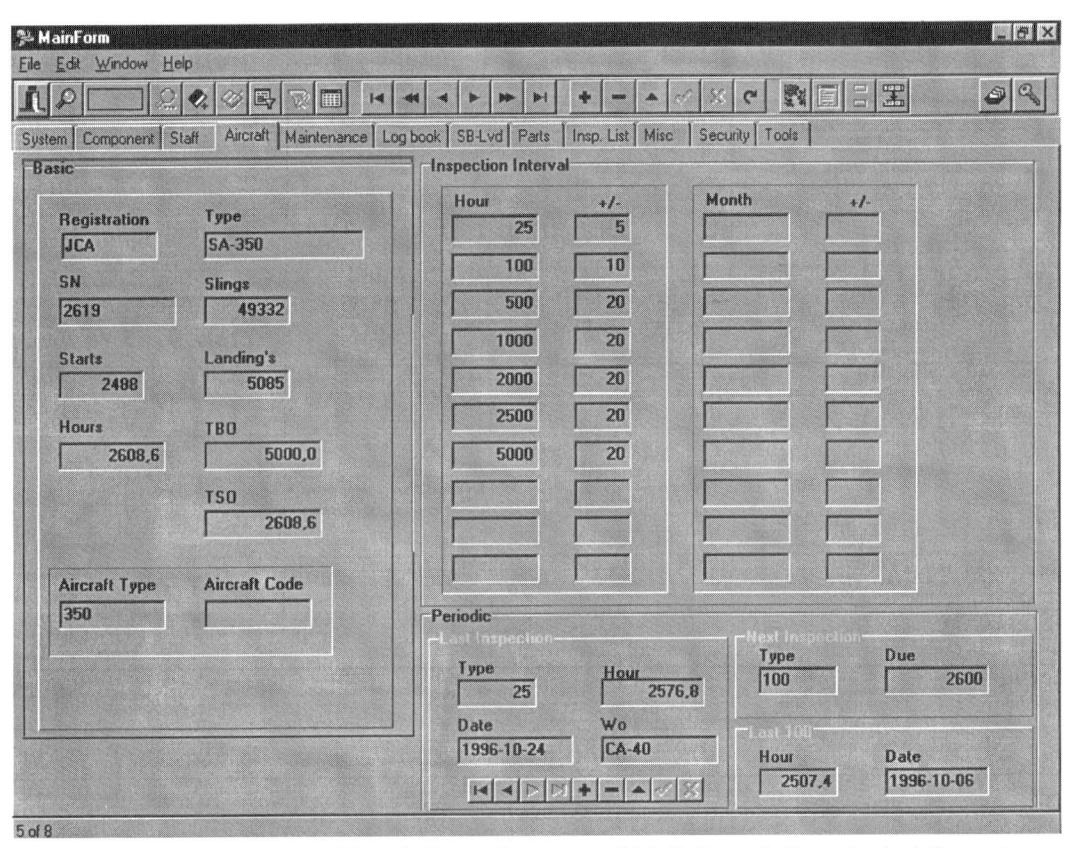

Figure 8-11. SaSiMS Basic Aircraft Data. Courtesy of ULF Cassel, President of Cassel Aero.

SaSiMS
Safety by Simplicity Maintenance Management System

Maintenance

Under the tab **Maintenance** [Aircraft | Maintenance | Log book] you create Work Orders, insert remarks and scheduled works.

Under the tab **WorkOrder** [WorkOrder | Snag/Remark | Work] you insert Work Orders.

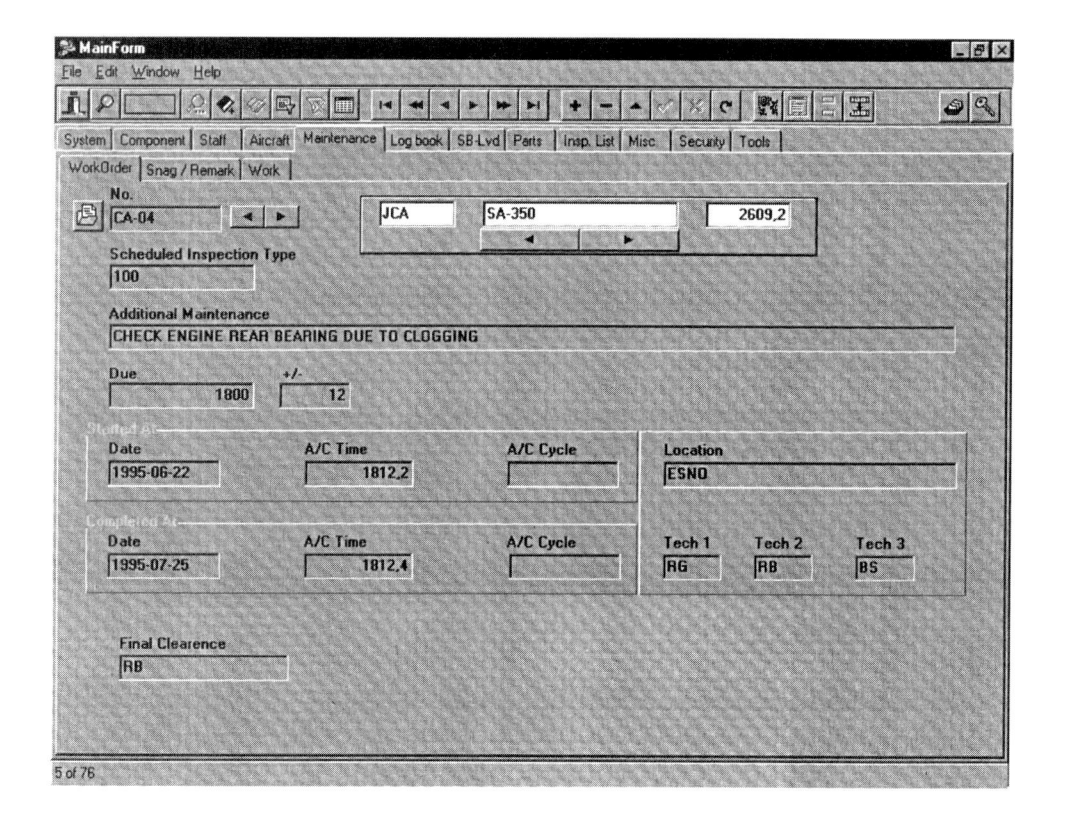

The current aircraft is shown in the window
Use the arrowkeys to scroll through the aircraft.

Figure 8-12. SaSiMS Maintenance. Courtesy of ULF Cassel, President of Cassel Aero.

252 *Chapter 8*

SaSiMS
Safety by Simplicity Maintenance Management System

Service Bulletins / Airworthiness Directives

Basic data of SBs/ADs and corresponding documents are inserted under this tab. The data inserted is at this stage not connected to a/c or components. The connections are made under the **Aircraft**- tab or the **Component**-tab.

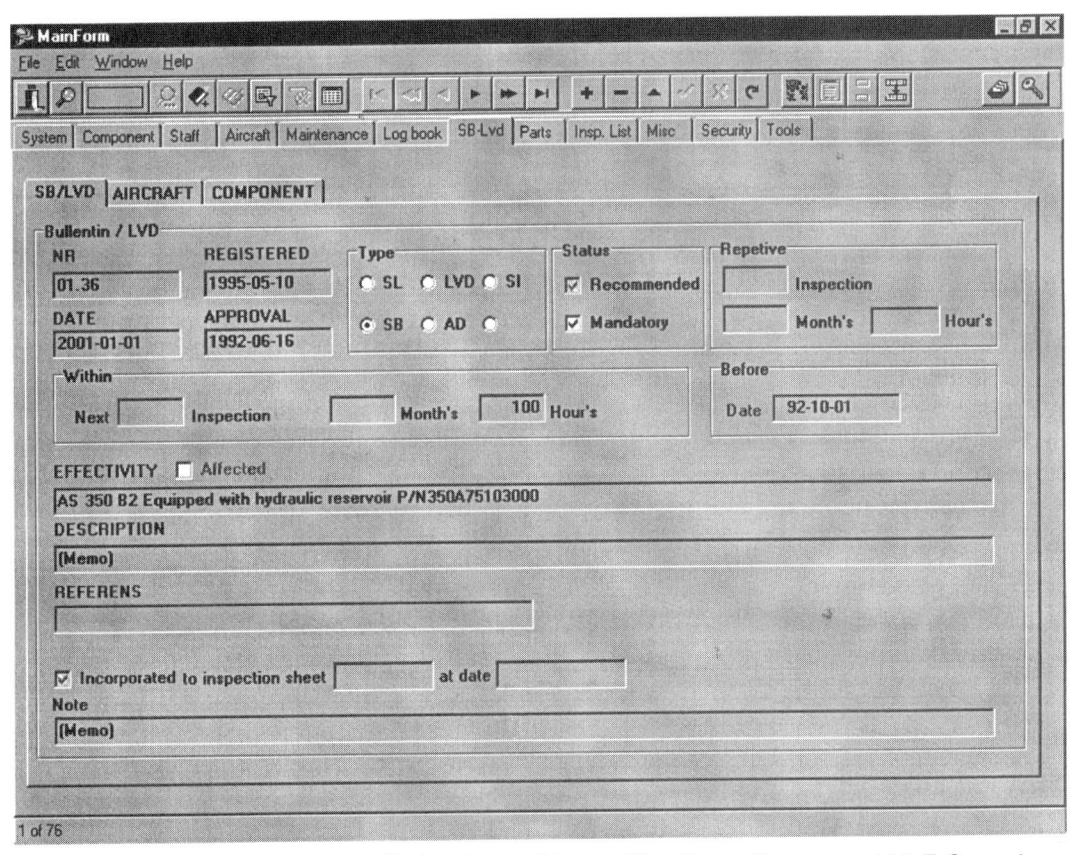

Figure 8-13. SaSiMS Service Bulletins/Airworthiness Directives. Courtesy of ULF Cassel, President of Cassel Aero.

8: *Aircraft Maintenance* 253

Activities

1. Talk to an aviation maintenance manager or technician about aircraft maintenance requirements.
2. Review the FARs, in particular FAR Parts 43 and 91, for aircraft maintenance requirements.
3. Consider the advantages and disadvantages for each of the three aircraft maintenance methods presented. Can you add other pros and cons for each method?
4. Visit an aircraft maintenance operation and determine the extent of maintenance performed. Does this appear to be reasonable?
5. Communicate with an FBO operator or other agency that provides aircraft maintenance. Determine the services provided and their costs.

Chapter Questions

1. Identify the primary purpose of the aviation maintenance department and explain its significance to the company.
2. Explain why the federal government, beginning with the Airmail Act of 1934, separated the aircraft manufacturers from the aircraft users and how this affected aircraft maintenance.
3. Discuss the differences among the three aircraft maintenance programs (annual, continuous, and progressive) presented in the chapter.
4. What should an aircraft operator base his/her aircraft maintenance on in selecting one of the three maintenance programs available?
5. Discuss the differences in using an in-house and contract aircraft maintenance method.
6. Why would a combination aircraft maintenance method be best for a company?
7. Discuss the pros and cons for using a crew-chief versus job-shop method for aircraft maintenance.
8. Provide an example of determining the number of aircraft technicians required for an in-house maintenance function responsible for a Learjet flown 650 hours annually.
9. How does an aircraft operator obtain a MEL list for an aircraft?
10. Discuss the CAMP Systems or the SaSIMS program.

CHAPTER 9

Safety and Security

Overview

Throughout this text there has been an emphasis on safety of flight operations; however, safety and security actions relate to all aspects of an aviation department. Everyone must be confident that safety is paramount. Security goes hand-in-hand with safety because through proper security measures, safety results. In this chapter, a brief review of some important safety statistics provided by governmental and aviation associations allows one to consider how various actions provide not only personal, but company safety and security. Preventative steps are necessary, but when accidents/incidents do occur, proper procedures must be followed. Information about safety precautions and procedures for general and specific items is presented to help understanding and promote effective performance.

Goals

- ✈ Define safety and security relative to corporate and business aviation.
- ✈ Discuss safety issues applicable to corporate aircraft operators.
- ✈ Understand the significance of safety/security statistics for business aviation.
- ✈ Develop an awareness of attention to safety/security.
- ✈ Formulate good management practices for safety/security.
- ✈ Discuss safety/security relative to specific activities in corporate/business aviation.
- ✈ Discuss safety programs; for example, ASRS and NTSB programs.
- ✈ Know aircraft accident reporting procedures.

Introduction

Since the earliest days of aviation, people have been concerned for flight safety. Some of the very first safety rules for flight came about because of French farmers' concern about early balloonists. When early flyers came down in fields unexpectedly, residents became upset over the damage to their crops and damages resulting from tracking around their fields. Police had to be notified of flight attempts. Stories of these early flight attempts and disasters can be obtained from the National Aeronautics and Space Administration (NASA), along with many accident/incident stories of heavier-than-air flight attempts and space mission successes and disasters. The very first man to fly, Jean de Rozier, was also identified as the first man to die from flying. Rosier fell to his death in June 1783 while attempting a balloon crossing of the English Channel (Kane, 1996). From the early aviators to the flyers and support people today, there is a primary concern for safety. Local, state, and federal governments have passed numerous, strict regulations to promote flight safety.

In the United States, "big brother," or the federal government, was convinced by popular opinion to manage and control aviation (review Chapter 1) beginning in 1926. First, the Aeronautics Branch, then the Civil Aeronautics Administration and later the Federal Aviation Administration developed and controlled flight activities through governmental regulations. Today, there are not only numerous FARs, but many state and local rules for aviation control and support. Aviation can be considered the most heavily regulated mode of transportation in the United States.

Though aviation is not the least safe mode of transportation in the U.S., it receives everyone's attention when an aircraft accident occurs. The news media and other informants quickly provide readers and listeners with information about aviation accidents. Annually within the U.S., almost 50,000 people are killed on the nation's highways. Why, then, is so much attention given to aviation when its statistics account for a fraction of this number? The answer is that even one death that could have been prevented is sufficient reason to practice safe procedures, whether involving a car or an airplane. An airline crash makes the headlines because of the thought of so many people involved in such a disastrous event. In this chapter, several of the flight operations procedures will be reviewed, along with data to help people understand the scope of safety in flight. The issues of safety and security should remain in the forefront of managerial concern, and everyone should act as if a rule or procedure affects him or her directly. Part of the vision of top management must concern safety.

Security and safety needs mean managers must provide effective operating policies to protect property and employees and to prevent accidents. Many managers identify their policies in operations manuals or company rules. There is a tendency by some managers, however, not to put in writing specific safety and security rules because they go without saying. Not so! Too often reports are made about a poor operating procedure or that no safety policy had been identified and the result was a serious aircraft accident or incident.

For indirect aircraft flight operations or those preflight and postflight activities such as ground transportation, reporting accidents and incidents follow general safety rules. For flight operations, the National Transportation Safety Board (NTSB) is directly responsible. To meet its responsibility, the NTSB provides NTSB Part 830, Rules Pertaining to the Notification and Reporting of Aircraft Accidents or Incidents and Overdue Aircraft, and Preservation or Aircraft Wreckage, Mail, Cargo, and Records.

The NTSB defines an aircraft **accident** as "an occurrence associated with the operation of an aircraft which takes place between the time any person boards the aircraft with the intention of flight and all such persons have disembarked, and in which any person suffers death or serious injury, or in which the aircraft receives substantial damage." Further, an **incident** is defined as "an occurrence other than an accident, associated with the operation of an aircraft, which affects or could affect the safety of operations" (National Transportation Safety Board [NTSB], 1997).

Security

Security is always a concern for traveling executives and has been heightened by increased terrorism, corporate espionage, and a general attitude of public awareness. Other company personnel are also concerned about having a safe work environment. Basically, security falls into two categories: information and physical. **Information security** refers to protecting company people and property from financial damage or loss of image. **Physical security** involves protecting people from injury and distress.

Terrorism is a real threat to all company employees, but more so for corporate executives who travel to foreign locations, especially to countries with unstable governments. This threat is *political*. NBAA suggestions for avoiding terrorists' acts are listed in Table 9-1.

Casual threats from angry employees or unstable people must be considered just as serious as political threats and may be more probable. Security of company property from internal and/or external theft and damage is also a major managerial concern.

For companies with small flight operations, perhaps security procedures are emphasized through self-awareness only; in other words, everyone keeps an open eye and mind for the unusual. For large flight operations, however, security policies and procedures may be identified in the operations manual or in other company directives. The size of a firm's *budget* may be the deciding factor on how specific and concrete security measures are. Large firms may have a security officer who provides training to company employees and develops and manages an effective security program. Outside security consultants may also be contracted to provide security training. Managers must provide security for company equipment, facilities, and other material, as well as personnel security. Providing information and physical security means *taking time and making an investment.*

News reports of disgruntled employees who have flown on corporate missions and caused serious damage are occasionally publicized. News media coverage of various incidents involving upset or angry employees is unwelcome whether it involves a company aircraft or other situation. Perhaps a quick check of passenger baggage by a flight manager may prevent an incident and a report ever being made. When employees know that they and their baggage will not be checked, opportunities exist for situational smuggling or other negative actions.

Table 9-1. Suggestions to Avoid Terrorists' Acts
Review all sources of media information for current acts. Maintain company awareness for security by newsprint. Study preventative measures publicized by government and other authorities. Review aircrew notes/records from previous trips. Consider removing company logos and identifications. Make no controversial statements. Cancel or reschedule trips to potential trouble areas. Ensure local hiring of guards and other support personnel is effective. Use passenger and load manifests. Use secured aircraft parking. Provide aircrew security training.
Note: Adapted from NBAA (1994). *NBAA management guide*. Washington, DC: National Business Aircraft Association. Used with permission.

The aviation department manager or chief pilot is responsible for providing flight support to ensure a safe flight. In this regard, whatever reasonable actions necessary for ensuring people, property, and equipment are secure and safe should be taken.

Information/Physical Security

The NBAA and Mescon Group (1989) developed various training programs for corporate aircraft operators. One of these programs pertains to flight operations and security training. Various aspects of their program follow; however, contact the NBAA for current details and other programs.

Information compromises result from (a) unintentional events that create adverse conditions, (b) loss or unintended sharing of company information, (c) arrest and incarceration, (d) adverse publicity, (e) flight delays, (f) exposed personal relationships, (g) possible bribery and extortion, and (h) possible illegal setups to obtain company information. What about company passengers who have been asked to carry something for someone else and think nothing of doing so or are too frightened or unsure to say no? A firm's aircraft can be seized because of third-party smuggling. Customs' regulations apply in every country and corporate flight managers and pilots must make themselves aware of their details.

Items such as alcohol, drugs, tobacco, jewelry, art, guns, plants, pornography, gifts, and illegal funds are just a few of the items that air crews should consider as being security risks. *Internal theft* involves company employees, usually only taking small items such as tools, small pieces of aircraft equipment, or miscellaneous items. Every company employee must conduct himself or herself with loyalty and integrity. Taking small company items home because they are there and no one would mind is not a valid excuse at any time. Resources are limited for every company; even the military (with the biggest budget) can not afford internal thefts. In one instance, a USAF aircraft mechanic at RAF Mildenhall took home a C-141 engine. What the person would have done with it cannot be determined, but the fact is that this person committed a major theft. Such large thefts are not impossible in corporate firms.

External theft results from breaking and entering a firm's premises with the purpose of taking property illegally. This property may be of substance or be informational only, for example, company records.

An ADM or flight department manager must be proactive and not wait for security failures to occur. Systems and company resources must be reviewed for effective and efficient use, but be properly secured at all times. Company personnel must be provided the protection needed to keep everyone safe, equipment usable, and information where and when required.

A list of questions to ask yourself about company security is provided in Table 9-2. See how many of these you can answer. Review your company's operations manual or security policies. Add to this list as needed to bring security awareness to everyone's attention.

For company personnel, security protection may include the use of photo identification cards, security keys, and uniforms. Effective communication systems must be used to monitor and control company employees so that they are where they are supposed to be when needed.

Knowing who is in charge of personnel admittance to a work place is important at all times. For company facilities, locked doors, closed circuit television, anti-intrusion devices, good lighting, and designated locations are helpful. A security guard at the hangar or central entrance door with lock-control devices is also useful. The best means of security is knowing who is supposed to be in an area and questioning anyone who is unknown or unescorted.

Table 9-2. Security Awareness Questions

Who is the firm's security manager?
Who should be contacted first when there is a security concern?
What is the phone number to use for a security alert?
Who should be given any suggestions for security awareness items?
Who is in charge when there is a security alert?
Who handles security decisions when an aircraft is in the air?
Who handles security decisions when an aircraft is on the ground?
What security equipment is available for use?
Who releases company information outside the firm when there is a problem or a question?
What company security training is available?

Aircraft Security

How about a firm's aircraft? There are several issues to consider here. Security of company aircraft at home base, enroute, and at locations other than home base must be provided. At *home base*, aircraft security will probably be at its highest because of the company's available resources; for example, hangars, guards, owned facilities, and watchful company employees. Having a secured aircraft parking area, with appropriate tiedown equipment and guards is one of the best security methods. Parking at an FBO location allows for good security when the aircraft parking area is separated and controlled by the agent. Unattended parking areas leave aircraft operators open for criticism and possible nonpayment of any insurance claim. The aircraft need to be locked and microswitches or other devices/methods used when feasible; for example, taping doors and hatches.

Preparing for a flight means fueling and cargo/baggage loading. Security during these times is also vital. All baggage should be checked and properly tagged. A load manifest must include all passengers, baggage, cargo, and/or mail. Passenger and baggage loading need to be controlled and guarded until flight departure. Using a ramp controller or duty officer to monitor and manage flight preparations is advisable. During preparation for a flight departure, faxing or calling flight information to others should be limited and controlled. Changes to flight schedules should only be given to those who have a need to know. In the competitive business world, being at the right place at the right time might mean a major contract, and if a firm's business is private, it should be kept that way. Not even dependents, friends, or relatives of the air crew going on a trip should feel free to discuss destinations and times with others.

For in-flight security, communications during flight involve also only talking to those who have a need to know. Radio frequencies are open. The appropriate radio licenses are required, along with common spares needed (fuses, batteries, etc.). Bomb threats, terrorist actions, and other flight disruptions should be handled by using well-thought-out company procedures and established flight procedures. Executing a distress radio call on 121.5 Mhz or 243.0 Mhz, setting the transponder to 7500, maintaining a true airspeed of no more than 400 knots, keeping an altitude between 10,000 and 25,000 feet mean sea level, and following forced instructions are actions identified in the Airman's Information Manual.

Training to handle the unexpected can only prepare the air crew to make on-the-spot decisions. The best help that a flight manager can give air crew members is pre-flight training. Air crew members who know company policies and company people can usually make good last minute decisions. Air crew personnel who haven't been trained or just never took the time to know the company people and rules will probably be uncertain during an unexpected security situation. Random company tests of security should be made to determine everyone's preparedness.

Carrying weapons, drugs, or other controlled items must be handled with the utmost care and in accordance with strict company rules/regulatory agencies. Pilots should know their passengers and be aware of any unrest within the labor force, company actions that have a negative effect on personnel, and the cargo/baggage being transported. Hazardous materials may be carried aboard aircraft, and when this occurs, the items and the procedures for handling the material should be made known to the aircrew. It is just not good enough for a corporate pilot to report that he or she just flies the aircraft and is not responsible for handling the cargo. Corporate flying requires greater attention to many aspects of air transportation movements that are handled by specialists in a commercial airline operation.

Air crew members should maintain contact with their home base throughout the flight and when any doubt exists about any procedure, a discussion with an appropriate company official may be helpful. At a destination, one of the first actions that should be taken is to communicate the flight's arrival to the company, followed by information about billeting of the passengers and air crew, and other support activities (aircraft parking, ground transportation, aircraft maintenance, etc.).

Postflight activities again include communication with everyone who has the need to know what happened on the flight and at the destination(s), especially about items that may affect others in the company and/or company equipment. If any incident or accident occurred, reports should be completed as required. Planning for the security and safety of the next flight should begin by providing important information about any aspect of previous flights being related to those who will be involved in future flights. Any item that involves security and safety should be reported to appropriate company officials who can decide on necessary policy changes or other steps to maintain an effective security and safety program.

Safety

Safety should remain a top priority for everyone, but what is the price for safety? In commercial airline operations, certain equipment can be used that should provide for increased safety. For example, smoke hoods have been tested which, when used in an air transportation emergency, will give extra precious minutes to passengers. Yet, smoke hoods are not required equipment. Smoke hoods and other equipment cost money and at one point or another, managers must make important decisions about how much money is to be spent on safety equipment. Better avionics, better aircraft, stronger and more powerful engines, and many other items can be purchased to improve safety for a corporate aircraft operator.

Safety of flight support activities and flight is the main concern. Management's budget must cover a flight department's operation to provide a reasonably safe service. Safety of personnel involves accomplishing many of the security actions identified above, but attention to flight safety appears to be the overall concern of managers.

Flight operations involves people, equipment, facilities, and procedures. ADMs must make a risk assessment of their functions to determine their acceptability. Improved technology, increased communications, and skilled people have resulted in better safety records for corporate aircraft operations.

Aviation Safety Statistics

Aviation safety statistics are used to measure the extent of safety for flight operations. The Federal Aviation Administration's main goal is safety, and it reports various statistics involving aircraft accidents and incidents. In turn, each of these statistics is divided into specific indicators for large air carriers, commuters, air taxi operators, general aviation, and rotorcraft. Midair collision indicators are also reviewed for accidents and incidents. Additional categories for aircraft incidents include pilot deviation rates, operational errors, runway incursion rates, and vehicle/pedestrian deviations. The latter involves movement on an airport by a vehicle or pedestrian that has not been authorized by traffic control (Federal Aviation Administration [FAA], 1996). The basic FAA statistical standard for the commercial airlines and general aviation is fatalities/accidents per 100,000 flight hours. Additional standards based on aircraft miles flown and passenger miles flown are also reported by the FAA.

NBAA (1997) presented a summary of aircraft accident rates (Figure 9-1) with subcategories for corporate/executive and business aircraft operations. In Table 9-3, statistics for rotorcraft accidents are provided for comparison.

According to Flight Safety Foundation, Breiling Associates reported that 1995 "was the worst year for corporate/executive aircraft accident involvement since 1987, with the highest number of fatalities since 1991" (1996, n.p.). In 9 of 12 aircraft accidents, business jets were involved. Overall, the accident rate for the corporate/executive category was 0.25 accidents per 100,000 flight hours. Corporate and business aircraft operators compiled a very good accident rate record, compared to the airlines and better than the rest of general aviation, the commuters, and air taxi operators. In the first half of 1997, Robert E. Breiling Associates reported 5 corporate U.S. jet fatal accidents and 4 business jet fatal accidents of the 14 total (the remaining 5 were U.S. commuter/air taxi aircraft accidents). For U.S. turboprops fatal accidents, 3 involved corporate aircraft and 7 business aircraft of 21 total fatal accidents ("Turbine Business," 1997, p. 44).

It is difficult to determine rotorcraft accident categories because the FAA and the Helicopter Association International do not divide rotorcraft operations into specific sub categories. Bradley (1997) reported that corporate transport helicopter operations emerged as the "safest of all helicopter operations" (p. 43) for the period 1982 to 1995. During this period, he reported 173 accidents/incidents for U.S.-registered twin-turbine helicopters. From 1988 through 1995, 15 helicopter accidents were reported by corporate operators, and 3 accidents were reported by business helicopter operators.

Specific accident/incident reports may be obtained from the FAA's Office of Accident Investigation as well as from the NTSB. Using NTSB's internet path of "iirform.htm" will provide a listing of 17 categories of accident/incident reports. These categories include all aircraft accidents; all general aviation accidents; all helicopters; fatal and serious injuries; and specific manufacturers' names like Cessna, Beech, Bell, Fairchild, and so forth. Write directly to the NTSB to obtain specific information on accidents at the following address: Public Inquiries Section, National Transportation Safety Board, 490 L'Enfant Plaza East, SW, Washington, DC 20594.

Aircraft Accident Rates, 1984-1996 (per 100,000 Flight Hours)

	General Aviation*	Air Taxi**	Commuter Air Carriers+	Airlines++	Corporate/ Executive#	Business##
	TOTAL/FATAL	TOTAL/FATAL	TOTAL/FATAL	TOTAL/FATAL	TOTAL/FATAL	TOTAL/FATAL
1984	10.36/1.87	5.14/0.81	1.260/0.401	0.208/0.012	0.52/0.08	3.96/1.24
1985	9.66/1.75	5.99/1.36	1.209/0.403	0.253/0.080	0.88/0.31	3.93/0.89
1986	9.54/1.75	4.35/1.15	0.870/0.116	0.231/0.020	0.53/0.08	3.11/0.93
1987	9.25/1.65	3.65/1.13	1.644/0.514	0.329/0.038	0.56/0.12	3.06/0.75
1988	8.69/1.68	3.84/1.06	0.908/0.096	0.251/0.018	0.27/0.05	3.18/1.01
1989	7.96/1.53	3.68/0.83	0.803/0.223	0.248/0.098	0.29/0.11	3.46/1.27
1990	7.78/1.55	4.71/1.24	0.642/0.128	0.198/0.049	0.21/0.09	3.71/0.96
1991	7.98/1.58	3.88/1.20	1.013/0.368	0.218/0.034	0.23/0.08	3.08/0.82
1992	8.71/1.87	3.78/1.19	1.008/0.321	0.144/0.032	0.21/0.08	2.17/0.68
1993	9.05/1.77	3.81/1.06	0.660/0.165	0.178/0.008	0.23/0.07	2.02/0.52
1994	9.09/1.83	4.26/1.30	0.429/0.129	0.159/0.030	0.18/0.07	1.81/0.51
1995	10.27/2.06	3.75/1.20	0.465/0.078	0.266/0.022	0.25/0.11	2.04/0.67
1996(P)	8.06/1.51	4.63/1.42	0.40/0.040	0.300/0.039	0.14/0.06	1.71/0.34

* = All U.S.-registered civil aircraft not operating under FAR Part 121 or 135

** = FAR Part 135 non-scheduled air carriers

+ = FAR Part 135 scheduled air carriers

++ = FAR Part 121 scheduled and non-scheduled air carriers

= Aircraft owned or leased and operated by a corporation or business firm for the transportation of personnel or cargo in furtherance of the corporation's or firm's business and which are flown by professional pilots receiving a direct salary or compensation for piloting.

= The use of aircraft by pilots (those not receiving direct salary or compensation for piloting) in conjunction with their occupation or in the furtherance of a business.

P = Preliminary

Source: National Transportation Safety Board and Robert E. Breiling Associates (shaded columns), 1997

Figure 9-1. Aircraft Accident Rates, 1984–1996 (per 100,000 Flight Hours). Source: NBAA *Business Aviation Fact Book 1997*, p. 25.

Accident Causes

The main cause of aircraft accidents is human based. Numerous factors, but not all, affecting air crews have been reported by NBAA and Mescon Group (1989) officials as being a cause of aircraft accidents and incidents. In Table 9-4, typical human error items apply whether an aircraft or a car is being operated. However, certain items such as "someone not

Table 9-3. Rotorcraft Accident Rate (100,000 flight hours)

Calendar Year	Number of Accidents	Number of Flight Hours	Accident Rate
1989	210	2,613,340	8.04
1990	216	2,440,119	8.85
1991	190	2,355,402	8.07
1992	194	2,074,207	9.35
1993	176	1,940,796	9.07
1994	207	1,950,586	10.61
1995	161	1,833,532	8.78

Source: Federal Aviation Administration. (1996). *Aviation system indicators*. Washington, DC; U.S. Government Printing Office. pp. 2–14.

Table 9-4. Human-Factor Causes of Aircraft Accidents

Anxiety	Distractions
Alcohol	Drugs
Complacency	Conversations
Ignorance	Inexperience
Judgmental errors	Passenger pressure
Peer pressure	Get-home-itis
Crew overfamiliarity	No one flying the aircraft
Thoughts not on flying	Operational distractions
Routine operations	Daydreaming
Stress	Incapacitation

Source: Adapted from NBAA's unpublished course leader's guide. National Business Aircraft Association. (1989). *Flight operations security and safety*. Washington, DC: Author. Used with permission.

flying the aircraft" and "crew overfamiliarity" apply specifically to aviation. Others like inexperience, passenger pressure, and daydreaming affect pilots, car drivers, truck drivers, and many other transportation operators. The key is personal attention, both in skills and mental attitude. For further research into this subject, contact the NTSB and FAA for accident and incident information. Various Internet addresses are listed in Appendix C, for example, FAA's Aviation Safety Links (links.htm) includes a listing of many aviation safety offices and agencies from whom requesters can obtain safety information. These include the FAA, state aviation organizations, NTSB, NASA, military, international, aviation safety organizations, universities, and other aviation safety sites.

Air crew incapacitation has been reported to result from gastrointestinal and ear problems, vertigo, headache, faintness, and other sudden medical problems. Fatigue is a major problem for all air crew members, and because there is no federal flight duty limitations for corporate pilots, it may become a more serious issue. Corporate managers who jeopardize their people and missions because they decide to fly the crew beyond practical duty limits are simply placing the budget or power management over common sense. Loss of life, aircraft, and equipment should never result from inadequate flight-crew rest. It is also important to realize that fatigue is cumulative. Proper air-crew rest, exercise, and food are equally as important as good flying skills.

Other aircraft accident causes are bad weather, inoperative or malfunctioning equipment, mechanical failures, airport conditions, and other miscellaneous items. The International Society of Air Safety Investigators (ISASI) reported that the predominance of corporate business jet accidents during the 1980s occurred in the landing phase (53.4 percent). Critical factors such as runway length, conditions, and environment were major concerns. The second most critical phase was the takeoff, involving 14.7 percent of accidents (Griffin, 1992). Together, these activities involved only 1 percent of the total flight time, but accounted for 68 percent of the accidents.

The approach phase accounted for 12.5 percent of the accidents, yet it is one of the most demanding phases of flight. Application of cockpit management training has been reported responsible for improvements in this flight phase. Lastly, Griffin (1992) reported that the cruise phase of flight accounted for 55 percent of the flight time, but involved the smallest percentage of accidents.

Bradley (1997) reported phases of flight based on multi-engine and single-engine turbine helicopters. These percentages and phases for the period 1982 through 1995 are identified in Table 9-5. From these statistics, aviation managers can understand the need for quality training for the air crew and support personnel.

Safety Programs

Many public and private agencies and activities promote safety through programs that recognize individuals and groups for their safety efforts. Whether flying an aircraft, working in the dispatch section, serving as a flight attendant, fixing the aircraft, or any of the many activities necessary to support flight operations, safety is needed above all. Recognizing those who work hard and make the effort to comply with the rules, who suggest improvements to the rules that need to be changed, and who do their jobs safely is very important to encouraging continued success and promoting further safe operations. Universities, training agencies, governmental offices (federal, state, and local), and aviation associations provide recognition for safety in aviation. Two such programs that promote safety are identified next.

Table 9-5. Phases of Flight in which Helicopter Accidents Occurred (1982–1995)

Flight Phase	Multi-engine Turbine (%)	Single-engine Turbine (%)
Cruise	29.3	26.8
Descent	1.5	2.6
Approach	6.6	5.6
Maneuver	16.7	32.2[a]
Landing	14.1	13.1
Takeoff	13.1	12.2
Hover	8.1	—
Climb	4.0	3.5

Note: a = combined hover and maneuver for single-engine. Percentages do not add to 100% because of other miscellaneous reasons for accidents. Extracted from Bradley, P. (1997, February). Helicopter safety statistics. *Business & Commercial Aviation*, 46.

ASRS Program

The FAA and NTSB established the Aviation Safety Reporting System (ASRS) in 1975. While the FAA provides most of the funding, NASA administers the program through a contractor selected via competitive bidding. In 1997, the contractor was Battelle Memorial Institute.

ASRS collects, analyzes, and responds to voluntarily submitted aviation safety incident reports to help decrease aviation accidents. ASRS data are used to:

1. Identify deficiencies and discrepancies in the National Aviation System (NAS) to facilitate corrective actions,
2. Support policy formulation and planning by NAS, and
3. Strengthen aviation human factors research (Federal Aviation Administration. (1997).

Anyone involved in aviation operations (pilots, air traffic controllers, flight attendants, mechanics, ground support personnel, and others) can voluntarily submit a report to ASRS when they are involved in, or observe, an aviation incident. The reports are held in strict confidence; there has been no breach of confidence in over 300,000 reports. Experienced aviation personnel read the reports and analyze them to identify aviation hazards. If any hazard is found, an FAA office or aviation authority is notified. The FAA identifies its waiver of penalties under FAR Part 91.25 and AC 00-46C, and basically the alleged violation must be:

1. Inadvertent and not deliberate,
2. Not reveal any event subject to Section 609 of the FAA Act,
3. The reporter must not have been guilty of a FAR violation during the preceding five years, and
4. The report has to be submitted within 10 days of the event.

Some of the program outputs include:

1. Alert messages. These take various forms but involve relaying safety information to persons in authority.
2. Callback. This is a monthly safety bulletin that contains excerpts from ASRS incident reports.
3. *ASRS Directline.* This is published periodically to inform operators and flight crews of significant items.
4. Database-search requests. When the need exists and a request for information is received, ASRS will provide reports.
5. Operational support. ASRS supports the FAA and NTSB during rulemaking, accident investigations, and other actions by providing information as requested.
6. Topical research. Many topics of interest are researched by ASRS, including wake turbulence, digital avionics, cockpit interactions, crew performance, and others.

NTSB Safety Programs

Established in 1953, the NBAA Flying Safety Awards program enables those operators who have achieved safety in flight to gain recognition for professionalism. The program promotes the safe use of business aircraft. Only member companies are recognized during the fall NBAA convention. The program includes:

1. *Corporate Flying Safety Award.* NBAA members whose aircraft have been flown in excess of three or more consecutive accident-free years.
2. *Commercial Flying Safety Award.* NBAA members whose aircraft have been flown in excess of three or more consecutive accident-free years in a nonscheduled, revenue-producing capacity.
3. *Pilot Safety Award.* NBAA member-company pilots who have flown corporate aircraft in excess of 1,500 accident-free hours.

4. *Aviation Maintenance Department Safety Award.* NBAA member companies that qualify for a Corporate or Commercial Safety Award and perform their own maintenance.
5. *Maintenance/Avionics Technician Safety Award.* NBAA member-company technicians who have been employed three or more accident-free years in support of corporate/business flight operations.
6. *Aviation Support Services Safety Award.* NBAA member-company support-services personnel who have been employed three or more accident-free years in support of corporate/business flight operations (National Business Aircraft Association, 1997).

Accident Investigations

The NTSB is responsible for determining the probable cause of transportation accidents and promoting safety through its recommendations. To meet this requirement, a small staff of governmental workers (approximately 400) manage various programs. Many experts from industry, universities, and other sources help to collect the facts and data surrounding each accident. The NTSB delegates to certain authorities the responsibility for accident investigation; for example, the FAA's Office of Aviation Safety reviews experimental, agricultural, and home-built aircraft accidents and the military services investigate their own accidents.

Aviation department managers, chief pilots, and chiefs of maintenance must know and understand the NTSB rules and many other company, state, and local regulations that apply when an aircraft accident occurs. A firm's operations manual should have clearly defined steps to follow, and everyone within the firm should understand and comply with them.

Normally, a pilot immediately notifies the NTSB through his or her aviation department manager or chief pilot when an accident occurs. The manager will, in turn, notify the nearest NTSB field office of any accident or incident under NTSB Part 830.5. NTSB field offices are listed under *U.S. Government* in telephone books and contact numbers should be maintained in a firm's flight dispatch/scheduling section. The NTSB accident/incident reporting requirements are provided in Table 9-6.

Information required to be provided to the NTSB includes:

1. type, nationality, and registration marks of the aircraft;
2. name of owner and operator of the aircraft;
3. name of pilot-in-command;
4. date and time of accident;
5. last point of departure and point of intended landing of the aircraft;
6. position of the aircraft with reference to some easily defined geographical point;
7. number of persons on board, number killed, and number seriously injured; and
8. description of any explosives, radioactive materials, or other dangerous articles (National Transportation Safety Board [NTSB], 1997).

Table 9-6. NTSB Accident/Incident Reporting Requirements

Flight control system malfunction or failure.
Inability of any required flight crew member to perform normal flight duties as a result of injury or illness.
Failure of structural components of a turbine engine excluding compressor and turbine blades and vanes.
In-flight fire.
Aircraft collision.
Damage to property, other than aircraft, estimated to exceed $25,000 for repair.
For large, multi-engine aircraft (over 12,500 pounds):
    In-flight failure of electrical systems.
    In-flight failure of hydraulic systems.
    Sustained loss of power or thrust produced by two or more engines.
    Evacuation of an aircraft in which an emergency egress system is utilized.
Aircraft is overdue and is believed involved in an accident.

Source: Adapted from NTSB Part 830.5. (1997). Rules pertaining to the notification and reporting of aircraft accidents or incidents and overdue aircraft, and preservation or aircraft wreckage, mail, cargo, and records. [On-line]. URL http://www/ntsb.gov.

Depending on the severity of the accident, the NTSB sends a "go-team" to the accident site. Various experts may be asked to be part of the team, which is headed by an NTSB member. These experts involve various specialties: avionics, air traffic control, operations, weather, human factors, engineering, medical, metallurgists, and others. Paperwork is extensive, but the overriding concern of the investigators is to determine probable cause in order to prevent future accidents.

Air crews must familiarize themselves with these procedures, as well as the procedures needed at the accident scene. Safety of the passengers comes first by removing them from any further possible harm. Although people need information about an accident, air crew members should not make any statements. It is best to leave any statements to the company's spokesperson or public affairs officer. Compliance with the NTSB rules does not mean that any air crew member should jeopardize him- or herself by making any statements unless capable guidance is given. It is extremely important that company managers develop clear procedures for accident/incident reporting and provide them in an operations manual or other company document.

Miscellaneous

Many agencies are working hard to improve the safety awareness of all business aircraft users. John Olcott, president of NBAA, presented a talk about the characteristics of safety cultures to the Flight Safety Foundation on May 1, 1997. He emphasized that complacency (one of the items listed in Table 9-4) was a significant factor in aviation safety. Examples of various corporations that developed excellent safety cultures were provided, for example, DuPont and Amoco corporations. Overall, Olcott's main goal was to identify common elements of safety cultures for the business aircraft user. In doing this, he listed 10 aspects of culture conditioning:

1. Unqualified commitment to safety as a way of life;
2. Unambiguous expectation by each manager that employees develop and follow safe life patterns;
3. Availability of equipment to do the work;
4. Clear, easily understood standard operating procedures;
5. Inclusive system of communications;
6. Nonretribution for submission of incident data;
7. Retraining without penalty or stigma;
8. System for tracking incident/accident data, analysis, and feedback;
9. Peer acceptance that accidents are preventable; and
10. Peer acceptance that safety is a lifestyle—a matter of culture (NBAA, 1997).

Safety cannot be taken for granted, and for anyone involved in aviation matters, it should be the top priority. Olcott and many others can only suggest and recommend courses of action, but it is left to each person to do the right thing at the right time. It is easy to use a shortcut in an operation, but if the procedure requires specific action, then that is the right thing to do, without hesitation or possible harmful intervention by outsiders. That is living safety!

Summary

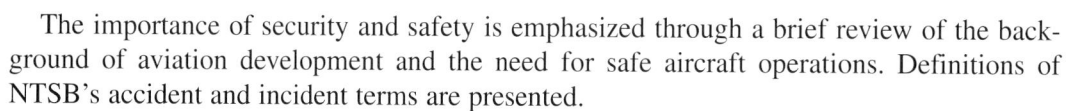

The importance of security and safety is emphasized through a brief review of the background of aviation development and the need for safe aircraft operations. Definitions of NTSB's accident and incident terms are presented.

To prevent accidents and/or incidents, several suggestions from the NBAA, FAA, and others are presented. Security programs can be categorized into information and physical security, with the latter involving political and casual threats. The size of a company's aircraft operation will have an effect on the size and scope of the company's security program. Smaller activities tend to rely more on personal awareness and understanding while larger operations will probably use formal security and safety programs managed by a security/safety officer.

Aircraft security involves actions prior to flight, enroute, and postflight. Numerous suggestions to maintain good security and safety are provided, especially for activities prior to any flight. These include aircraft preparation, loading of passengers and cargo, and scheduling/communications. Enroute security activities are limited, but specific actions (flight speed, radio use, and others) must be known and understood by all air crew members. Postflight security and safety procedures are important to prevent future accidents/incidents; however, required steps must be followed in the unfortunate event an accident or incident occurred.

Various aviation safety statistics are presented relative to causes of aircraft accidents/incidents and phases of flight in which they occurred. Human error remains the major cause for aircraft accidents/incidents, while landing/takeoff operations were the major phases of flight in which accidents occurred. Corporate/business aviation accident rates fall well below others in general aviation and compare quite favorably with the commercial airlines.

Rotorcraft statistics cannot be completely presented because the FAA does not separate helicopter operations in a manner similar to fixed-wing operations. Rotorcraft accident rates were presented in Table 9-5.

Two of the many safety programs made available by government, aviation associations, universities, and others are presented—ASRS and NBAA safety programs. Lastly, NTSB Part 830 should be reviewed and fully understood by air crew members and people involved in corporate/business flights. In NTSB Part 830, the type of accidents/incidents to be reported to the NTSB and the actions to be followed in the event of an occurrence are identified.

Activities

1. Conduct a search of accident data reported by the FAA, NTSB, or other agencies by using the Internet listing provided in Appendix C.
2. Discuss with corporate air crew members their main concerns for safety in flight operations. Which phase of flight is reported to be most serious?
3. Review a specific aircraft accident report provided by the NTSB and determine how the accident could have been prevented.
4. If working for a company having a flight department, conduct a critical analysis of its security/safety program. Can improvements be made?
5. For a corporate flight department, make an informal review of actions employees take to maintain good security. Any recommendations?
6. Review a firm's aircraft accident and incident rate and determine the major reasons for its occurrence.
7. Make a comparison of aircraft accident/incident rates between commercial airlines and general aviation. Attempt to break the rates into specific categories in order to identify corporate and business aircraft operators by type of aircraft operated.

Chapter Questions

1. Differentiate between an aircraft accident and incident as defined by the NTSB.
2. Discuss the major reasons for corporate aircraft accidents.
3. Identify the major phases of flight in which aircraft accidents occur and specify their percentage differences.
4. List and discuss various actions to prevent terrorist acts against corporate flight activities.
5. Explain the differences between informational and physical threats to security for a corporate aircraft operator.
6. Explain internal and external thefts as they relate to corporate flight activities.
7. Discuss any differences between a security program of a large and small corporate aircraft operator.
8. Identify and explain actions necessary to provide good aircraft security at home base and away from home base.
9. How should flight department personnel be trained for security/safety awareness?
10. When new technology or other advanced equipment is available to improve safety within an aircraft operation, why do some managers not use them?
11. Discuss the differences in accident rates between the commercial airlines and general aviation.
12. How do rotorcraft and corporate/business aircraft operators differ in accident percentages by phase of flight?
13. Why is human error considered to be the main cause for aircraft accidents?
14. Identify other causes for aircraft accidents and/or incidents besides human error.
15. Present an established safety program, public or private, that promotes safety awareness and action.
16. Identify several types of aircraft accidents that must be reported by pilots to the NTSB.
17. What is the "go-team?"
18. What steps should be taken by a person involved in an aircraft accident?
19. Discuss who should report aircraft accidents when a corporation is involved.

CHAPTER 10

Current/Future Issues

Overview

Many of today's issues involving corporate and business aviation will also be the industry's concerns of tomorrow. Various selected topics, by no means all-inclusive, are presented in this chapter in order to focus attention on the value and benefit of corporate/business aviation. To realize value from using business aircraft, aviation personnel must keep abreast of the many issues that affect aircraft operations.

Throughout this text, a prevailing concept has been that managers conduct their business by using their limited resources effectively. How aviation resources are used has evolved into a serious endeavor. When managing aircraft, managers must comply with many federal, state, and local regulations. Additionally, globalization of the marketplace means that important international regulations must also be followed, not only in conducting business but in using business aircraft. Safety remains the primary concern, but attention needs to be given to other important issues. This chapter covers topical areas including management, aircraft, maintenance, finance, people, technology, operations, and others. Information concerning these topics is now more readily available than before through the use of global communications and especially through the use of personal computers and the Internet (see Appendix C). Importantly, various aviation associations provided information about their programs and issues they are working on.

Goals

✈ Discuss major issues facing corporate and business aviation managers.
✈ Identify and discuss three major issues of international corporate aviation.
✈ Discuss airport and airway access as it relates to corporate/business aviation.
✈ Explain the impact of the MNPS/RVSM rules affecting corporate aviation.
✈ Identify and explain the effect on corporate aviation of major issues relating to operations, maintenance, and management.

- ✈ Describe any differences of security and safety for a corporate flight activity versus a commercial airline operation.
- ✈ Discuss the difficulty of justifying corporate aviation operations to stockholders.
- ✈ Explain major regulatory actions that affect corporate aviation operations.

Introduction

Most people do not thoroughly understand general aviation, much less corporate and business aviation. These same people, however, do understand how the commercial airlines contribute to society and their personal demands because they use them. Phil Boyer, president of the Aircraft Owners & Pilots Association, presented an interesting slide presentation on the perception of Americans about general aviation at the 6th Annual FAA General Aviation Forecast Conference in March 1996. One informational slide emphasized an important point that relates to corporate and business aviation. Boyer presented that one out of ten individuals has an unfavorable view of general aviation when understanding of the industry exists (Federal Aviation Administration [FAA], 1996). Another way of considering this statement is that nine of ten people view general aviation as favorable when they understand what it is and what it does.

Corporate and business aviation managers' *primary effort* should be to let others know what their activity is and how it contributes not only to specific companies, but to society and to the nation. Efforts to do this are being made by many aviation groups, for example, NBAA, GAMA, AOPA, HAI, IBAC, FAA, and others. But for corporate managers who have established corporate flight departments, the task becomes more central. The aviation department manager must justify the continued operation of the normally non-revenue earning function to the top company managers. At the same time, the ADM must meet many federal, state, and local issues affecting flight operations. Only by letting the top managers know what value corporate aviation brings to the company will others understand the significance of the industry. John Olcott, NBAA President, emphasized a proactive approach to the industry. Others followed by publicizing how aircraft operations contribute to company growth, local jobs, increased commerce, and general business.

To meet this challenge of letting people know what the industry is and how business aviation benefits not only a company but many others, those directly involved have to aggressively promote their activities and address any deficiencies. They must fight to be heard by regulators, corporate officials, and others. Unsatisfactory and unsafe aviation operations must be corrected by continual efforts to improve efficiency of flight operations and support the company's mission safely.

An informal personal survey taken in May 1997 of ten aviation graduate students resulted in the identification of topics in Table 10-1. When asked to identify the top ten items that would have an effect on corporate aviation in the future, these pilots and aviation-employed people listed 48 different items in the 100 items reported. The major repeat items were tax

Table 10-1. Items Affecting Corporate Aviation in the Future (May 1997 Survey Results)

User fees	Airport issues*	Global politics*
New pilot starts/labor supply	Economy*	Economic inflation
Airport and airway congestion*	ATC funding*	Regulations
Global business travel	Parts prices	Cost of entry
New travel technologies*	Noise rules*	Tax rules*
Fraudulent parts	Military reduction	Fuel prices*
FAA*	JARs	Free flight
Product liability*	Fractional ownership*	GPS*
FAR Part 135 rules	Qualified personnel	Terrorism
Work rules	Budgeting	Labor
Training*	Maintenance	Planning*
Environment	Flight planning	National funding
Safety*	Security*	Video conferencing
Commercial airlines' efficiencies	ATC congestion	Rule changes
Bottom lining	Administration	Hub and spoke changes
Spacemen and alien rides	Operation efficiencies	Value of time

Note: Items affecting the aviation industry identified by 10 Embry-Riddle Aeronautical University students taking Corporate Aviation Operations (MAS 622), May 1997. Of the 100 items listed, many were repeat items (denoted by *).

laws, liability, airport and airway congestion, global politics, noise rules, economy, fractional ownership, GPS, fuel prices, new technology, air traffic control, and airport issues.

Each item reported in Table 10-1 is important (perhaps disregarding the spacemen/alien rides at this time), and aviation concerns were reported basically by the type of work the person was doing. For example, pilots were more attentive to flight procedures, flying regulations, and aircraft technology. Administrators were more concerned with budgeting and planning. Most corporate aviation workers were concerned with changing tax rules, safety, security, training, and the state of the economy. *Bottom lining,* by the way, was defined as operators working to achieve absolute independence from middlemen and shops, or reaching the lowest cost achievable. Some of these major issues facing aviation managers are addressed in this chapter under major headings such as **s**afety/security, **m**anagement, **a**ircraft, **r**egulation, **t**echnology, **o**perations, **p**lanning/procedures, and **s**ystems (**SMART OPS**).

Before reviewing SMART OPS, chapter information previously presented in the tables should be looked at again. General aviation aircraft sales are increasing in the latter part of the 1990s, and corporate international flying is established as a business means. Using business aircraft is based on mission requirements, and new aircraft ownership programs, like fractional ownership, are available. Aviation regulations are changing, and national and international events are affecting business aircraft usage. Aircraft associations like NBAA, IBAC, CBAA, and others are promoting the use of aircraft for business and are fighting hard to obtain fair support from governments. Business aircraft operators are not trying to compete with the airline operators, but are attempting to supplement their services when and where necessary to meet company mission requirements. Many business aircraft operators still use the commercial airlines, but time savings and other value-added factors are being obtained with business aircraft.

SMART OPS

Safety/Security

In the last chapter, safety and security issues were presented. Above all, business flight activities must remain safe and secure. Aviation, as a mode of transport, affects everyone when accidents/incidents occur. Although only a few people may be directly involved in an aircraft accident, the outcome will cause people to consider the continued use of air travel, no matter if it is commercial or business flying. People are wary of aircraft coming down anywhere, of terrorist activities, and of aviation-related accidents and incidents.

The National Transportation Safety Board recommends corrective actions to other regulatory bodies like the FAA, but these actions are post hoc activities. The damage has usually been done. Governments can make flying more safe, but then people wouldn't be able to afford the *costs* of flying. Good judgment has to balance the costs of safety with the risks of harm. As a mode of passenger transport, air transportation remains the safest mode of transport. The human-error causes of aircraft accidents are being addressed by many governmental agencies, by the aircraft associations, and others. Aircraft manufacturers are producing new equipment to make aircraft easier and safer to fly. Training schools are developing new courses for piloting and for flight operations; for example, cockpit resource management training.

Security awareness has increased because of terrorist activities and the ease of international travel. Additionally, business activities are becoming more global because of new world markets, lower labor costs, cheaper raw materials, and new managerial strategies. The need for security is not temporary, but will exist for all time. IBAC, NBAA, and the other business aircraft associations provide excellent guidance to aircraft operators and should be consulted when any concern for flying exists. If there is any sincere doubt that a flight is not safe or there might be a security problem, then it should be canceled.

A *safe attitude* does not cost anything, but it can be the foundation for an effective aviation operation. Training helps to reduce risks in aviation operations, and such training involves not only the pilots, but the dispatchers, the mechanics, the flight attendants, fuelers, and other support people contributing to the flight operation. Certification standards for aviation specialists must be continually reviewed and updated.

Management

Having the right person to do the job comes first. Whether an aviation department manager is a pilot or not should not be the main concern. The concern is whether the person can do the job most effectively with the resources available; whether the person is the right type of manager with the best qualifications for the job. Once selected, the person chosen to manage the flight activities then has many responsibilities to meet. Planning becomes a major activity for future successes! Duties involving budgeting, scheduling, managing people, controlling operations, communicating with authorities, representing the company, and many others are challenging to any good manager, but not impossible to do and do well.

Perhaps one of the greatest managerial issues today and in the future is the existence of the flight operation itself. Whoever decides to begin a flight activity wants to know that there is a real need for the operation and that the aircraft operation provides value to the business. If it is a personal flight operation, then that person makes the final decision to keep the flight operation. Justifying the aviation function to top company managers, however, may be the major concern for the aviation manager. Supporting how the use of a business aircraft contributes to employee productivity, to increased company sales, or to other tangible variables may require significant record keeping. Presenting the intangible benefits is much harder to do, but may be the reason for the continued existence of the department. Having the people who use the aircraft report its value is the best method in justifying the operation.

Future attention must be given to developing excellent managerial skills and abilities. Good management remains an art and a science, and while a pilot must maintain technical proficiency through training, a good manager increases personal qualifications through studies and learning programs.

Aviation managers must attend to four major items:

1. Mission,
2. People,
3. Equipment, and
4. Facilities.

Knowing what one's mission is and analyzing the resources available to meet this mission means success.

Many issues fall under management, and one of the main issues is budgeting. Aircraft operations involves many fixed and variable costs, but there are other costs. Labor sources are not as plentiful today as in the past. Programs like GA Team 2000 are helpful to interest new pilots. GAMA has an active role in this project to increase the number of student pilots to 100,000 per year in the United States. It is estimated that over one million young people are interested in flying, but the commitment to undertake flight training takes money and time—both of which are at a premium.

Various agency programs have been initiated and will be helpful to attract new pilots. For example, in 1992 the Experimental Aircraft Association began its *Young Eagles* program to introduce one million young people to aviation through airplane rides. The Aircraft Owners and Pilots Association *Project Pilot* was designed to encourage young people to obtain a private pilot's license. A similar program, *Learn to Fly,* helped to encourage young people in fixed-base operations and flight schools, and Women in Aviation, International promotes the role of women in aviation.

Paying employees low wages, but spending large amounts on training, means that the likelihood of long-term employment decreases. To counter the effect of low wages and possible unhappiness in employees, managers need to encourage additional benefits. Providing medical insurance, uniforms, dental care, life insurance, disability insurance, and loss-of-license insurance will go a long way in keeping good employees.

During high employment, it stands to reason that personal qualifications will be lower and just the opposite during low employment. The highest pilot certification, ATP, is recommended for all pilots to provide a greater safety margin, but realistically, not everyone will have the time or money to achieve this qualification. The major factors for higher salaries are the type of aircraft used and the size of the operation, but aviation managers should provide acceptable compensation to all employees.

The severe downsizing of the U.S. military has affected the labor pool of pilots, mechanics, dispatchers, air traffic controllers, and others. To offset this, the use of simulator training is on the rise, and ab initio flight training is increasing. There is a *distinct work experience difference* between flying or working for a commercial airline and a corporate flight department. Flying and other technical skills may be comparable, but for work in smaller aviation departments, teamwork, a positive attitude, and a willingness to do more for less pay prevails in the corporate arena. *Customer service* is important for the airlines, but for the corporation, it *is the major concern.*

Various aircraft acquisition programs are available today, for example, fractional ownership, leasing, buying new or used aircraft, chartering, joint ownership, and partnership. Tax issues are likewise extremely important to corporate aircraft operators; therefore, the many tax rules that apply have to be reviewed by an aviation manager as well as a company tax expert.

Being a public or a private corporation does not change management's goal; it does affect the obligations to be met. Aircraft management firms also provide the services required to support a firm's travel, and this results in external management. These management firms accomplish outstanding service, for a price, and should be reviewed by top management.

Aircraft

Obtaining the right aircraft is essential to any firm's mission. Helicopters are being used more than before because of their technical improvements and increased services to business users. Perhaps the tilt-rotor aircraft will be a major aircraft for future operators—combining the speed of a fixed-wing with the agility of the rotorcraft. Various new and used aircraft are presented in the major periodicals such as *Business & Commercial Aviation, Professional Pilot,* and various aviation trade journals. Aircraft manufacturers, aircraft brokers, and other sales representatives provide those interested with a variety of aircraft for consideration. Jet Solutions, Executive Aviation, and others provide guidance and support for aviation users. Choosing the right aircraft will result in optimizing costs.

Deciding upon a replacement aircraft after years of service with the present aircraft begins with an intensive effort of keeping the firm's mission and needs foremost in mind. Planning for the future requires a review of past activities, relative to the mission and needs identified. A new aircraft decision follows.

To obtain a new or used aircraft becomes another consideration. Management activities involve evaluating aircraft use and suitability, accomplishing cost analyses, understanding maintenance requirements, determining systems control, and other actions required in aircraft operations.

Various general aviation aircraft are available from U.S. and foreign manufacturers, and the variety of piston-engined, turboprop, turbojet, and rotorcraft come in all shapes and sizes. Additionally, Boeing has available the corporate jetliner. Actually, any aircraft, foreign or domestic, can be used by the business aircraft operator as long as it meets the needs of the operator.

Current and future aircraft issues include not only the costs of these aircraft, but their specifications, performance, equipment requirements, materials, and suitability. Bell, Bombardier, Cessna, Gulfstream, Learjet, Raytheon, Sikorsky, and others have excellent products available for the business aircraft operator; the concern is fitting the right aircraft to the right user. Good aviation management, therefore, requires knowing the rules, keeping abreast of changing regulations, supporting appropriate programs, participating in aircraft association activities, and keeping an open and watchful eye for any event or activity that may affect a firm's aviation function.

Regulations

New and changing rules governing aircraft ownership and operations continue to affect aircraft operators. MNPS, RVSM, taxes, operating, noise, air traffic control, navigation equipment, and other rules are being passed by the federal government. State and local regulations also must be followed.

For example, obtaining flight control equipment to meet the RVSM communication and navigation rules to reduce risk of vertical collisions requires thousands of dollars for the equipment and the training that is necessary. Stage 3 noise rules dictate that aircraft operators must not exceed specified noise level requirements for aircraft engines. Technical amendments to the FARs mean that operators must comply with changes such as keeping manuals in electronic form or complying with new definitions of aircraft operations. Airworthiness directives and advisory circulars provide new changes and/or requirements. Any of the above activities are magnified when a corporate aircraft operator decides to operate under FAR Part 135 in order to gain revenue. Operating manuals, flight-time limitations, and other regulatory restrictions apply to those who carry passengers for compensation.

More regulators are looking at the various aviation categories and considering that safety pertains to each equally; that is, they are questioning differences in Part 91, 135, and 121 operations. A major concern of corporate managers is when it costs a corporate operator more time and money to meet flight rules designed for the commercial airlines. ETOPS, or extended-range, twin-engine operations, require airline managers to meet time restrictions for their flights. If applied to business jets by European Joint Aviation Authorities (JAA), not only would the one-engine inoperative diversionary rules increase flight time, the precedent would be set to make corporate aircraft operators more in line with the commercial aircraft operators. International flight rules may apply to more U.S. corporate aircraft operators because of the changing global marketplace. Harmonizing the JAA regulations with FAA rules continues to be a major concern with regulatory officials for international aircraft operations.

The General Aviation Revitalization Act (GARA) helped to reduce liability costs for some owners/operators and aircraft manufacturers, but it in no way eliminated them. Concern for liability costs will remain a serious issue for many, and no doubt lawyers will challenge case law in every provision of the act.

Changes to the FAA and to the National Airspace Plan direct attention at corporate aircraft operators more than in the past. New FARs pertaining to pilot certification, to maintenance activities, to aircraft parts, and to other items will need continued review by the business aircraft associations and individual operators. Overregulation is stifling and is considered counterproductive by many aircraft operators. At the same time, safety remains a top priority and requires constant attention to flight operations and equipment. Possibly a major future regulatory impact will be the reform of the FAA. It could become an independent agency or be retained under the Department of Transportation.

Technology

Previously, reference was made to the many types of aircraft available today. In the future, still more aircraft will be available. The tilt-rotor has not yet been produced for corporate use, but its availability may provide operators opportunities to use the new technology to corporate advantages. Perhaps new forms of dirigibles or small homebuilt craft may be used for business activities. New aircraft communications equipment will advance control and reporting systems.

Since weather is a major factor in general aviation aircraft accidents, the FAA has improved its programs through the Automated Surface Observation System and uses real-time weather information. This occurs through an aeronautical data link.

Major technological improvements in communication systems and tools have helped training schools like Flight Safety, SimulFlite, and other aviation-oriented programs. Embry-Riddle Aeronautical University, for example, uses its advanced communications to teach students through coordinated efforts. Student pilots prepare flight plans, weather students provide meteorological briefings, dispatchers coordinate movements, air traffic controllers monitor flights, and on it goes. These students work individually, yet as a coordinated group, to effect real-time actions that dramatically improve their aeronautical learning. Students have reviewed aircraft accidents and developed software to project flight problems. The FAA and NTSB have used such programs in their accident analyses and in developing new rules.

Innovative products will not only be realized in systems, but in new airframes, engines, and avionics. The Global Express and Gulfstream V aircraft will provide intercontinental services at very high speeds. The supersonic business jet may be a reality. Williams and BMW-Rolls Royce may develop new aircraft engines, while avionics improvements will be made in global positioning satellites, terrain collision avoidance systems (TCAS), and others. The use of computer aided design and computer aided manufacturing (CAD/CAM) in aircraft manufacturing will provide a modular approach to designing and partnering. Technological improvements will continue to affect corporate aircraft operators, but advantages of their use may be offset by the added costs. Each improvement must be reviewed and evaluated.

Operations

The heart of business aviation is operations. Herein, the business flight is conducted to meet the mission requirement; however, operational concerns exist. Airspace and airway congestion has prevented some corporate pilots from making their scheduled times and even from making their destinations at all. Commercial airline hub-and-spoke activities have increased to the detriment of general aviation operators; conversely, this has resulted in interest for corporate and business aviation. Certain airport authorities would rather have the larger aircraft that can carry more passengers land at their airports than the smaller general aviation aircraft. The *greatest threat* to the use of business aircraft is airport and airway access restrictions for business aircraft operators.

Pilots have identified many concerns affecting their operations. Some of these included fuel costs, noise rules, flight planning, customs, maintenance, air traffic control limitations, landing fees, and others. Most pilots report that the high costs associated with aircraft operations is the major concern of their bosses, while their concern is for training, good pay, and a safe overall flight activity.

Corporate pilots cannot leave the worries of keeping costs down to the department managers; they must actively work to save costs through efficient flight operations. When away from home station, the pilot is responsible for the care and safety of the aircraft and its passengers/cargo. Keeping the aircraft secured, obtaining necessary maintenance, obtaining low-cost fuel, and many other actions will require air crew members who understand their duties and know their limits. Terrorists' activities are real, and they must be considered at all times. Above all, a pilot's main concern is his or her flying proficiency.

Planning and Systems

Planning has often been reported in this text as a continuing process to meet the firm's mission and satisfy the customer. The customer in an aviation department is not just the passenger who gets to ride the aircraft, it is anyone and everyone who needs attention throughout the vertical and horizontal elements in the corporate structure. By this, it is meant that everyone serves a customer in some form. A mechanic fixes an aircraft for the pilot. A caterer provides needed meals. A scheduler plans a flight program and obtains necessary flight clearances, and on it goes.

For aviation managers, software programs like NBAA's Travel$ense program may be quite helpful. For aircrews, Mentor Plus' FliteStar software can be very helpful. For aircraft maintenance, CAMP Systems is great to use. Other software are available for flight planning, inventory control, weather charting, budgeting, maintenance, training, and just about every function in an aviation activity.

Planning is vital to every business activity because planning means readiness. Of course, unexpected events do occur and even the best plans may go wrong. Even though, people can be prepared to deal with the unexpected, such as generally knowing what to do when a terrorist strikes or when there is an unfortunate accident. Flight operations in a foreign country may require obtaining additional information from the State Department or doing research into cultural norms and other subjects. Knowing one's mission begins the planning process, but the end of planning never is reached.

Miscellaneous

The current and future concerns for corporate and business aviation operators are many. IBAC and the regional business aircraft associations; GAMA, PAMA, and other professional groups; and individual users actively promote business aviation through legislative support, training, and informational programs.

Future concerns for business aviation are many—the use of the Airport and Airway Trust Fund monies, the effect of the fringe-benefit rule, the future development of joint aviation rules, and others continue to affect the industry. Review the current periodicals and aviation association publications for issues of the day and plans for tomorrow. One thing is certain—change is constant.

The environment in which we live and its social, economic, and political impacts affect everyone. The personal skills and knowledge we use will have an enormous effect on the business aviation industry. Business activities have increased across the globe, and with this the need for international travel expands. Governmental interaction will affect more managers, and, in turn, the corporate and business aviation industry will change. Whether the industry is affected by government, by internal management, or by outside forces, change will occur.

Summary

Understanding what general aviation is helps people to appreciate its use. Aviation managers need to let others, at least their corporate officials, know of their activities and their importance and value, at least their corporate officials. One technique suggested is to obtain from their customers, the passengers, what value was gained through the use of the firm's aircraft.

A table is presented of 48 different items affecting the future of corporate aviation as reported by 10 aviation students, and this, in itself, alludes to the many concerns those involved in the industry have about their activities and events that affect them. SMART OPS presents various issues within eight specific categories: security/safety, management, aircraft, regulations, technology, operations, planning, and systems. Overall, the major threat to business aircraft users is perceived as airport and airway access for their business aircraft.

Important to the business aviation manager and everyone who works within a firm's aviation unit are the mission, people, equipment, and facilities. To meet a firm's mission, three indispensable components of a mission statement must be presented. These are the product, quality of service, and customer.

Lastly, the industry itself is faced with changes, and these are being reviewed by individuals, company managers, and various aviation associations so that the future of the industry is bright. The mood of the people in the industry is positive and proactive because of the value and benefits realized. By working together, those in the industry will remain a voice to be heard in business aviation's progress.

Activities

1. Review the current literature and prepare a list of 10 issues most important to the business aircraft industry.
2. Discuss with a friend or colleague the most important issue facing corporate aviation operators today.
3. Research the Airport and Airway Trust Fund and determine the amount and use of the monies within.
4. Obtain information from a business aviation group, for example, GAMA, NBAA, or PAMA, to determine its position on an issue affecting the industry.
5. Review statistics provided by the NBAA, GAMA, FAA, and NTSB to determine their significance for general aviation, with specific attention to business aviation.

Chapter Questions

1. Discuss why airport and airway access is a major concern for corporate and business aircraft operators.
2. List 10 major issues affecting corporate aviation and discuss how the top three affect the industry.
3. Explain how product liability remains a major concern for business aircraft owners/operators.
4. Discuss the importance of aircraft noise and how it affects the corporate aircraft operator.
5. How has the airline hub-and-spoke method affected business aviation?
6. Discuss the effect of MNPS/RVSM on corporate aviation operators.

7. For international air transportation operators, what are some of the major issues that they must face.
8. Why is it important for an aviation department manager to obtain information from the aircraft users about the value of flights provided, and what information should be obtained?
9. Identify and explain two of the current federal, state, or local regulations that are having an effect on corporate aviation operators.
10. Why is security/safety a major concern for aviation managers. Describe methods used by aviation managers to maintain safe and secure operations.

Appendix A

Glossary

AAAE - American Association of Airport Executives.

AACC - Airports' Associations Co-ordinating Council, located in Geneva, Switzerland, to improve airport activities.

AC - Advisory Circular. FAA publications which provide users with information about policy, guidance, or other informational items.

ACDO - Air Carrier District Office. A FAA field office staffed to help on matters related to the certification and operation of scheduled air carriers and other large certificated aircraft.

AD - Airworthiness Directive. FAA regulatory notice informing aircraft owners of a condition which must be corrected. ADs are prescribed under FAR Part 39.

ADIZ - Air Defense Identification Zone. See FAR Part 99.

AECMA - European Association of Aerospace Manufacturers.

AFSS - Automated Flight Service Station. Non-air traffic control FAA facility providing weather briefings and flight plan filing. Monitors flight plans for overdue aircraft and intiates search and rescue services.

ALPA - Airline Pilots Association.

A & P - Airframe and Powerplant Maintenance Technician. An aircraft technician who has met the experience and knowledge requirements required by the FAA and FAR Part 65.

Acquisition Costs - The costs involved in acquiring an asset—the purchase price.

Administrative Action - Enforcement action used by the FAA under FAR Section 13.11 if an alleged violation meets the following criteria: (a) no significant unsafe condition existed, (b) lack of competency or qualification was not involved, (c) violation was not deliberate, and (d) alleged violator has a constructive attitude toward complying with the regulations and has no similar violations. Action takes the form of a Warning Notice or a Letter of Correction (FAA Order 2150.3A).

Administrator - Federal Aviation Administrator or any person who has been delegated authority to act for the Administrator.

AGL - Above Ground Level. Refers to altitude expressed as feet above terrain or airport elevation.

AIM - Airman's Information Manual. FAA publication informing airmen about operating in the National Airspace of the U.S. Basic flight information, ATC procedures, health and medical facts, accident reporting, aeronautical charting, and other general information are included in the AIM.

Air Carrier - A person who undertakes directly by lease or other arrangement to engage in air transportation (FAR Section 1.1). The commercial system of U.S. air transportation consisting of the certificated air carriers, air taxis (including commuters), supplemental air carriers, commercial operators of large aircraft, and air travel clubs.

Air Service Agreement - An agreement between two or more nations contracting for reciprocal rights for air service. Mainly, bilateral, or two nation, agreements are referred to when international air services are presented. These agreements may include provisions for capacity, tariffs, negotiation procedures, aircraft airworthiness, aircrew certification requirements, and many other points of contention between two nations.

Air Taxi - A classification of air carriers which transports persons, property, and mail using small aircraft (under 30 seats or a maximum payload capacity of less than 7,500 pounds) in accordance with FAR Part 135 (FAA Statistical Handbook of Aviation, 1991).

Air Traffic Hub - Cities and Metropolitan Statistical Areas requiring aviation services. May include more than one airport. Hubs fall into four classes as determined by the community's percentage of the total enplaned passengers by scheduled air carriers in the 50 United States, District of Columbia, and the other U.S. areas designated by the FAA (FAA Statistical Handbook of Aviation, 1991).

Large Hub - 1.00 percent.
Medium Hub - 0.25 percent to 0.999 percent.
Small Hub - 0.05 percent to 0.249 percent.
Nonhub - Less than 0.05 percent of total passengers.

Aircraft - A device that is used or intended to be used for flight in the air (FAA Part 1).

Aircraft Accident - As defined by the National Transportation Safety Board, "an occurrence associated with the operation of an aircraft which takes place between the time any person boards the aircraft with the intention of flight until such time as all such persons have disembarked, and in which any person suffers death or serious injury as a result of being in or upon the aircraft or by direct contact with the aircraft or anything attached thereto, or in which the aircraft receives substantial damage."

Aircraft Incident - An occurrence, other than an accident, associated with the operation of an aircraft that affects or could affect the safety of operations and that is investigated and reported on FAA form 8020-5.

Airline Deregulation Act - This is Public Law 94-504, or the Airline Deregulation Act of 1978. It is an amendment to the Federal Aviation Act of 1958 and provides for many economic rule changes for passenger airline operations. Some of its key rulings apply to federal preemption, tariffs, mergers, interlocking agreements, mutual aid agreements, antitrust exemptions, small community air service, and the sunset provisions.

Airline Transport Pilot - The highest commercial airline pilot rating meeting the requirements of FAR Part 61, Subpart F. May act as a pilot-in-command of an aircraft engaged in air carrier services. Applicant must be at least 23 years old; be of good moral character; able to read read, write, and speak English; be a high school graduate; and pass a first class medical exam.

Airman - A pilot, maintenance technician, or other licensed aviation technician. Certification standards are found in FAR Part 61, Certification: Pilots and Flight Instructors; FAR Part 63, Certification: Flight crew members other than pilots; and FAR Part 65, Certification: Airmen other than flight crew members. Medical standards are found in FAR Part 67, Medical Standards and Certification (NBAA Management Guide, 1991).

Airport - An area of land or water that is used or intended to be used for the landing or takeoff of aircraft, and includes its buildings and facilities (FAA Part 1).

Airport and Airway Trust Fund - Established under Section II of the Airport and Airway Development Act of 1970 to collect monies from various airline charges, e.g., fuel, domestic ticket taxes, international ticket surcharge, tires, registration fees, and others.

Airport Operations - An aircraft takeoff or landing. There are two types of operations—local and itinerant. Local operations are performed by aircraft which:

a. Operate in the local traffic pattern or within sight of the airport.
b. Are known to be departing for, or arriving from, flight in local practice areas within a 20-mile radius of the airport.
c. Execute simulated instrument approaches or low passes at the airport.
Itinerant operations are all airport operations other than local.
(FAA Statistical Handbook of Aviation, 1991).

Airworthiness Certificate - A FAA certificate awarded to all civil aircraft which have met certification standards.

Annual Inspection - A complete inspection of an aircraft and engine required by FAR Part 91.409 to be accomplished every 12 calendar months on all certificated aircraft.

ARSA - Regulatory airspace area around the busiest U.S. airports (other than TCAs). Radio contact with approach control is mandatory for all traffic from the surface to 4,000 feet AGL out to 10 miles and from an intermediate altitude to 4,000 feet AGL to 20 miles from the airport. Class "C" airspace.

Assets - Generally, this is an accounting term and refers to as all the resources of a firm, tangible and intangible.

ATA - Airline Transport Association.

ATA 100 Aircraft System Code - The Airline Transport Association aircraft maintenance codification of maintenance numbers to aircraft systems in order to standardize and simplify working procedures.

Available Seat Miles/Kilometers - The number of available seats per mile/kilometer. For example, a 100 seat aircraft which is flown 200 miles provides for 2,000 available seat miles.

Average Stage Length - Total number of travel miles divided by the total landings or takeoffs and generally expressed in nautical miles for operational purposes and statute miles for business purposes.

Aviation Flight Department - Within corporate activities, this is the establishment of a flight activity involving company aircraft and aircrew. Normally, an aviation flight manager controls flight operations, of which many may be unscheduled air travel support flights for company executives. An aviation flight department may be small (only one or two aircraft), medium (three to five aircraft), or large (having over five aircraft).

Barnstorming - The term given to the activities of early aviators (1920–1940) who flew their aircraft during airshows, carnivals, aviation fairs, or other air-oriented public gatherings to earn monies and promote aviation.

Bilateral Agreement - An agreement between two nations to establish international air services. See *Air Services Agreement.*

Block Time - The time from the moment the aircraft first moves under its own power for the purpose of flight until it comes to rest at the next point of landing; also known as block-to-block time (FAA Part 1).

Brain Power - A term used during the 1940s to label people who had intellectual skills and were considered a product in itself. Corporate executives sought to employ these people for their company development.

Business Aviation - use of an aircraft by a pilot not receiving a direct salary or compensation for piloting in connection with his/her occupation or in furtherance of a private business.

Cabotage - Eighth freedom of the air. Right of a country A aircraft to be flown between points in country B.

Carrier Group - A grouping of certificated air carriers determined by annual operating revenues as indicated below:

    Majors - over $1 billion.
    National - $100,000,000 to $1 billion.
    Large Regionals - $20,000,000 to $99,999,999.
    Medium Regionals - 0 - $19,999,999 or that operate aircraft with 60 or less seats, or a maximum payload capacity of 18,000 pounds (FAA Statistical Handbook of Aviation, 1991).

FAR Part 241, Section 04 Groupings:
    I $0 - $100,000,000
    II $100,000,001 - $1,000,000,000
    III $1,000,000,001+
    Group I further divides into two subgroups: $20,000,000 to $100,000,000 and below $20,000,000.

Center - An FAA Air route Traffic control center providing radar surveillance and traffic separation to participating en route traffic above airspace handled by Approach and departure control—generally above 6,000–10,000 feet.

CEO - Chief Executive Officer.

Certificated Air Carrier - An air carrier holding a Certificate of Public Convenience and Necessity issued by DOT to conduct scheduled services interstate. Nonscheduled or charter operations may also be conducted by these carriers. These carriers operate large aircraft (30 seats or more or a maximum payload capacity of 7,500 or more) in accordance with FAR Part 121 (FAA Statistical Handbook of Aviation, 1991).

Certificated Flight Instructor - CFI. A certificated pilot authorized by the FAA to give flight training in accordance with FAR Part 61.

Chief Pilot - Designated pilot whose experience as a leader and competence as a pilot are qualifications to direct other pilots. Requirements for a chief pilot for FAR 135 operations fall under FAR 135, Section 135.39. No regulations exist for operations under FAR Part 91 (NBAA Management Guide, 1994).

City-Pair - Designation given to two different cities for which air transportation is being provided or considered (e.g., New York–Cleveland or Chicago–Washington).

Class A Airspace - Former positive control area above 18,000 feet over the conterminous U.S.

Class B Airspace - Former Terminal Control Area (TCA).

Class C Airspace - Former Airport Radar Service Area (ARSA).

Class D Airspace - Former Airport Traffic Area (ATA) combined with a control zone.

Class E Airspace - General controlled airspace formerly comprising control areas, transition areas, Victor areas, the Continental U.S., etc.

Class F Airspace - International airspace designation not used in the U.S.

Class G Airspace - Uncontrolled airspace, generally the airspace below 700 or 1,200 feet in most of the U.S., extending up to 14,500 feet in some western and sparsely populated areas.

Clearance - Formal instructions from air traffic control authorizing a specific route or action.

Commerce - Trade or business (buying and/or selling of goods and services) conducted among and between peoples.

Commercial Operator - A person who engages in the intrastate carriage of persons or property for compensation or for hire and is not otherwise an air carrier or foreign air carrier (FAA Statistical Handbook of Aviation, 1991).

Commercial Operators of Large Aircraft - Commercial operator operating aircraft with 30 or more seats or a maximum payload capacity of 7,500 pounds or more (FAA Statistical Handbook of Aviation, 1991).

Commercial Pilot - May act as pilot-in-command of an aircraft carrying passengers for compensation or hire and act as pilot-in-command in an aircraft that is being operated for compensation or hire e.g., one that has been hired to do pipeline patrol but carries no passengers (FAA Statistical Handbook of Aviation, 1991). Must be at least 18 years old; be able to read, write, and speak English; hold at least a valid second-class medical; and meet other requirements under FAR Part 61, Subpart E.

Common Carrier - A scheduled, certificated transportation carrier which holds itself out to the general public and has a certificate of public convenience and necessity.

Communications - General term used to convey a message, conduct business, or interchange between and among people and businesses.

Commuter Air Carrier - An air taxi operator which performs at least five round trips per week between two or more points and publishes flight schedules which specify the times, days of the week, and points between which such flights are performed (FAA Statistical Handbook of Aviation, 1991).

Contract Carrier - A transportation carrier which negotiates a contract for transportation services and receives compensation for its services.

Corporate Aviation - Use of an aircraft, owned or leased, which is operated by a company or business for the transportation of people, cargo, or mail in furtherance of a firm's business, and which is flown by a professional pilot who receives a direct salary or compensation for piloting.

Crewmember - A person assigned to perform duties in an aircraft during flight.

CVR - Cockpit voice recorder.

Decibel - A logarithmic measurement used to define the relative intensity or ratio of sound. Noted by symbol dB.

DME - Distance Measuring Equipment.

Dry Lease - Lease of an aircraft only. No aircrew or other provisions are contracted.

DUAT - Direct User Access terminal. A personal computer system used to retrieve weather/NOTAMs and to file flight plans.

Duty Time - The total amount of time a flightcrew member is on duty, beginning when the person reports for an assignment and ending when the person is released from that assignment. All flight and non-related flight tasks are included (NBAA Management Guide, 1994).

Economic - Having to do with business or the distribution and production of wealth. Economic rules involving rates, routes, fares, and other factors affecting monetary amounts.

Environmentalists - Relevant to aviation, the people who show major concern with noise and air pollution created by aircraft or airport operations, among others.

Equal Time Point (ETP) - The point at which the time to return to point of origin and the time required to reach the destination is equal.

Executive Aviation - See Corporate Aviation. This term was used early during the development of corporate aviation, but has generally been replaced with the term corporate aviation. The FAA still uses the term to designate corporate aviation.

Extended Range Operations (ETOPS) - With respect to other than helicopter operations, operations over water at a horizontal distance of more than 50 nautical miles from nearest shoreline, and with respect to helicopters, operations at a horizontal distance of more than 50 nautical miles from nearest shoreline and more than 50 nautical miles from off-shore heliport or structure (FAR Part 1).

FALPA - Federation of Airline Pilots Association.

FANS - Future Air Navigation System (ICAO).

FATCA - Federal Air Traffic Controllers' Association.

Fatal Injury - Means any injury which results in death within seven days of the accident.

FDA - Flight Data Recorder.

Ferry Flight - A flight to move an aircraft to a location where necessary work may be accomplished. When aircraft operations require it, a one-time FAA approval may be needed to make the flight.

FIFO - Flight Inspection Field Office.

Fixed Base Operator - An aviation activity located at an airport which provides aviation-related services for compensation (e.g., fueling, aircraft cleaning, catering, transportation, maintenance, and others).

Flight Attendant - An aircrew member who provides support to passengers during flight. An attendant is required based on aircraft seating capacity (FAR 91.533, FAR 135.107 and FAR 121.391).

Flight Information Center (FIC) - Established to provide flight information and alerting service.

Flight Plan - Means specified information, relating to the intended flight of an aircraft, that is filed orally or in writing with air traffic control (FAR Part 1).

Flight Standards District Office - FAA field office serving an assigned geographical area and staffed with Flight Standards personnel who serve the aviation industry and the general public on matters relating to the certification and operation of air carrier and general aviation aircraft.

Activities include general surveillance of operational safety, certification of airmen and aircraft, accident prevention, investigation, enforcement, and others.

Flight Time - See Block Time.

FMS - Flight Management System. An on-board computer system that integrates multiple sources of navigation input from a database and real time entries.

Fringe Benefit Tax Rule - Name given to the Internal Revenue regulation which requires employees who receive a "free" ride on a corporate aircraft to file an income tax return for the fair market value of the flight. The rules are complex and company legal or tax people should be consulted.

FSDO - Flight Standards District Office. FAA filed office serving an assigned geographical area.

FSS - Flight Service Station. Air traffic facilities that (a) provide pilot briefing, en route communications, and VFR search and rescue services; (b) assist lost aircraft and aircraft in emergency situations; (c) relay ATC clearances; (d) originate notices to airmen; (e) broadcast aviation weather and national airspace system information; (f) receive and process IFR flight plans; and (g) monitor navigation aids.

General Aviation - Involves all flying activities except those as an air carrier under FARs 121, 123, 127, and 135.

Global Marketplace - The term given to the world at large as a place where commerce may be conducted.

GNP - Gross National Product—the value of all goods and services produced by a nation.

GPS (NAVSTAR) - Global Positioning Satellite Navigation System which is operated by the U.S. Department of Defense.

Gross Navigation Error (GNE) - Reported when an aircraft is 25 nautical miles off course.

Helicopter Association International (HAI) - Trade association of the civil helicopter industry worldwide.

Hub - See Air Traffic Hub.

Hub-and-Spoke/Hubbing - The term given to the commercial airline system by which commercial flights are scheduled to arrive at a major point to consolidate small passenger loads into larger ones and to make more efficient use of available aircraft. "Operational system, whereby flights from numerous points arrive at and then depart from a common point within a short time frame so that passengers arriving from any given point can connect to flights departing to all other points" (Secretary's Task Force, 1990).

HUD - Head-up Display. Presents electronic flight information in the pilot's field of vision while the pilot looks throught the windshield.

IA - Inspection Authorization. FAA-issued certificate to an individual under FAR Section 65.91 granting authority to certify that specific maintenance inspections have been completed in order to return aircraft to service after maintenance, alterations, or repairs.

IACA - International Civil Airports Association.

IAOPA - International Council of Aircraft Owner and Pilot Association.

IATA - International Air Transport Association.

ICAA - International Civil Airports Association.

ICAO - International Civil Aviation Organization.

IFALPA - International Federation Airline Pilots Association.

IFATCA - International Federation of Air Traffic Controllers' Association.

IFIM - International Flight Information Manual.

IFR - Instrument Flight Rules. Rules of the road for flights permitted to penetrate clouds and low visibility conditions by reference to cockpit flight instrumnents and radio navigation (AOPA, 1995).

IFSS - International Flight Service Station. Central operations facility in the flight advisory system to control aeronautical point-to-point telecommunications and air/ground telecommunications with pilots operating over international territory or waters which provide flight plan following, weather information, search and rescue action, and other flight assistance operations.

INOTAMS - International Notices to Airmen.

ITC - Investment Tax Credit. U.S. ruling which allowed a tax credit for the purchase of new equipment (e.g., an airplane). This rule was repealed, but is being strongly advocated for renewal by general aviation manufacturers.

Incident - An occurrence other than an accident associated with the operation of an aircraft, that affects or could affect the safety of operations (NTSB Sec 830.2).

Industrial Revolution - The time period when agricultural production gave way to advances in technology, and more people went to work in factories than being on farms. A period of social and economic change, beginning in England about 1760 and later in the United States and other countries.

Informational Security - The minimizing of personal or organizational risk of financial harm or embarrassment (NBAA Foundation and The Mescon Group, 1989).

Intangible Benefits - Benefits which cannot be materially identified or which have intrinsic value. These benefits are subjectively determined by the user.

Interchange Agreement - Arrangement by which a person leases an airplane to another person in exchange for equal time, when needed, on the others' airplane, and no charge, assessment, or fee is made, except that a charge may be made not to exceed the difference between the cost of owning, operating, and maintaining the two airplanes (FAR Part 91.501 (c)).

International Handler - A management and/or aviation service activity which provides international flight operations services, which may include flight planning, flight documentation, fuel operations, maintenance, or other necessary flight activities.

Joint Aviation Regulation (JAR) - A regulation developed by joint agreement among European countries, similar to Federal Aviation Regulations in format, but applicable to European aviation operations.

Joint Ownership - Means an arrangement whereby one of the registered joint owners of an airplane employs and furnishes the flight crew for that airplane and each of the registered joint owners pays a share of the charge specified in the agreement (FAR Part 91.503).

Knot (Nautical Mile) - Approximately 1.15 statute miles equals one nautical mile.

Large Aircraft - Within the U.S. the FAA defines this as an aircraft of more than 12,500 pounds, maximum certificated takeoff weight.

Liability - A legal or responsible obligation to make good for any damage or loss. Product liability is often heard in aviation, and this means that an aircraft manufacturer is liable or responsible for the aircraft produced.

Load Factor - For this text, load equals revenue passenger miles divided by available seat miles (revenue passenger kilometers/available passenger kilometers internationally) or revenue ton miles/kilometers divided by available ton miles/kilometers.

Loran - Long Range Navigation. Electronic navigation system by which hyperbolic lines of position are determined by measuring the difference in the time of reception of synochronized pulse signals from two fixed transmitters.

Maintenance - Means the inspection, overhaul, repair, preservation, and the replacement of parts, but excludes preventive maintenance (FAR 1).

Major Alteration - Alteration not listed in the aircraft, aircraft engine, or propeller specifications that might appreciably affect weight, balance, structural strength, performance, powerplant operation, flight characteristics, or other qualities affecting airworthiness; or that is not done according to accepted practices or cannot be done by elementary operations (FAR Section 1.1).

Major Repair - A repair if properly done, might appreciably affect weight, balance, structural strength, performance, powerplant operation, flight characteristics, or other qualities affecting airworthiness; or that is not done according to accepted practices or cannot be done by elementary operations (FAR Section 1.1).

Matrix - A tabular form having rows and columns within which terms or numbers can be placed for informational/planning purposes.

Megacarrier - A term used to designate one of the major airlines which has developed a global system of passenger and cargo service. It is generally accepted that such a carrier has a very large fleet of aircraft, has interline agreements with other airlines, has a code-sharing program with other airlines, and has many thousands of employees.

MEL - Minimum Equipment List. A list of aircraft appliances and equipment that must be functioning for an aircraft to be flown. Approval requirements are found in FAR Section 1.1. (AC 91-67 and NBAA Management Guide, 1994).

Minor Alteration - An alteration that is not major.

Minor Repair - A repair that is not major.

MIS - Management Information System.

MLS - Microwave Landing System. A precision instrument approach system operating in the microwave spectrum that consists of an azimuth station, an elevation station, and precision distance measuring equipment.

MMEL - Master Minimum Equipment List. The master FAA listing of aircraft appliances and equipment from which an owner/operator develops an MEL (AC 91-67).

MNPS - Minimum Navigation Performance Specifications. A specified set of minimum navigation performance standards that aircraft must meet in order to operate in MNPS-designated airspace. The aircraft must be certified by the State of Registry for MNPS operation. The objective of MNPS is to ensure the safe operation of aircraft and to derive maximum benefit, generally through reduced separation standards, from the improvement in accuracy of navigation equipment developed in recent years (NBAA Management Guide, 1994).

Mode C - Altitude-reporting mode of secondary radar used with ATCRBS transponders.

Mode S - Discrete, addressable secondary radar system that may also include data link.

Modes of Transportation - There are five modes of transportation: Water, Motor, Rail, Pipeline, and Air.

MSL - Mean Sea Level. Altitude expressed as feet above sea level irrespective of local terrain or airport elevation.

National Airspace System - Common network of U.S. airspace, navigation aids, communication facilities, aeronautical charts and information, rules, regulations and procedures, technical information, and FAA manpower and material.

Nautical Mile - See Knot.

Noneconomic - A term used to designate any rule, item, or issue which does not have economic meanings. For example, safety, security, and licensing rules which fall within FARs 1 to 199 are considered non-economic regulations.

NOTAM - Notice to Airmen. Contains information concerning the establishment, condition, or change in any component of, or hazard in the National Airspace System, the timely knowledge of which is essential to people concerned with flight operations.

OAC - Oceanic Area Control Center. Any Area Control Center (ACC) with jurisdiction over oceanic airspace for the purpose of providing air traffic services.

Objective - In decision-making, a measurement based on actual data (e.g., answers to a multiple choice test, true-false test, or clear choice which is not subject to interpretation).

Operator - Any person who causes or authorizes the operations of an aircraft, such as the owner, leasee, or bailee of an aircraft (NTSB Section 830.2).

Overbooked - Within commercial aviation, this means that an airline official has deliberately allowed more passengers to purchase a ticket than the number of seats available to occupy. If 200 tickets were sold and all these passengers showed for the flight for an aircraft with only 180 seats, then 20 people were overbooked.

PANS - Procedures for Air Navigation Services. ICAO advisory material which is similar in nature to U.S. Advisory Circulars—the difference is that PANS apply to international aviation matters.

Payload - The weight an aircraft can carry including passengers, cargo, mail, and baggage.

Physical Security - Minimizing the risk of injury, fright, or physical damage or harm (NBAA Foundation and The Mescon Group, 1989).

PIC - Pilot in Command. The pilot responsible for the operation and safety of an aircraft during flight time (FAA Part 1).

Pilot Deviation - Actions of a pilot that result in the violation of an FAR or a North American Aerospace Defense Command (NORAD) Air Defense Identifiaction Zone (ADIZ) tolerance.

Policy - A plan of action.

Possession Costs - The costs involved in maintaining an item. This may include interest payments, maintenance costs, insurance, fuels and lubricants, employee salaries, and other fixed and variable expenses.

Preventive Maintenance - Means simple or minor preservation operations and the replacement of small standard parts not involving complex assembly operations (FAR Part 1). See Appendix A, para (c) of FAR Part 43 for listing of preventive maintenance work items.

Primary Business Test - Means the review of the income generation items for a company. If more than 50 percent of the revenue earned by a company comes from transportation services, then that company must apply for a transportation certificate under federal regulations. If less than 50 percent of its revenues comes from transportation services, then the company is not considered a transport carrier.

Primary Use - The use category in which an aircraft flew the most hours. The 10 use categories are defined below:

Aerial Application - Any use of an aircraft for work purposes which concerns the production of food, fibers, and health control in which the aircraft is used in lieu of farm implements or ground vehicles for the particular task accomplished. This includes fire fighting operations, the distribution of chemicals or seeds in agriculture, reforestation, or insect control.

Aerial Observation - Any use of an aircraft for aerial mapping/photography, survey, patrol, fish spotting, search and rescue, hunting, highway traffic advisory, or sightseeing; not included under Part 135.

Business Transportation - Use of an aircraft not for compensation or hire by individuals for the purposes of transportation required by business in which they are engaged.

Commuter Air Carrier - An air taxi that performs at least five scheduled round trips per week between two or more points or carriers mail. See Air Carrier.

Demand Air Taxi - Use of an aircraft operating under Federal Aviation Regulation Part 135, Passenger and cargo operations, including charter and excluding commuter air carrier. See Air Carrier.

Executive/Corporate Transportation - Any use of an aircraft by a corporation, company, or other organization for the purposes of transporting its employees and/or property not for compensation or hire, and employing professional pilots for the operation of the aircraft.

Instructional Flying - Any use of an aircraft for the purpose of formal instruction with the flying instructor aboard, or with the maneuvers on the particular flight(s) specified by the flight instructor; excludes proficiency flying.

Personal Flying - Any use of an aircraft for personal purposes not associated with a business or profession, and not for hire. This includes maintenance of pilot proficiency.

Other Work Use - Any aircraft used for construction work (not included under Part 135), helicopter, hoist, towing gliders, or parachuting.

Other - Any other use of an aircraft not included above. (Example: experimentation, R & D, testing, demonstration, government) (Office of Aviation Policy and Plans, 1994).

Private Pilot - May not act as pilot-in-command of aircraft that is carrying passengers for compensation or hire nor act as pilot-in- command in an aircraft that is being operated for compensation or hire; e.g., one that has been hired to do pipeline patrol but carries no passengers (FAA Statistical Handbook of Aviation, 1991).

Productivity - Output over input. Can be measured partially, as a multi-factor, or as a total.

Proficiency Check - Check given by a check airman to flightcrew members to test the ability of the pilot in specific situations and maneuvers.

Prohibited Area - An airspace area where flight is prohibited except by prior arrangement or by the controlling agency.

Progressive Inspection - A continuing sequential airworthiness inspection of an aircraft and its various components and systems at scheduled intervals in accordance with procedures approved by the FAA. See FAR Part 91.409. (NBAA Management Guide, 1994).

Public Convenience and Necessity - A certificate issued under the Federal Aviation Act of 1958, as amended, Section 401, and which an air carrier must obtain to engage in certificated air transportation. Basically, an air carrier must prove fitness, willingness, and ability.

Public-Use Airport - An airport open for public use without prior permission and without restrictions.

Registered Aircraft - Aircraft registered with the Federal Aviation Administration.

Revenue Ton-Mile/Kilometer - One ton of revenue traffic transported one mile/kilometer.

Rotorcraft - Means a heavier-than-air aircraft that depends principally for its support in flight on the lift generated by one or more rotors (FAR Part 1).

Royal Barge - The term applied to early corporate aircraft when only the chief executives were transported by company flight departments on company aircraft.

RVR - Runway visual range.

SARPS - Standards and Recommended Practices. ICAO rules which are mandatory international flight rules.

Second-in-Command - Means a pilot who is designated to be second in command of an aircraft during flight time (FAR Part 1).

Serious Injury - means any injury which (1) requires hospitalization for more than 48 hours, commencing within seven days from the date the injury was received; (2) results in a fracture of any bone (except simple fractures of fingers, toes, or nose); (3) involves lacerations which cause severe hemorrhages, nerve, muscle, or tendon damage; (4) involves injury to any internal organ; (5) involves second-or third-degree burns, or any burns affecting more than five percent of the body surface (FAA Statistical Handbook of Aviation, 1991).

SID - Standard Instrument Departure.

SIFL - Standrad Industry Fare Level.

Small Aircraft - Aircraft less than 12,500 pounds, maximum certificated takeoff weight.

Small Regional - A term used to designate all other commercial air operations which do not include any of the majors, nationals, large regionals, or medium regionals. The latter are categorized by revenues, while the small regionals are not-exempt under FAR Part 298.

SOP - Standard Operating Procedure.

Spread Sheet - A planning tool used in accomplishing a travel analysis. The number of trips, mileage between origins and destinations, and locations of travel are used in preparing a spread sheet.

Squawk - Radio transmission of the radar transponder onboard an aircraft. Also, ATC instruction to the pilot to set one of 4,096 possible codes to identify the aircraft on controller radar.

Stage 3 Aircraft - An aircraft that has been shown to comply with the quieter noise requirements under FAR Part 36.

Statute Mile - Unit of linear measure of 5,280 feet.

Stockholders' Equity - An accounting term used to indicate that portion of a company's net worth which belong to its stockholders; shares of stock.

STOL - Short takeoff and landing.

Subjective - An individual's judgement based on feelings, thought, or other non-objective measures.

Supplemental Air Carrier (Charter) - An air carrier which holds a Certificate of Public Convenience and Necessity issued by DOT, authorizing performance of passenger and cargo interstate charter services supplementing the scheduled service of the certificated scheduled air carriers (FAA Statistical Handbook of Aviation, 1991).

Tangible - Something that has intrinsic value, substance, or which can be measured; e.g., time and money can be counted.

TBO - Time between overhaul.

TCA - Terminal Control Area. Regulatory airspace area around the busiest U.S. airports, typically to a radius of 20 nautical miles and 7,000–10,000 feet above ground level.

TCAS - Traffic Alert and Collision Avoidance System. A cockpit system to detect other transponder-equipped aircraft, alert pilots, and command/coordinate evasive action between TCAS aircraft.

Time Share - Means an arrangement whereby a person leases an airplane with flight crew to another person, and no charge is made for the flights conducted under that arrangement other than those specified under FAR Part 91.501 (d) (FAR Part 91).

Trade - The buying and selling of goods and/or services. See Commerce.

Transponder - A special onboard 1090 Mhz radio transmitter to enhance and code an aircraft's radar return. When interrogated, transmits a return signal which controllers use to identify the flight on radar.

TRSA - Terminal Radar Service Area.

Travel Analysis - Action which involves objective and subjective information and judgment to determine the travel needs for company personnel/cargo movements. Used to determine whether company aircraft may be of value and to determine the number and type of aircraft useful.

Travel Pattern Graph - Used in accomplishing a travel analysis—a decision planning tool. The number of individual trips, distance of each trip, and aircraft type and ranges are used to accomplish a travel pattern graph.

Travel Pattern Map - Used in accomplishing a travel analysis—a decision planning tool. Selected radii of aircraft ranges are plotted on a map to indicate to a planner the origin and destinations of travel with consideration to the range of selected aircraft.

Type Rating - An additional rating to a pilot's certification that authorizes the pilot to act as pilot in command of a specific aircraft.

UDF - Unducted propfan.

Utility - Usefulness. For example, time and place utility means being in the right place at the right time.

Value - The quality of something which provides usefulness, importance, or other desirable meaning.

Visual Flight Rules (VFR) - Rules that govern the procedures for conducting flight under visual conditions. Also used in the U.S. to indicate weather conditions that are equal to or greater than minimum VFR requirements. Used by pilots and controllers to indicate type of flight plan (FAA Aviation Forecasts, 1993).

VOR - Very high frequency omnidirectional radio range.

VORATC - Integrated VOR and Tactical Air Navigation (TACAN), navigation devices that provide azimuth and distance-measuring capability.

Wet Lease - The lease of an aircraft with aircrew, fuel, maintenance and/or other service (See AC 91-37A).

Windshear - Large changes in wind speed and direction with altitude. Can cause sudden gain or loss of airspeed on takeoff or during landing approaches (AOPA, 1995).

Wingtip Vortices - Small tornado-like whirlwinds trailing an aircraft wingtips.

Appendix B

U.S. Government Printing Offices

The U.S. Government Printing Office (GPO) operates U.S. Government bookstores all around the United States. Any item not in inventory can be ordered and sent to you. All the bookstores accept VISA, MasterCard, and Superintendent of Documents deposit account orders. For more information, contact Superintendent of Government Printing Office, Washington, DC, 20402, (202) 783-3238, or contact any of the following GPO Bookstores. Addresses are listed for ordering purposes.

Atlanta, GA
275 Peachtree Street, NE, Room 100
P.O. Box 56445
Atlanta, GA 30343
(404) 331-6947

Birmingham, AL
O'Neill Building
2021 Third Avenue, North
Birmingham, AL 35203
(205) 731-1056

Boston, MA
Thomas P. O'Neill Building
10 Causeway Street, Room 169
Boston, MA 02222
(617) 720-4180

Chicago, IL
Room 1365, Federal Building
219 S. Dearborn Street
Chicago, IL 60604
(312) 353-5133

Cleveland, OH
Room 1653, Federal Building
1240 E. 9th Street
Cleveland, OH 44199
(216) 522-4922

Columbus, OH
Room 207, Federal Building
200 N. High Street
Columbus, OH 43215
(614) 469-6956

Dallas, TX
Room 1C46, Federal Building
1100 Commerce Street
Dallas, TX 75242
(214) 767-0076

Denver, CO
Room 117, Federal Building
1961 Stout Street
Denver, CO 80294
(303) 844-3964

Detroit, MI
Suite 160, Federal Building
477 Michigan Avenue
Detroit, MI 48226
(313) 226-7816

Houston, TX
Texas Crude Building
801 Travis Street, Suite 120
Houston, TX 77002
(713) 228-1187

Jacksonville, FL
Room 158, Federal Building
400 W. Bay Street
P.O. Box 35089
Jacksonville, FL 32202
(904) 353-0569

Kansas City, MO
120 Bannister Mall
5600 E. Bannister Road
Kansas City, MO 64137
(816) 765-2256

Laurel, MD
Warehouse Sales Outlet
8660 Cherry Lane
Laurel, MD 20707
(301) 953-7974

Los Angeles, CA
ARCO Plaza, C-Level
505 South Flower Street
Los Angeles, CA 90071
(213) 239 9844

Milwaukee, WI
Room 190, Federal Building
517 E. Wisconsin Avenue
Milwaukee, WI 53202
(414) 297-1304

New York, NY
Room 110, 26 Federal Plaza
New York, NY 10278
(212) 264-3825

Philadelphia, PA
Robert Morris Building
100 North 17th Street
Philadelphia, PA 19103
(215) 597-0677

Pittsburg, PA
Room 118, Federal bulding
1000 Liberty Avenue
Pittsburg, PA 15222
(412) 644-2721

Portland, OR
1305 S.W. First Avenue
Portland, OR 97201
(503) 221-6217

Pueblo, CO
World Savings Building
720 North Main Street
Pueblo, CO 81003
(719) 544-3142

San Francisco, CA
Room 1023
Federal Office Building
450 Golden Gate Avenue
San Francisco, CA 94102
(415) 252-5334

Seattle, WA
Room 194, Federal Building
915 Second Street
Seattle, WA 98174
(206) 442-4270

Washington, DC
U.S. Government Printing Office
10 North Capitol Street, NW
Washington, DC 20402
202) 275-2091

Washington, DC
1510 H Street, NW
Washington, DC 20005
(202) 653-5075

Appendix C

Aviation Internet Addresses

Air Traffic Control:
http://www.natca.org

Air Transportation-Government:
http://www.dot.gov/
http://www.faa.gov/
http://www.nasa.gov/
http://www.ntsb.gov/

Aircraft Accident Investigation:
http://www.ntsb.gov/

Aircraft Design:
ftp://rascal.ics.utexas.edu
http://www.arc.nasa.gov
http://www.techreports.larc.nasa.gov/cgi-bin/NTRS
http://www.tc.faa.gov

Aircraft Manufacturers:

Boeing - http://www.boeing.com/
Douglas - http://www.dac.mdc.com/
Learjet - http://www.learjet.com
Lockheed-Martin - http://www.lockheed.com/

Cessna - http://www.cessna.com
Falcon - http://www.falcon.com
Gulfstream - http://www.gulfstream.com
Raytheon - http://www.raytheon.com

Aircraft Sales/Parts/etc:
http:/www.mcgraw-hill.com/acfon-line/
http://www.sonic.net/aso/aso.html
http://www.airparts.com/airparts

Airports:
http://w3.one.net/-flypba/AIRLINES/airports/

Associations:
Air Transport Association of America (ATA) -http://www.air-transport.org/
Aircraft Owners & Pilots Association - http://www.aopa.org/
American Association of Airport Executives & International Association of Airport Executives - http://airportnet.org
Amercian Institute of Aeronautics and Astronautics (AIAA) - http://www.lainet.com/~rich/aiaaonline/
American Psychological Association (APA) - http://www.apa.org
Experimental Aircraft Association - http://atlantic.austin.apple.com/people.pages.jhovan/esa/eaa.html

General Aviation Manufacturers Association - http://generalaviation.org
Helicopter Association International (HAI) - http://www.rotor.com/
International Air Transport Association - http://www.iata.org
International Civil Aviation Organization (ICAO) - http://www.CAM.ORG/~icao/
National Business Aircraft Association - http://www.nbaa.org
National Safety Council - http://www.cais.com/nsc/
The Ninety-Nines Inc (99s) - http://acro.harvard.edu/99/HomePage.html

Aviation Careers:
http://www.aeac.com/aeac/images/navbar.htm

Aviation General:
http://www.guides.com/acg/nlet1295.htm
http://www.avweb.com/

Aviation Human Factors:
http://olias.arc.nasa.gov/publications/reference-publications.html

Aviation Law & Insurance:
http://www.law.cornell.edu/topics/aviation.html
http://guide-p.infoseek.com/Titles?qt=Aviation+insurance&col=WW&sv-N2

Aviation Maintenance:
http://www.hfskyway.com/
http://www.faa.gov/avr/afshome.htm
http://www.calm-systems.com/default.htm

Aviation Safety:
http://www.faa.gov/avr/news/ASPHome.htm
http://www.ntsb.gov/
http://www.aviation.org/
http://206.31.38.97/ASI/ASIrpt95.html
http://www.portal.com/~stevep/avsafetypres.html

Business News:
CNN - http://www.cnn.com/
Economist - http://www.economist.com/
Financial Times - http://www.ft.com/
Fortune 500, 1000, & Global 500 - http://www.fortune.com
Management Review - http://www.enews.com/magazines/mr/
New York Times - http://nytimesfax.com/
Reuters - http://www.reuters.com/
Time - http://pathfinder.com/time/international/
US News & World Report - http://www.usnews.com/usnews/main.htm
USA Today - http://www.usatoday.com/
Wall Street Journal - http://wsj.com/
Washingtom Post - http://washingtonpost.com/

Corporations:
Corporate Angel Network - http://oncolink.upenn.edu/psycho_stuff/corp_angel.html
Jet Aviation - http://www.jetaviation.com
Raytheon - http://raytheon.com/rac/

Directories:
   Business Directories
   http://www.commerce.net/directories/
   Commercial Services on the Net
   http://www.directory.net/
   Global Business Directory
   http://www.pronett.com
   Internet University Directory
   http://www.caso.com/

   Schools Libraries of the Web Directory
   http://www.libertynet.org/~bertland/libs.html
   Web Directory
   http://www.webdirectory.com
   WWW Library Directory
   http://www/albany.net/~ms0669/cra/libs.html

FAA Library:
ftp://fwux.fedworld.gov/pub/faa/faa.htm

FARs:
http://www.faa.gov/avr/AFS/FARS/far_idx.htm
http://www.access.gpo.gov/cgi-bin/cfrassemble.cgi

Flight Safety Systems, Inc:
http://www.avweb.com/sponsors/fssi

Government:
FAA AFS Web - ftp://www.opspecs.com/afs/data/mmels
U.S. Department of Education - http://www.ed.gov

Great Circle Distance Calculator:
http://www.sonic.net/aso/pointers.html

Libraries:
Internet Public Library - http://www.ipl.org
Library of Congress - http://www.loc.gov/homepage/lchp.html
Smithsonian Institute Libraries - http://www.sil.si.edu/
WWW Virtual Library-   http://www.w3.org/pub/DataSources/bySubject/Overview2.html
Yahoo! Reference Library - http://www.yahoo.com/Reference/Libraries/

Meteorology:
http://www.comet.vcar.edu
http://www.weather.com/
http://http.ucar.edu/metapage.html

Search Form for N Numbers:
http://acro.harvard.edu/GA/search_nnr.html

Standard Industry Fare Level (SIFL):
http://www.nbaa.org/nonmember/library/digest/sepdi96/sepdi96.htm

Thomas - Legislative Information on the Net:
http://thomas.loc.gov

Used Aircraft:
http://www.avweb.com/toc/usedacft.html

U.S. Government Printing Office:
http://www.access.gov.gpo/su_docs/aces/aaces/001.html

Weather World:
http://http.rad.ucar.edu/staff/knight/
http://www.atmos.uiuc.edu/wxworld/html/general.html
http://www.atmos.uiuc.edu/wxworld/html/satimg.html
http://www.weather.com/
http://rad.ucar.edu/Staff/Gthomson/cur_wx/wx_index.htm\#ruc_model

Worldwide Web Virtual Library: Subject Catalogue:
http://info.cern.ch/hypertext/DataSources/by Subject/Overview.html

NOTE: Names and addresses are subject to change. Data as of April 1998. Please send any corretions or new addresses to Ken J. Kovach, (Compuserve) - 74321,135, or INTERNET:74321.135@compuserve.com.

Appendix D

Federal Aviation Regulations

(Chapter I, Title 14, Code of Federal Regulations)

NON-ECONOMIC REGULATIONS

SUBCHAPTER A - DEFINITIONS

Part
1   Definitions and abbreviations

SUBCHAPTER B - PROCEDURAL RULES
11 General rule-making procedures
13 Investigative and enforcement procedures

SUBCHAPTER C - AIRCRAFT
21 Certification procedures for product and parts
23 Airworthiness standards: Normal, utility, and acrobatic category airplanes
25 Airworthiness standards: Transport category airplanes
27 Airworthiness standards: Normal category rotorcraft
29 Airworthiness standards: Transport category rotorcraft
31 Airworthiness standards: Manned free balloons
33 Airworthiness standards: Aircraft engines
35 Airworthiness standards: Propellers
36 Noise standards: Aircraft type and airworthiness certification
39 Airworthiness directives
43 Maintenance, preventive maintenance, rebuilding, and alteration
45 Identification and registration marking
47 Aircraft registration
49 Recording of aircraft titles and security documents
    50-59 (Reserved)

SUBCHAPTER D - AIRMEN
60 (Reserved)
61 Certification: Pilots and flight instructors
63 Certification: Flight crew members other than pilots
65 Certification: Airmen other than flight crewmembers
67 Medical standards and certification

SUBCHAPTER E - AIRSPACE
71 Designation of federal airways, area low routes, controlled airspace, and reporting points
73 Special use airspace
75 Establishment of jet routes and area high routes
77 Objects affecting navigable airspace

SUBCHAPTER F - AIR TRAFFIC AND GENERAL OPERATING RULES
General operating and flight rules

- 93 Special air traffic rules and airport traffic patterns
- 95 IFR altitudes
- 97 Standard instrument approach procedures
- 99 Security control of air traffic
- 101 Moored balloons, kites, unmanned rockets, and unmanned free balloons
- 103 Ultralight vehicles
- 105 Parachute jumping
- 107 Airport security
- 109 Indirect air carrier security

SUBCHAPTER G - AIR CARRIERS, AIR TRAVEL CLUBS, AND OPERATORS FOR COMPENSATION OR HIRE: CERTIFICATION AND OPERATIONS

- 121 Certification and operations: Domestic, flag, and supplemental air carriers and commercial operators of large aircraft
- 125 Certification and operations: Airplanes having a seating capacity of 20 or more passengers or a maximum payload capacity of 6,000 pounds or more
- 127 Certification and operations of scheduled air carriers with helicopters
- 129 Operations of foreign air carriers
- 133 Rotorcraft external-load operations
- 135 Air taxi operators and commercial operators
- 137 Agricultural aircraft operations
- 139 Certification and operations: Land airports serving CAB-certificated air carriers

SUBCHAPTER H - SCHOOLS AND OTHER CERTIFICATED AGENCIES

- 141 Pilot schools
- 143 Ground instructors
- 145 Repair stations
- 147 Aviation maintenance technician schools
- 149 Parachute lofts

SUBCHAPTER I - AIRPORTS

- 150 Airports noise compatibility planning
- 151 Federal aid to airports
- 152 Airport aid program
- 153 Acquisition of U.S. land for public airports
- 154 Acquisition of U.S. land for public airports under the Airport and Airway Development Act of 1970
- 155 Release of airport property from surplus property disposal restrictions
- 157 Notice of construction, alteration, activation, and deactivation of airports
- 159 National capital airports
- 169 Expenditure of federal funds for nonmilitary airports or air navigation facilities thereon

SUBCHAPTER J - NAVIGATIONAL FACILITIES

- 171 Non-federal navigation facilities

SUBCHAPTER K - ADMINISTRATIVE REGULATIONS

183 Representatives of the Administrator
185 Testimony by employees and production of records in legal proceedings, and service of legal process and pleadings
187 Fees
189 Use of Federal Aviation Administration communications system
191 Withholding security information from disclosure under the Air Transportation Security Act of 1974

SUBCHAPTERS L-M (RESERVED)

SUBCHAPTER N - WAR RISK INSURANCE

198 War risk insurance

SUBCHAPTER O - AIRCRAFT LOAN GUARANTEE PROGRAM

199 Aircraft loan guarantee program

ECONOMIC REGULATIONS

SUBCHAPTER A - ECONOMIC REGULATIONS

Part
200 Definitions and instructions
201 Applications for certificates of public convenience and necessity
202 Certificates authorizing scheduled route service: Terms, conditions, and limitations
203 Waiver of Warsaw Convention liability limits and defenses
204 Data to support fitness determinations
205 Aircraft accident liability insurance
206 Certificates of public convenience and necessity: Special authorizations
207 Charter trips and special services
208 Terms, conditions,and limitations of certificates to engage in charter air transportation
211 Applications for permits to foreign air carriers
212 Charter trips by foreign air carriers
213 Terms, conditions, and limitations of foreign air carrier permits
214 Terms, conditions, and limitations of foreign air carrier permits authorizing charter transportation only
215 Names of air carriers and foreign air carriers
216 Commingling of blind sector traffic by foreign air carriers
217 Reporting data pertaining to civil aircraft charters performed by U.S. certificated and foreign air carriers
218 Lease by foreign air carrier or other foreign person of aircraft with crew
221 Tariffs
222 Intermodal cargo services by foreign air carriers
223 Free and reduced-rate transportation
228 Embargoes on property
231 Exemption from schedule filing
232 Transportation of mail, review of orders of postmaster general
235 Reinvestment of gains derived from the sale or other disposition of flight equipment

240	Inspection of accounts and property
241	Uniform system of accounts and reports for certificated air carriers
247	Direct airport-to-airport mileage records
248	Submission of audit reports
249	Preservation of air carrier records
250	Oversales
251	Prohibited interests; interlocking relationships
252	Smoking aboard aircraft
253	Notice of terms of contract of carriage
254	Domestic baggage liability
255	Carrier-owned computer reservation systems
256	Display of joint operations in carrier-owned computer reservation systems
261	Filing of agreements
263	Participation of air carrier associations in board proceedings
270	Criteria for designating eligible points
271	Guidelines for subsidizing air carriers providing essential air service air transportation
287	Exemption and approval of certain interlocking relationships
288	Exemption of air carriers for military transportation
291	Domestic cargo transportation
292	Classification and exemption of Alaskan air carriers
294	Canadian charter air taxi operators
296	Indirect air transportation of property
297	Foreign air freight forwarders and foreign cooperative shippers associations
298	Exemptions for air taxi operations
299	Exemption from section 408 for aircraft acquisitions

SUBCHAPTER B - PROCEDURAL

300	Rules of conduct in DOT proceedings under this chapter
302	Rules of practice in proceedings
303	Review of air carrier agreements and 408 applications
305	Rules of practice in informal nonpublic investigations
310	Inspection and copying of DOT opinions, orders, and records
311	Classification and declassification of national security information and material
313	Implementation of the Energy Policy and Conservation Act
314	Employee Protection Program
316	Collection of claims owed the United States
320	Procedures for awarding Japanese charter authorizations
323	Terminations, suspensions, and reductions of service
324	Procedures for compensating all carriers for losses
325	Essential air service procedures
326	Procedures for bumping subsidized air carriers from eligible points

SUBCHAPTER C - (RESERVED)

SUBCHAPTER D - SPECIAL REGULATIONS

372 Overseas military personnel charters
373 Implementation of the Equal Access to Justice Act
374 Implementation of the Consumer Credit Protection Act with respect to air carriers and foreign air carriers
374a Extension of credit by airlines to federal political candidates
375 Navigation of foreign civil aircraft within the United States
377 Continuance of expired authorizations by operation of law pending final determination of applications for renewal thereof
379 Nondiscrimination in federally assisted programs of the board- Effectuation of Title VI of the Civil Rights Act of 1964
380 Public charters
382 Nondiscrimination on the basis of handicap

SUBCHAPTER E - ORGANIZATION

384 Statement of organization, delegation of authority, and availability of records and information
385 Staff assignments and review of action under assignments
387 Organization and operation during emergency conditions
389 Fees and charges for special services

SUBCHAPTER F - POLICY STATEMENTS

398 Guidelines for individual determinations of essential air transportation
399 Statements of general policy
400-1199 (Reserved)

Source: Office of the Federal Register. *Code of Federal Regulations*.
Title 14 Aeronautics and Space. Washington, DC: U.S. Government Printing Office. 1994.

Appendix E

State Aviation Offices

Alabama Department of Aeronautics
770 Washington Avenue, Suite 544
Montgomery, AL 36130
(205) 242-4480

Arizona Division of Aeronautics
Department of Transportation
2612 South 46th Street
Phoenix, AZ 85034
(602) 255-7691

California Division of Aeronautics
Department of Transportation
P.O. Box 942873
Sacramento, CA 94273-0001
(916) 322-3090

Connecticut Bureau of Aviation
and Ports
Department of Transportation
P.O. Drawer A, 24 Wolcott Hill Road
Wethersfield, CT 06129-0801
(203) 566-4417

Florida Aviation Office
Department of Transportation
605 Suwannee Street, M.S. 46
Tallahassee, FL 32399-0450
(904) 488-8444

Guam Airport Authority
P.O. Box 8770
Tamuning, Guam 96931
(011)-(671) 646-0300

Idaho-Bureau of Aeronautics
Idaho Transportation Department
3483 Rickenbacker Street
Boise, ID 83705
(208) 334-8775/6

Alaska Department of Transportation
and Public Utilities
P.O. Box 196900
Anchorage, AL 99519-6900
(907) 266-1666

Arkansas Department of Aeronautics
Regional Airport terminal, 3rd Floor
1 Airport Drive
Little Rock, AR 72202
(501) 376-6781

Colorado Division of Aeronautics
Department of Transportation
6848 South Revere Parkway, Suite 3-101
Englewood, CO 80112-6703
(303) 397-3039

Delaware Aeronautics Administration
Transportation Authority
Department of Transportation
P.O. Box 778
Dover, DE 19903
(302) 739-3264

Georgia Department of Transportation
Office of Intermodal Programs
276 Memorial Drive, S.W.
Atlanta, GA 30303-3743
(404) 651-9201

Hawaii-Airports Division
Hawaii Department of Transportation
Honolulu International Airport
Honolulu, HI 96819-1898
(808) 836-6432

Illinois-Division of Aeronautics
Illinois Department of Transportation
Capital Airport, One Langhorne Bond Dr.
Springfield, IL 62707-8415
(217) 785-8500

Indiana-Division of Aeronautics
Indiana Department of Transportation
143 West Market Street, Suite 300
Indianapolis, IN 46204
(317) 232-1496

Kansas-Division of Aviation
Kansas Department of Transportation
Docking State Office Building
Topeka, KS 66612-1568
(913) 296-2553

Louisiana-Aviation Division
Department of Transportation and Development
P.O. Box 94245
Baton Rouge, LA
(504) 379-1242

Maryland Aviation Administration
Department of Transportation
P.O. Box 8766
Baltimore-Washington International Airport, MD 21240

(410) 859-7100

Michigan-Bureau of Aeronautics
Michigan Department of Transportation
2nd Floor Terminal Building
Capital City Airport
Lansing, MI 48906
(517) 373-1834

Mississippi Aeronautics Bureau
Department of Economic and Community Development
P.O. Box 5
Jackson, MS 39205
(601) 354-6970

Nebraska Department of Transportation
P.O. Box 82088
Lincoln, NE 68501
(402) 471-2371

New Hampshire-Division of Aeronautics
New Hampshire department of Transportation
Municipal Airport, 65 Airport Road
Concord, NH 03301-5298
(603) 271-2551

Iowa-Office of Aeronautics
Air and Transit Division
Iowa Department of Transportation
International Airport
Des Moines, IA 50321

Kentucky-Office of Aeronautics
Kentucky Transportation Cabinet
421 Ann Street
Frankfort, KY 40622
(502) 564-4480

Maine-Air Transportation Division
Maine Department of Transportation
State House Station #16
Augusta, ME 04333
(207) 289-3186

Massachusetts Aeronautics Commission
10 Park Plaza, Room 6620
Boston, MA 02116-3966
(617) 973-7350

Minnesota-Aeronautics Office
Minnesota Department of Transportation
State Transportation Building, Room 417
395 John Ireland Boulevard
St. Paul, MN 55155
(612) 296-8202

Montana-Aeronautics Division
Montana Department of Transportation
P.O. Box 5178
Helena, MT 59604
(4406) 444-2506

Nevada Department of Transportation
1263 South Stewart Street
Carson City, NV 89712
(702) 687-5440

New Jersey-Office of Aviation
New Jersey Department of Transportation
1035 Parkway Avenue, CN 610
Trenton, NJ 08625
(609) 530-2900

New Mexico-Aviation Division
P.O. Box 1149
Sante Fe, NM 87504-1149
(505) 827-0332

North Carolina-Division of Aeronautics
North Carolina Department of Transportation
P.O.Box 25201
Raleigh, NC 27611
(919) 787-9618

Ohio Department of Transportation
Division of Aviation
2829 West Dublin-Granville Road
Columbus, OH 43235
(614) 466-7120

Oregon-Division of Aeronautics
Oregon Department of Transportation
3040 25th Street, S.E.
Salem, OR 97310
(503) 378-4880

Puerto Rico Ports Authority
P.O. Box 362829
San Juan, PR 00936-2829
(809) 723-2260

South Carolina Aeronautics Commission
P.O. Box 280068
Columbia, SC 29228-0068
(803) 822-5400

Tennessee-Office of Aeronautics
Tennessee Department of Transportation
P.O. Box 17326
Nashville, TN 37217
(615) 741-3208

New York-Aviation Division
New York State Department of Transportation
1220 Washington Avenue
Albany, NY 12232
(518) 457-2820

North Dakota Aeronautics Commission
Box 5020
Bismark, ND 58502
(701) 224-2748

Oklahoma Aeronautics Commission
Department of Transportation Building
200 N. E. 21st Street, Room B-7
1st Floor
Oklahoma City, OK 73105
(405) 521-2377

Pennsylvania-Bureau of Aviation
Pennsylvania Department of Transportation
208 Airport Drive
Harrisburg International Airport
Middletown, PA 17057
(717) 948-3915

Rhode Island Department of Transportation
Division of Airports
Theodore Francis Green State Airport
Warwick, RI 02886
(401) 737-4000

South Dakota-South Dakota Office of Aeronautics
700 Broadway Avenue East
Pierre, SD 57501-2586
(605) 773-3574

Texas Department of Transportation
Division of Aviation
P.O. Box 12607
Austin, TX 78711-2607
(512) 476-9262

Utah-Aeronautical Operations Division
Utah Department of Transportation
135 North 2400 West
Salt Lake City, UT 84116
(801) 533-5057

Virginia-Department of Aviation
4508 S. Laburnum Avenue
Richmond, VA 23231-2422
(804) 786-1364

West Virginia-Department of Transportation
Building 5, Room A-109
West Virginia State Capitol
(304) 348-0444

Wyoming Department of Transportation
P.O. Box 1708
Cheyenne, WY 82002-9019
(307) 777-7481

Vermont-Agency of Transportation
133 State Street
Montpelier, VT 05633
(802) 828-2093

Washington-Division of Aeronautics
Washington Department of Transportation
8600 Perimeter Road, Boeing Field
Seattle, WA 98108-3885
(206) 764-4131

Wisconsin-Bureau of Aeronautics
Division of Transportation Assistance
P.O. Box 7914
Madison, WI 53707-7914
(608) 266-3351

Appendix F

Aviation Associations

Aeronautical Radio, Inc.
2551 Riva Rd.
Annapolis, MD 21401
(410) 266-4000

Aeronautical Repair Station Association
121 N. Henry St
Alexandria, VA 22314
(703) 739-9543

Aerospace Industries Association of America
Suite 1200, 1250 Eye St. N.W.
Washington, DC 20005
(202) 371-8400

Aerospace Industries Association of Canada
Suite 1200
60 Queen Street
Ottawa, ON, Canada K1P 5Y7
(613) 232-4297

**Aircraft Appraisal Association
of America, Inc.**
1105 Sovereign Row
Oklahoma City, OK 73108
(405) 942-8225

Aircraft Operations Group Association
Suite 330
130 Slater St
6E2
(613) 230-2668

Aircraft Owners & Pilots Association
421 Aviation Way
Frederick, MD 21701
(301) 695-2000

Airline Dispatchers Federation
Suite 950
700 13th St N.W.
Washington, DC
(800) 676-2685

Airport Consultant Council
Suite 100
908 King St.
Alexandria, VA 22314
703-683-5900

**Airport Management
Conference of Ontario**
P.O. Box 179
Perkinsfield, ON, Canada L0L 2J0
(705) 526-8086

Airports Council International
World Headquarters
P.O. Box 16, CH-1215
Geneva 15 Airport
Switzerland
(22) 798-4141

Air Traffic Control Association, Inc.
Suite 711
Clarendon Blvd
Arlington, VA 22201
(703) 522-5717

Alpha Eta Rho
1615 Gamble Lane
Escondido, CA 92029
Ottawa, ON, Canada K1P
(619) 740-1159

**American Association of
Airport Executives**
4212 King St.
Alexandria, VA 22302
(703) 824-0500

Air Line Pilots Association
1625 Massachusetts Ave. N.W
Washington, DC 20036
(202) 797-4010

American Helicopter Society, Inc.
217 N. Washington St.
Alexandria, VA 22314-2538
(703) 684-6777

Association of Aviation Psychologists
NASA Ames Research Center
M/S 257-1
Moffet Field, CA 94035
(415) 604-6737

Association of German Aviation Professionals
Hohenstaufenning 43,
D-50674
Koln, Germany
(44) 2219 218290

Aviation Distributors & Manufacturers Association
1900 Arch St
Philadelphia, PA 19103
(215) 564-3484

Aviation Safety Institute
Suite 316
6797 N. High St.
Worthington, OH 43085
(614) 885-4242

General Aviation Manufacturers Association
Suite 801
1400 K St. N.W.
Washington, DC 20005-2485
(202) 393-1500

International Council of Aircraft Owner & Pilot Associations
421 Aviation Way
Frederick, MD 21701
(301) 695-2221

International Federation of Airworthiness
IFA Sectretariat Ave.
58 Whiteeheath
Ruislip, Middx,
England HA4 7PW
(44) 1895-672504

International Society of Air Safety Investigators
Technology Trading Park
Five Export Dr.
Sterling, VA 20164-4421
(703) 430-9668

International Society of Aviation Maintenance Professionals
1008 Russell Lane
West Chester, PA 19382
(610) 399-9034

Joint Aviation Authorities
P.O. Box 3000
Saturnsstr. 8-10, NL-2130KA
Hoofddorp, Netherlands
(31) 2503-79700

Mooney Aircraft Pilots Association
P.O. Box 460607
San Antonio, TX 78246
(210) 525-8008

Helicopter Association International
1635 Prince St.
Alexandria, VA 22314-2818
(703) 683-4646

International Civil Aviation Organization
Suite 652
1000 Sherbrooke St. W.
Montreal, PQ, Canada H3A 2R2
(514) 285-8219

National Aviation Associations
Coalition (c/o NBAA)
Suite 400
1200 18th St. N.W.
Washington, DC 20036
(202) 783-9000

National Aviation Club
No 104, 1500 N. Beauregard St.
Alexandria, VA 22311
(703) 379-1506

National Aviation Hall of Fame, Inc.
One Chamber Plaza
Dayton, OH 45402
(513) 226-0800

National Business Travel Association
Suite 401
1650 King St.
Alexandria, VA 22314
(703) 684-0836

National Safety Council
1121 Spring Lake Dr.
Itasca, IL 60143
(708) 285-1121

Pilots Intl. Association, Inc.
4000 Olson Memorial Highway
Minneapolis, MN 55422
(612) 588-5175

National Aircraft Finance Association
P.O. Box 85
Pooleville, MD 20837
(301) 349-2070

National Air Transportation Association, Inc.
4226 King St.
Alexandria, VA 22302
(703) 845-9000

Regional Airline Association
Suite 300
1200 19th St. N.W.
Washington, DC 20036-2422
(202) 857-1170

SAFE Association
Suite 112
107 Music City Circle
Nashville, TN 37214
(615) 902-0056

Seaplane Pilots Association
421 Aviation Way
Frederick, MD 21701-4756
(301) 695-2083

Small Aircraft Manufacturers Association
4226 King St.
Alexandria, VA 22302-1507
(703) 379-1800

Tennessee Aviation Association, Inc.
P.O. Box 2157
Hendersonville, TN 37077
(615) 824-9411

United States Pilots Association
Suite 10
483 S. Kirkwood Rd.
St. Louis, MO 63122
(314) 849-8772

Professional Aviation Maintenance Association
Suite 401
1200 18th St. NW
Washington, DC 20036-2598
(202) 296-0545

Professional Helicopter Pilot Association
P.O. Box 9558
Glendale, CA 91206
(213) 891-3636

Wright Flight
1300 Valencia, #300
Tucson, AZ 85706
(520) 294-0404

University Aviation Association
3410 Skyway Dr.
Auburn, AL 36830
(334) 844-2434

The Wings Club, Inc.
52 Vanderbilt Ave
New York, NY 10017-3885
(212) 867-1770

Women in Aviation, Intl.
P.O. Box 188
Dayton, OH 45402
(513) 225-9440

Appendix G

Aviation Manufacturers

Aerospatiale
37, blvd de Montmorency, F-75781 Paris
Cedex 16, France
42-24-24-24

Astra Jet Corp.
4 Independence Way
Princeton, NJ 0845-6620
Tel 609-987-1125

Avtek Corp
2200 Teal Club Rd.
Oxnard, CA 93030
Tel 805-482-2700

Beech Aircraft Corp.
PO Box 85
Wichita, KS 67201-0085

BMW/Rolls-Royce GMBH
D-6171
Oberursel, Germany
Tel 6171-90-0

Bombardier Aerospace Group
400 Ch de la Cote-Vertu
Dorval, PQ Canada H4S 1Y9
Tel 514-855-5000

British Aerospace PLC
Jetstream Aircraft Ltd
Prestwick Intl Airport
Prestwick, Ayrshire
Scotland KA9 2RW
Tel 1292-479888

Augusta Aerospace Corp.
P.O. Box 16002
3050 Red Lion Rd.
Philadelphia, PA 19114
Tel 215-281-1400

Aviat, Inc
The Airport
PO Box 1149
S. Washington St.
Afton, WY 83110-1149
Tel 307-886-3151

Bell Helicopter Textron Inc.
Box 482
Ft Worth, TX 76101
Tel 817-280-2011

Boeing Commercial Airplane
GroupHohenmarkstr. 60-70 P.O. Box 3707
Seattle, WA 98124-2207
Tel 206-655-2121

Brantly Intl, Inc
12399 Airport Dr
Wilbarger County Airport
Vernon, TX 76384
Tel 81-552-5451

Canadair
P.O. Box 6087
Station Centre-Ville
Montreal, PQ, Canada
Tel 514-855-5000

Cessna Aircraft Co.
Box 7704
Wichita, KS 67277
Tel 316-941-6000

C.A.S.A.
Head office
Avenida de Aragon, 404
E-228022
Madrid, Spain
Tel 585-7000

Commander Aircraft Co.
Wiley post Airport
7200 N.W. 63rd St.
Bethany, OK 73008
Tel 405-495-8080

Enstrom Helicopter Corp.
P.O. Box 490
Twin County Airport
Menominee, MI 49858
Tel 906-863-1200

Fairchild Aircraft
Box 790490
San Antonio, TX 78279-0490
Tel 210-824-9421

Gulfstream Aerospace Corp.
Box 2206
500 Gulfstream Rd.
Savannah, GA 31402-2206
Tel 912-965-3000

Learjet Inc.
P.O. Box 7707
One Learjet Way
Wichita, KS 67277-7707
Tel 316-946-2000

Mooney Aircraft Corp.
Louis Schreiner Field
Kerrville, TX 78028
Tel 210-896-6000

SAAB AB
S-581 88
Linkoping
Sweden
Tel 113-187000

Dassault Falcon Jet
East 15 Midland Ave.
Paramus, NJ 07652-2937
Tel 201-262-0800

Embraer-Empress Brasilera de Aeronautica S.A.
Box 343
Av. Brig. Faria Lima, 2170
Sao Jose dos Campos, SP
Brazil 12227-901
Tel 123 251000

Eurocopter Intl
American Eurocopter Corp.
2701 Forum Dr.
Grand Prairie
TX 75052-7099
Tel 214-641-0000

GKN Westland Helicopters Ltd
Yeovil
Somerset, England BA20 2YB
Tel 1935-75222

Israel Aircraft Industries Ltd.
Commercial Aircraft Group
Ben Gurion Intl Airport 70100
Israel
Tel 3-935-3337

Lockheed Martin Corp.
6801 Rockledge Dr.
Bethesda, MD 20817
Tel 301-897-6000

Raytheon Co.
141 Spring St.
Lexington, MA 02173
Tel 617-862-6600

Sabreliner Corp.
Suite 1500
7733 Forsyth Blvd
St. Louis, MO 63105-1821
Tel 314-863-6880

Short Brothers PLC
P.O. Box 241
Airport Rd.
Belfast, Northern Ireland BT3 9DZ
Tel 1232-458444

Sino Swearingen Aircraft Co.
1770 Sky Place Blvd.
San Antonio, TX 78216
Tel 210-824-2053

Sikorsky Aircraft Corp.
P.O. Box 9729
6900 Main St.
Stratford, CT 06497-9129
Tel 203-386-4000

Textron Inc.
40 Westminister St.
Providence, RI 02903
Tel 401-421-2800

Appendix H

Company Operations Manual Extracts

Two samples (extracts only) of Operations Manuals are provided. The first is for a FAR Part 91 and FAR Part 135 combined operation, and the second is for a FAR Part 91 operation only. Names are withheld upon request.

For the **first sample** (labeled XYZ), notice that an organizational chart is presented (Section II) and that the FAA representative has a block to sign. Additionally, revisions are noted at the bottom of the page. Section numbering allows for easy referencing. Section III and VI extracts are included.

The **second sample** of an actual Operations Manual includes more details in order to provide a greater understanding of the many items which may be required by management. After a title page, the Table of Contents is presented; eight Sections are identified in the manual.

The Introduction is not labeled by paragraph; however, each Section item is numbered for easy referencing. The contents of each section are self-explanatory, with sample data presented for each section. Lastly, a Flight Operations Notice is included as an recent addendum to the manual.

To help prepare an Operations Manual, contact your regional business aircraft association. The NBAA Management Guide, for example, provides information about a manual. It is strongly suggested that aviation department managers develop and use a well-developed operations manual to identify company requirements and standardize effective procedures.

OPERATIONS MANUAL

First Sample (FAR 91/135 Operations)

Section II.
COMPANY ORGANIZATION

2-1. <u>COMPANY NAME</u>

    XYZ INCORPORATED

2-3. <u>ORGANIZATIONAL CHART</u> - Flight Operations

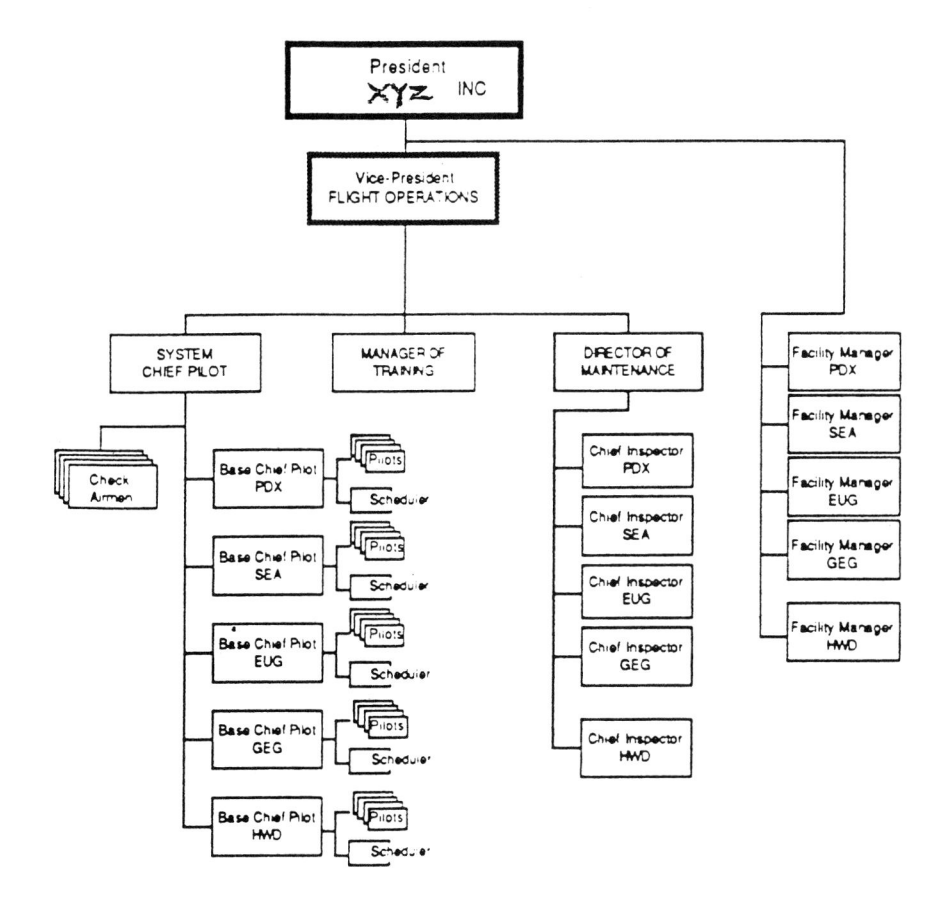

```
                        President
                         XYZ INC
                            |
                     Vice-President
                    FLIGHT OPERATIONS
   _____|_____
   |              |              |                          |
 SYSTEM       MANAGER OF      DIRECTOR OF              Facility Manager
CHIEF PILOT    TRAINING      MAINTENANCE                    PDX
   |              |              |                    Facility Manager
 Check       Base Chief Pilot  Chief Inspector              SEA
 Airmen         PDX              PDX                 Facility Manager
                 — Pilots                                  EUG
                 — Scheduler   Chief Inspector       Facility Manager
             Base Chief Pilot   SEA                       GEG
                SEA                                  Facility Manager
                 — Pilots      Chief Inspector            HWD
                 — Scheduler    EUG
             Base Chief Pilot
                EUG            Chief Inspector
                 — Pilots       GEG
                 — Scheduler
             Base Chief Pilot  Chief Inspector
                GEG             HWD
                 — Pilots
                 — Scheduler
             Base Chief Pilot
                HWD
                 — Pilots
                 — Scheduler
```

326 *Appendix H*

FLIGHT OPERATIONS MANUAL

TABLE OF CONTENTS

Table of Contents i
List of Effective Pages .... iii - iv
Introduction .... v - ix

Duties and Responsibilities .... SECTION 1
Personnel Policies .... SECTION 2
General Operating Procedures .... SECTION 3
Standard Operating Procedures .... SECTION 4
Standard Callouts .... SECTION 5
Emergency Procedures .... SECTION 6
International Procedures .... SECTION 7
Administrative .... SECTION 8

FLIGHT OPERATIONS MANUAL

INTRODUCTION

Purpose of Manual:

The primary purpose of this Flight Operations Manual is to promote the highest level of safety for our flight operations. One of the best ways to accomplish this is by ensuring clear communications between all members of our flight operations.

Another important aim of this manual is to reduce the cockpit workload and to streamline our flight operations so that crew members may handle the often busy cockpit workload with greater efficiency. This is best achieved by ensuring that each flight deck crew member knows precisely what to expect from all other crew members during the conduct of normal, abnormal, and emergency procedures.

All United Technologies Corporate Aircraft Department flight deck crew members are expected to be familiar with and adhere to the policies, procedures, and guidelines set forth in this manual.

Recommendations for Revisions Form:

Our Flight Operations Manual is one of our department's key resources for maintaining a safe and efficient operation. Like any "living document," it needs to be continually fine-tuned and updated in order to reflect our current philosophies.

Feedback from a variety of sources is needed for this process to work. You are encouraged to participate in the continual improvement of this manual using the "Recommendations for Revisions Form" provided in this manual.

Appendix H   **327**

Section III.
COMPANY POLICIES

3-1. **GENERAL**

All FAR 91 and 135 Air Carrier operations conducted by flight, ground, and maintenance personnel shall be conducted in accordance with Federal Aviation Regulations, State and Local laws, and XYZ Company Policies outli in this manual and the XYZ Company Personnel Manual.

3-3. **SAFETY**

The XYZ operating policies are based on the concept that safety com first. Essential elements of safety include quality condition of equipment, meticu pre-flight inspections, thorough training and high motivation of pilots, crewmem and support personnel. It is essential that all personnel are devoted to duty, good judgement, sound operational planning, and efficient use of available resources.

3-5. **RELIABILITY**

Reliability is important in both the equipment we operate and the service we provi to our customers. Every effort should be made—within the limits of safety concerns—to meet scheduled departure and arrival times of our flights and to consistently be counted upon to provide a high quality of flight services.

OUR CORPORATE OBJECTIVE IS TO CREATE A SATISFIED CUSTOMER.

3-7. **COMFORT AND CONVENIENCE**

a. GENERAL

Comfort and convenience of the passenger must be a consideration at all times in t ground and flight operations. Make maneuvers as smooth and gentle as possible restrict climb and descent rates to angles and rates comfortable to passengers and t safety of flight. Abrupt maneuvers should be avoided except in emergencies whe necessity demands.

b. SPECIFIC REQUESTS

Specific requests from passengers or customers shall be honored to the extent that it is not illegal and does not conflict or interfere with safe flight operations. This could include, but not limited to, special food or drink requests.

3-9. **PERSONAL APPEARANCE AND CONDUCT**

a. GENERAL

All personnel of XYZ are expected to present a neat and proper appearance, and to conduct themselves in a manner to reflect favorably upon themselves and the Company. No one shall use loud, offensive, boisterous, or profane language in the presence of passengers or customers. Your activities both on and off duty can reflect on the Company public image, and the well-being of your fellow employees.

b. FLIGHT CREWMEMBERS

Pilots should remember that as aircrew members they are traditionally held in high regard by the customers and should always conduct themselves to preserve that trust and confidence.

(1) If a uniform is specified for the operation then it must be worn for the flight, otherwise crewmembers are expected to dress in a neat and conservative business attire for all flights.

(2) No pilot should wear any article of clothing which would interfere with any pilot flight or ground operation.

(3) Ties for men should be worn on flights when weather conditions do not cause discomfort.

(4) Certain flights, such as transporting forest fighters or equipment, may indicate that the crewmembers should dress in a work uniform or clothing appropriate to the circumstances of the mission.

(5) Crewmembers on over-night trips shall take along adequate clean clothes and toilet articles.

328 Appendix H

3-11. SMOKING POLICY

(1) Smoking is prohibited by crewmembers during any flight operation.

(2) Smoking by anyone in or around aircraft servicing or fueling operations is strictly forbidden, and if observed, the person should be removed from the site immediately.

(3) Smoking is not allowed by passengers on any aeromedical flight at anytime.

(4) Smoking is not allowed during any take-off or landing and whenever the NO SMOKING sign (if equipped) is illuminated.

→ 3-13. MEDICAL EXAMS

a. Air Carrier crewmembers will maintain a current medical, at their expense, appropriate to the certificate and operational level of the flight being conducted:

First-Class within the past twelve months for:

(1) All Second-in-Command positions.
(2) Pilot-in-Command operations in:

   (a) FAR 91 operations of any aircraft.
   (b) Any flight training.
   (c) Air carrier piston and turbo prop aircraft.

b. First-Class within the past six months for PIC positions in:

(1) Air Carrier turbojet aircraft.

c. XYZ employees who are pilots not engaged in Air Carrier operations (such as sales, flight instructors, and service personnel) shall maintain at least a Second-Class medical certificate at their expense.

OPERATIONS MANUAL

Section VI.
FLIGHT PREPARATIONS

☞ Throughout this Operations Manual the XYZ management has tried to avoid having the pilot being required to reference extensive lists of FARs in order to make operational determinations. In this section that objective cannot be fully met due to the complexity of rules surrounding aircraft performance. Pilots should be certain that if a question of applicability arises that the specific regulations are consulted.

6-1. REPORTING FOR DUTY

Flight crewmembers should report for duty at least one hour prior to the planned departure time. Some occasions, such as adverse weather, may require a longer reporting period. It is the PIC's responsibility to determine the requirements of the flight and arrange for the crew to have adequate time to prepare for the flight.

6-3. PILOT'S READING FILE

Each base Chief Pilot keeps a Pilot's Reading File of updated items of interest concerning Company operations and other items of aviation interest in the areas of safety and FAA compliance. Pilots shall insure that they have read the file at least once each month.

6-5. AIRCRAFT ACCEPTANCE

a. Prior to accepting an aircraft for a flight the PIC shall insure that:

(1) The aircraft is authorized for the type of mission being flown.

(2) The aircraft is appropriate to the mission requirement.

(3) There are no outstanding maintenance discrepancies that would either render the aircraft unairworthy or cause it to not be able to complete the mission requirements. If there are deferred maintenance items the PIC shall consult the MEL for limitations and the MEL Use Policy in Section XIV for more guidance. The PIC must be able to comply with the limitations imposed either by the maintenance item or the MEL before the flight may depart.

Appendix H   329

Second Sample (FAR 91 Operations)

Manual Usage:

All flight operations conducted by United Technologies Corporate Aircraft Department personnel shall be carried out in accordance with:

- All applicable Federal Aviation Regulations (FAR's)
- Aircraft Flight Manuals (AFM's)
- Airman's Information Manual (AIM)
- All applicable ICAO procedures and rules
- Any other pertinent rules and regulations including this Flight Operations Manual (FOM)

Anytime in the judgment of a crew member a discrepancy exists in an interpretation between the FOM, FAR's, or/and AFM, the FAR's or AFM will take precedence.

As stated in FAR Part 91, a crew member may for reasons of safety deviate from any rule or regulation: "In an emergency requiring immediate action, the pilot in command may deviate from any rule of this Subpart or of Subpart B to the extent required to meet that emergency."

If a crew member does deviate from a FAR, AFM or FOM policy or procedure, the crew member(s) shall complete a Nonroutine Operations Report (NOR) and forward it to the base Chief Pilot.

Manual Description:

Copies of this manual will be assigned to:

- All flight deck crew members
- All company aircraft
- All company dispatch offices
- All training vendors
- Any other person(s) or locations as specified by the Director, Aviation Services

This Flight Operations Manual is the property of _____ Corporation and is restricted to use by _____ Corporate Aircraft Department personnel. It is not to be made available to persons or agencies not affiliated with _____ Corporation unless specifically authorized by the Director, Aviation Services.

Manual Revisions:

Revisions or replacements to the Flight Operations Manual will be accompanied by a Log of Revisions sheet. This sheet will contain instructions for incorporating the revisions into the manual.

Each manual holder is responsible for maintaining the FOM in a current and up-to-date condition.

Distribution of the Flight Operations Manual and its revisions will be through the Office of the Manager, Safety, Standards and Training.

Corporate Aircraft Department information which needs to be communicated to crew members at the various bases will be divided into one of three categories: Flight Operations Notices, Local Aircrew Notices, or Internal Correspondence. Below is a brief description of each.

Flight Operations Notices:

There may be times that either a new policy or procedure or a modification to an already existing policy or procedure needs to be distributed in a timely manner. Since both a manual revision or writing a new section may take a considerable amount of time, the policy or procedure will be distributed in the form of a Flight Operations Notice (FON).

FON's will contain information of a policy or procedural nature affecting flight operations. FON's may add, modify or delete information contained in the Flight Operations Manual.

FON's will have sequential numbers, effective dates, and expiration dates. Upon reaching the expiration date, these Notices will be discarded, reissued, or incorporated into the Flight Operations Manual.

FON's must be signed by the Director, Aviation Services.

FON's should be inserted in the appropriate tabbed sections in the back of your Flight Operations Manual. It is each manual holder's responsibility to continually update their manual and maintain a current set of FON's. See Appendices Section 1 (Forms), pages 1-2 and 1-3 for an example of a Flight Operations Notice.

FON and LAN Distribution:

Both FON's and LAN's have an "application" section that indicates to which group of crew members the Notice applies.

In the case of the FON, if the Notice is general, it will apply to all CAD crew members. If the FON is aircraft specific, it will apply to only those crew members that crew on that type of aircraft.

In the case of the LAN, if the Notice is general, it will apply to all crew members at that crew base. If the LAN is aircraft specific, it will apply to only those crew members that crew on that type of aircraft at that crew base.

Each <u>crew member</u> will have their own personal copy of the Flight Operations Manual which should include the following FON's and LAN's:

- FON's - GENERAL.
- FON's - The aircraft types flown by that crew member.
- LAN's - For that crew members crew base - GENERAL.
- LAN's - For that crew members crew bases - The aircraft types flown by that crew member.

Each <u>aircraft</u> shall have onboard a copy of the Flight Operations Manual which should include the following FON's and LAN's:

- FON's - GENERAL.
- FON's - That aircraft type.
- LAN's - For that crew base - GENERAL.
- LAN's - For that crew base - That aircraft type.

Each <u>dispatch/flight planning</u> area "master copy" of the Flight Operations Manual should include the following FON's and LAN's:

- FON's - GENERAL.
- FON's - All aircraft types at that crew base.
- LAN's - From all crew bases - GENERAL.
- LAN's - From that crew base - All aircraft types at that base.

Local Aircrew Notices:

At times base Chief Pilots may wish to distribute a policy or procedure that affects flight deck crew members at their local base. These written communications will be distributed as a Local Aircrew Notice (LAN).

LAN's contain information that pertain to operations at a particular crew base. This information is similar in content to FSS NOTAMS. Information contained in LAN's will not modify any policy or procedure contained in the Flight Operations Manual or in a Flight Operations Notice.

LAN's have sequential numbers, effective dates, and expiration dates. Upon reaching the expiration date, these NOTICES will be discarded or reissued.

LAN's must be signed by the base Chief Pilot. The base Chief Pilot will also distribute the LAN's.

LAN's should be inserted in the appropriate tabbed sections in the back of your Flight Operations Manual. It is each manual holder's responsibility to continually update their manual and maintain a current set of LAN's. See Appendices Section 1 (Forms), pages 1-4 and 1-5 for an example of a Local Aircrew Notice.

Internal Correspondence:

Internal correspondence will be all memos which contain information of an administrative nature.

They will be generated by any of the department managers within the Corporate Aircraft Department.

They will not contain information which effects the way we operate CAD aircraft.

They will not contain information which modifies the Flight Operations Manual.

Appendix H   331

FLIGHT OPERATIONS MANUAL

DUTIES AND RESPONSIBILITIES

SECTION 1.00

1.01 Director, Aviation Services 1-1
1.03 Administrative Assistant 1-1
1.05 Manager, Safety, Standards, and Training 1-1
1.07 Manager, Finance and Operations Analysis 1-1
1.09 Operations Assistant 1-1
1.11 Manager, Flight Support 1-2
1.13 Supervisor of Flight (SOF) 1-2
1.15 Flight Operations Coordinator 1-2
1.17 Chief Pilot 1-2
1.19 Supervisory Captains 1-2
1.21 Pilot, Executive Aircraft 1-3
1.23 CoPilot, Executive Aircraft 1-3
1.25 Flight Attendant Supervisor 1-3
1.27 Flight Attendant 1-3
1.29 Manager, Maintenance and Technical Services 1-4
1.31 Maintenance Planner 1-4
1.33 Specialist, Quality Assurance 1-4
1.35 Materials Coordinator 1-4

1.41 Chief Pilot 1-4
1.43 Pilot, Executive Aircraft 1-4
1.45 CoPilot, Executive Aircraft 1-5
1.47 Chief of Maintenance 1-5
1.49 Technician, Aircraft Maintenance 1-5

1.01 Director, Aviation Services

(Reports to Senior Vice President, Human Resources and Organization)

Responsible for the Aviation organization. Develops, implements, and administers operating policies and procedures for all executive flight operations. Studies new developments in aviation, including aircraft and equipment. Negotiates the cost of aircraft and equipment including purchase price, warranties and maintenance. Has financial responsibility for developing, administering, and monitoring the department operating budget.

1.03 Administrative Assistant

(Reports to Director, Aviation Services)

Provides administrative support for the Director of Aviation Services. Maintains personnel records, aircraft permits and licenses, contracts, and leases. Conducts a full range of administrative support functions for department managers and associates.

1.05 Manager, Safety, Standards, and Training

(Reports to Director, Aviation Services)

Responsible for implementing an accident prevention program to ensure safe and standardized flight operations of all UTC Corporate Aircraft Department flight personnel. Plan, monitor and evaluate flight crew vendor provided training programs to ensure a high degree of proficiency is maintained.

1.07 Manager, Finance and Operations Analysis

(Reports to Director, Aviation Services)

Responsible for the overall planning, measurement, analysis, reporting and control of financial policies and programs. Manages the Corporate Aircraft Department budget process. Performs administrative tasks and special projects to include audit functions, purchase and sale of aircraft, and operational analysis for senior management.

1.09 Operations Assistant

(Reports to Manager, Finance and Operations Analysis)

Performs administrative support functions to include audits, file management, local accounting and invoice processing, telecommunications, supply inventory management, budget tracking, aircraft scheduling and operational support.

332 Appendix H

1.11 Manager, Flight Support

(Reports to Director, Aviation Services)

Responsible for the planning and implementation of aircraft scheduling for executive transportation. Monitors and flight follows daily air operations to ensure safe and efficient use of aircraft. Facilitates the coordination, planning and scheduling of all international flights for senior management.

1.13 Supervisor of Flight (SOF)

Reports to Director, Aviation Services)

The SOF functions as the principal contact for the Corporate Aircraft Department beyond normal business hours and on weekends and holidays. Normally during the business work week, the Manager Flight Support is the SOF after business hours. During weekends and holidays, a rotating list of BDL Supervisory Captains and flying Staff Managers function as the SOF. When assigned SOF duty, the SOF is available and on call at all times.

1.15 Flight Operations Coordinator

(Reports to Manager, Flight Operations Support)

Schedule company aircraft for executive transportation in accordance with company policies and government regulations. Provide coordination and support services for such trips with all appropriate parties.

1.17 Chief Pilot

(Reports to Director, Aviation Services)

Manages flight crew personnel. Ensures operational compliance with FAA and all company regulations. Participates in the development and implementation of flight policies and guidelines. Maintains budgetary control of flight operations and approves flight related expenditures. Function as a Captain crew member on company owned or leased aircraft.

1.19 Supervisory Captains

(Reports to   Chief Pilot)

Develops and conducts programs to ensure standardization and maintains pilot-in-command proficiency for their designated aircraft type. Conducts training flights that prepare pilots for qualification check flights. Assists the Chief Pilot with supervision and development of pilot personnel. Serves as a technical advisor for both aircraft maintenance and operational matters on specific types of aircraft.

1.21 Pilot, Executive Aircraft

(Reports to Supervisory Captain)

Functions as the pilot-in-command on   owned or leased aircraft. Has direct responsibility for the safety of passengers and crew and the comfortable, timely operation of the aircraft. Supervises the flight crew. Ensures that all flight and ground operations comply with federal aviation regulations.

1.23 CoPilot, Executive Aircraft

(Reports to Supervisory Captain)

Act as Second-in-Command (CoPilot) on specified aircraft in the air transportation of   personnel and their business associates.

1.25 Flight Attendant Supervisor

(Reports to   Chief Pilot)

Supervises the activities of flight attendant personnel. Functions in a dedicated flight service position and is professionally trained in cabin safety and evacuation. Provides for the care, comfort, and well-being of passengers on assigned flights. Ensures that commissary supplies and food are properly stocked before every flight.

1.27 Flight Attendant

(Reports to Flight Attendant Supervisor)

Functions in a dedicated flight service position and is professionally trained in cabin safety and evacuation. Provides for the care, comfort, and well-being of passengers on assigned flights. Ensures that commissary supplies and food are properly stocked before every flight.

1.29 Manager, Maintenance and Technical Services

(Reports to Director, Aviation Services)

Responsible for overall aircraft maintenance, repair, modification, inspection and ground service activities at the   base. Ensure compliance with all FAA regulations and manufacturer service bulletins. Prepares specifications for bids on outside repairs, maintenance, and inspections. Reviews maintenance contracts and makes recommendations to the Director of Aviation Services. Functions as the Senior Environmental Health and Safety Coordinator for the Aviation Department. Provides direction and leadership to all associated maintenance personnel.

Appendix H   333

FLIGHT OPERATIONS MANUAL

PERSONNEL POLICIES

SECTION 2.00

2.01 Crew Member Qualifications for New Hires 2-1
2.03 Flight and Duty Time Limits 2-2
2.05 Pilot Assessment and Rating System 2-4
2.07 Medical Examinations 2-6
2.09 Pilot Illness 2-6
2.11 Use of Alcoholic Beverages 2-6
2.13 Drugs 2-7
2.15 Blood Donations 2-7
2.17 Crew Member Dress and Conduct 2-7

2.01 Crew Member Qualifications for New Hires

Staffing for aircrews is a cooperative effort involving the Corporate Headquarters Human Resources, Director, Aviation Services, and the Chief Pilot from the appropriate base. Human Resources lends the expertise necessary to ensure compliance with Federal and State regulations and to conform to Corporate standards while the Corporate Aircraft Department establishes the necessary requirements relating to aviation skills.

The following qualifications have been established as guidelines in initially selecting crew member personnel. The requirements are not binding. Qualifications may be reduced at the discretion of flight department management personnel.

PILOT-IN-COMMAND (CAPTAIN)

1. Meets all FAA requirements to serve as pilot-in-command.
2. Airline Transport Pilot Certificate in that category aircraft.
3. Type Rated in aircraft to be flown.
4. Minimum of 3,000 hours flight time.
5. Second Class FAA Medical Certificate.
6. FCC Radio Telephone Permit.
7. Minimum age, 23 years.

SECOND-IN-COMMAND (FIRST OFFICER/CO-PILOT)

1. Commercial Pilot Certificate in that category aircraft.
2. Multi-engine and Instrument Ratings.
3. Minimum of 2,000 hours flight time.
4. Second Class FAA Medical Certificate.
5. FCC Radio Telephone Permit.
6. Minimum age, 21 years.

FLIGHT ATTENDANT

1. Minimum age, 21 years.
2. Familiarity with aircraft operations.
3. Third Class FAA Medical Certificate.

2.03 Flight and Duty Time Limits

Fatigue is the main element to be considered when establishing flight and duty time limitations for any aircraft operation. Fatigue, divided into transient fatigue (temporary) and cumulative fatigue (accumulating), will seriously reduce reflexes and efficiency.

Because fatigue can be the result of many factors and will vary with the individual, a crew member should advise the Captain or Chief Pilot as appropriate when he or she feels they are nearing their natural limitations.

Since fatigue, at increased levels, is directly related to flight safety, the following scheduling limitations have been established for flight crew members:

LIMITATION	FIXED WING	ROTORCRAFT

BASIC:

	MORE THAN 6 CYCLES	MORE THAN 8 CYCLES
Flight Duty Period *	16	14
Flight Time	12	10
Minimum Crew Rest	10	10

Flight Duty Period	14	12
Flight Time	10	8
Minimum Crew Rest	10	10

AUGMENTED CREW:

Flight Duty Period	18	--
Flight Time	15	--
Minimum Crew Rest	12	--

May be extended by 2 hours if a rest period of 8 hours or more can be scheduled, during the flight duty period, and private sleeping facilities are available so that at least 6 of those 8 hours can be spent at such a facility.

2.03 Flight and Duty Time Limits (Cont'd)

If a situation arises in flight that could result in exceeding these limitations, the crew must make this decision. Proposed deviations which occur while the crew is on the ground require the approval of both the crew and the base Chief Pilot.

It is incumbent upon the Chief Pilot to determine the level of safety of the proposed deviation. Many factors must be considered when making this determination. How may hours or cycles has the crew flown today? How many consecutive days has the crew flown prior to today? What is the weather at the time?

In no case shall pressure be applied to a flight crew to cause that crew to continue a flight which they feel should be terminated.

If a decision is made to exceed the limitations, the crew will submit a Non-Routine Operation Report to the Chief Pilot.

2.05 Pilot Assessment and Rating System

This rating system places a numerical value on an individual's job performance covering several broad areas that are key in the performance of a pilot's duties. After these numerical values are totaled for each individual and then compared to all of the other CAD pilots, a numerical ranking will be assigned to each pilot.

This system will provide CAD management and pilots with as fair and objective a list as possible that can then be used when making decisions pertaining to aircraft assignments, promotions, transfers or manpower reductions.

All pilots will be evaluated using the Pilot Assessment and Rating System. This system is designed to be a part of a complete review process. The different parts of this process could include:

- Pilot Assessment and Rating System
- Corporate Office Performance Assessment Plan
- Personal Improvement Plan
- Corrective Action Plan
- CAD Composite Rating

Pilot Assessment and Rating System:

This review form is divided into two areas: Key Personal Competencies and Key Leadership Competencies. All pilots will be evaluated in the area of Key Personal Competencies, while the management pilots will also be evaluated in the area of Key Leadership Competencies. The form utilizes numerical values in assessing an individual pilot's performance in areas that are key to his/her performance of a pilot's duties.

Appendix H **335**

2.05 Pilot Assessment and Rating System (Cont'd)

Corporate Office Performance Assessment Plan:

The Corporate Office Performance Assessment Plan mirrors the Pilot Assessment and Rating System. This form will still be used and will be a part of an individual's yearly review.

Personal Improvement Plan (PIP):

This form is an integral part of an individual's review process. It enables both the employee and his/her manager to set "target goals" in order to better enable each to become more productive. The PIP is normally accomplished in the six month interval between an individual's yearly review. The employee initiates the PIP, it's use is optional.

Corrective Action Plan:

This form will be used whenever an employee receives an unsatisfactory rating in any area. It's purpose is to help map out a specific plan to help improve an employees performance and contribution to the organization.

CAD Composite Rating:

Additionally, each year the Director, Aviation Services will convene the Operations Council for the specific purpose of establishing a single composite ranking of all CAD pilots.

Utilizing the CAD Pilot Assessment Rating System, Operations Council Members will independently evaluate each pilot. All Council Member scores will then be averaged into a single numerical value for each pilot. This score will then be added to the Base Chief Pilot's numerical value that was established during the annual performance evaluation. The total of these two numerical values (Base Chief Pilot's + Operation's Council) will establish each pilot's position on a list that will be used by management. In the event that some pilot's ratings are within one point or less of each other, flight deck seniority will be used as a deciding factor to determine his or her position on the list. Flight deck seniority is defined as the date an individual was hired as a Pilot, CoPilot or Flight Engineer.

2.07 Medical Examinations

All flight deck crew members shall maintain at least a current FAA second-class medical certificate. A copy of the most recent medical form must be given to the Manager, SST at the crew members first opportunity.

2.09 Pilot Illness

No person may act as pilot-in-command, or in any capacity as a required pilot crew member while they have a known medical deficiency that would make them unable to meet the requirements of their current medical certificate.

If a pilot has been off flight duty because of sickness or an accident for two weeks or longer, a medical examination may be requested by the Director, Aviation Services before the pilot returns to flight duty. Pilots should not fly while under serious mental or emotional stress.

2.11 Use of Alcoholic Beverages

Flight crew members shall not consume alcoholic beverages while on duty, nor shall they consume such beverages within ten hours prior to acting as a flight crew member or standing by for the purpose of acting as a flight crew member.

Even though ten hours may have elapsed since your last alcoholic beverage consumption, you may still be affected by the residual alcohol in your bloodstream. The FAA regards a blood alcohol level of .04 as being under the influence. If you feel your performance as a flight crew member could be impaired as a result of prior alcohol use, you must remove yourself from flight status.

2.13 Drugs

There are certain drugs in common use that have a marked effect on the nervous system and which may temporarily affect the pilot's flying ability. They include sulfa, streptomycin, antihistamines, and may other drugs. Before using any of these drugs, all flight crews should make certain that such drugs will have no detrimental effect upon the nervous system. An Aviation Medical Examiner (AME) should be consulted if there is any doubt as to the immediate or residual effect of a particular drug.

2.15 Blood Donations

Corporate Aircraft Department employees who are assigned duty as a flight deck crew member will refrain from giving blood donations or transfusions except for life threatening situations in the employee's family. Employees who must give blood or make transfusions within 72 hours of a flight will be required to remove themselves from flight duty.

2.17 Crew Member Dress and Conduct

As flight crew members, we have a unique opportunity that most employees do not have, close accessibility to senior executives. In fact, it may become too easy for us to become comfortable with executives in day to day interaction so that we cross the line between what is appropriate conversation and what is not. Questions regarding   business issues, personal requests on behalf of the crew member and inappropriate familiarity are not acceptable.

We must be cautious to act professionally and with discretion. All crew members must demonstrate respect and sensitivity to their position and possible encroachment on their personal time.

All personnel shall respect the confidentiality of all Company business and shall not reveal to any unauthorized persons the nature or content of any form of communication, conferences or meetings pertaining to Company affairs.

2.17 Crew Member Dress and Conduct (Cont'd)

When on flight duty, all flight crew members will be properly dressed in the company supplied uniform and will maintain a neat and well groomed appearance that reflects self-respect and respect for the Company, and its customers.

The following are the only items approved to be worn as a part of your pilots uniform:

1. Uniform Suit (Jacket and Slacks)
2. White dress shirt
3. Necktie
4. Tie Chain,   Length of Service Award, or other company supplied emblems
5. Leather Belt, Black
6. Dress Socks, Black
7. Dress Shoes, Black Leather
8. Leather Flight Jacket (Helicopter only)
9. Burgundy Sweater
10. Silk pocket square
11. Topcoat/Raincoat

The above listed items excluding shoes, socks, and sweaters are company supplied and are the only pieces of clothing which may be worn as a part of the uniform. In cold weather, black gloves and a suitable scarf may be worn. They must be business attire (i.e. not ski gloves). These uniforms may only be worn when performing the duties of a flight crew member for United Technologies Corporation.

Pilots must adhere to the following additional items:

- The uniform jacket and slacks shall be clean and pressed.

- The white dress shirt will be clean and pressed. All buttons shall be buttoned at all times.

- Only company supplied ties may be worn with the uniform. The tie should be tied so as to be just even (or nearly so) with your belt.

- The belt should be replaced frequently enough so that it does not show extensive signs of wear.

- Shoes will be clean and freshly polished. Heels which show signs of excessive wear should be replaced.

Appendix H **337**

FLIGHT OPERATIONS MANUAL

PERSONNEL POLICIES

SECTION 2.00

2.01 Crew Member Qualifications for New Hires 2-1
2.03 Flight and Duty Time Limits 2-2
2.05 Pilot Assessment and Rating System 2-4
2.07 Medical Examinations 2-6
2.09 Pilot Illness 2-6
2.11 Use of Alcoholic Beverages 2-6
2.13 Drugs 2-7
2.15 Blood Donations 2-7
2.17 Crew Member Dress and Conduct 2-7

2.01 Crew Member Qualifications for New Hires

Staffing for aircrews is a cooperative effort involving the Corporate Headquarters Human Resources, Director, Aviation Services, and the Chief Pilot from the appropriate base. Human Resources lends the expertise necessary to ensure compliance with Federal and State regulations and to conform to Corporate standards while the Corporate Aircraft Department establishes the necessary requirements relating to aviation skills.

The following qualifications have been established as guidelines in initially selecting crew member personnel. The requirements are not binding. Qualifications may be reduced at the discretion of flight department management personnel.

PILOT-IN-COMMAND (CAPTAIN)

1. Meets all FAA requirements to serve as pilot-in-command.
2. Airline Transport Pilot Certificate in that category aircraft.
3. Type Rated in aircraft to be flown.
4. Minimum of 3,000 hours flight time.
5. Second Class FAA Medical Certificate.
6. FCC Radio Telephone Permit.
7. Minimum age, 23 years.

SECOND-IN-COMMAND (FIRST OFFICER/CO-PILOT)

1. Commercial Pilot Certificate in that category aircraft.
2. Multi-engine and Instrument Ratings.
3. Minimum of 2,000 hours flight time.
4. Second Class FAA Medical Certificate.
5. FCC Radio Telephone Permit.
6. Minimum age, 21 years.

FLIGHT ATTENDANT

1. Minimum age, 21 years.
2. Familiarity with aircraft operations.
3. Third Class FAA Medical Certificate.

3.01 Pilot-In-Command Authority

The Pilot-In-Command (PIC) assigned to a flight shall have exclusive and final authority as to whether or not the aircraft shall proceed to any destination or undertake any flight. The PIC shall not be overruled by any passenger or executive, nor disciplined for their decision having to do with weather, mechanical condition of the aircraft, or other hazards. The PIC has the final authority on all decisions relating to the operation of the aircraft.

3.03 Pilots At Controls

Except as necessary to perform duties in connection with the operation of the aircraft or in connection with physiological needs, crew members shall remain at their flight stations with seat belts fastened during take-off's, climbs, enroute, descents, and landings *(all aspects of flight)*. At least one qualified pilot shall remain at the flight controls at all times with their seat belt fastened.

No person other than a qualified pilot employed by   Corporate Aircraft Department will be authorized to sit at or manipulate the controls during passenger flights. Contract pilots and training vendor instructor pilots are considered "employed"   for this definition.

3.05 Flight Plans

The filing of a flight plan, IFR & VFR, is the responsibility of the pilot designated as the PIC for the trip. An FAA or ICAO flight plan must be filed for all Company flights. The home base call in number, or BDL flight operations call in number, (, must be on all flight plans to assist ATC or the contract flight following agency in notification in case of emergency.

Fixed Wing Aircraft:

All fixed wing flight operations are expected to file and use an IFR flight plan between the point of departure and the point of arrival. Exceptions for this requirement are for maintenance check flights and pilot currency flights that are conducted in the immediate area of the airport. For short flights (less than 30 minutes) where it may not be practical or efficient to file and fly an IFR flight plan, the crew may fly VFR *provided* they utilize positive radar control for traffic advisories and flight following services and flight safety is not compromised.

Helicopter VFR operations:

Helicopter VFR operations are not required to file a VFR flight plan for flights conducted within 300 NM of BDL flight operations if the crew can maintain VHF or flight phone contact with home base for flight following purposes. The route of intended flight must be conducted in accordance with the dispatch manifest, in a direct route between the point of departure and the point of arrival, otherwise a VFR flight plan must be filed.

3.12 Crew Resource Management

Crew Resource Management (CRM) is an essential element of the Department's aviation safety culture. The CAD's policy regarding CRM is that all   pilots will be trained in CRM initially and then periodically throughout their career. All crew members are expected to practice and adhere to the basic elements of CRM while conducting flight operations.

Crew members will be trained and are expected to be proficient in the following CRM skills;

- Communications and Decision Making
    Assertion, Inquiry and Advocacy statements
    Crew Self Critique
    Conflict Resolution

- Crew Coordination
    Leadership
    Planning
    Interpersonal Relationships

- Workload Management
    Preparation
    Planning
    Distraction Avoidance
    Automation Management

3.13 Sterile Cockpit Philosophy

Flight crews will maintain a sterile cockpit environment anytime the aircraft is in motion on the surface or airborne while operating at altitudes below 10,000 feet AGL.

If the cruise altitude is below 10,000 AGL, a sterile cockpit environment will be maintained until the aircraft has been stabilized in the cruise configuration and the AFTER TAKEOFF CHECKLIST has been completed.

A sterile cockpit environment will also be maintained during climbs and descents when within 1,000 feet of reaching an assigned altitude.

3.15 Two-Communications Rule

Whenever an aircraft is operating in a critical phase of flight, if the PNF makes a statement and the PF fails to respond after PNF repeats that statement, the PNF shall assume control of the aircraft by stating, "I HAVE THE CONTROLS."

A critical phase of flight is defined as any time the aircraft is operating below 500 AGL.

Appendix H **339**

3.21 Passenger Relations

In the routine approach to aircraft operation, it is sometimes easy to overlook a very important factor -- the SAFETY and COMFORT of the PASSENGERS.

Almost all passengers, even the most experienced, are sometime apprehensive about flying. The crew must alleviate these feelings and make every trip as comfortable and pleasant as possible.

The following factors are important in establishing good relationships with passengers:

- When crew members are in the presence of passengers, they should plan everything they say and do to promote comfort and peace of mind.

- During obvious business conferences or when guests are on board, crew members should use discretion in entering the cabin.

- How something is said is as important as what is said. Be reassuring. The "voice with a smile" technique may be an old one, but it is a good one.

- At the discretion of the PIC, passengers may be given the opportunity to visit the cockpit during cruise flight. Passengers may be allowed to ride in the cockpit for takeoffs and landings with the permission of the PIC provided a jump seat is available and they have been briefed on the use of the jump seat, oxygen mask location, communications and sterile cockpit procedures.

- If marginal weather exists at the destination and there is a possibility of going to more than one suitable alternate, consult your principal passenger to determine which alternate would be most convenient for them.

- By winning the confidence of the passengers, the pilot will enhance the probability of their cooperation in any situation that develops. This is particularly important should an emergency arise wherein prompt compliance on the part of the passengers is a safety factor. As soon as practical after an emergency arises, the Captain will ensure that the passengers are informed.

- When there is a delay because of weather or other unforeseen reasons, the Captain and/or dispatch will notify the principal passenger on board as soon as possible.

3.21 Passenger Relations (Cont'd)

- Flight personnel shall not disclose to anyone any conversation, discussion or any other information regarding company business that a crew member might overhear or have access to as a result of their assignments, whether on or off company aircraft. This rule shall be strictly adhered to by all flight department personnel.

3.23 Passenger Boarding and Briefing

Passengers will be greeted by a crew member who will ascertain the validity of the trip details including passenger names, company affiliation and itinerary. At least one crew member will supervise the loading of passengers and baggage and will ensure that all exterior doors are firmly secured and all pins and chocks are removed prior to boarding the aircraft.

The crew should provide the passengers with information concerning destination weather, time enroute, any unusual flight condition such as turbulence and information regarding catering or special passenger arrangements. In addition to this, as required by FAR Part 91.519, all passengers should be orally briefed on:

1. Smoking.
2. Use of safety belts.
3. Location and use of entry doors and emergency exits.
4. Location of survival equipment.
5. Ditching procedures and the use of flotation equipment for over-water flights.
6. The normal and emergency use of oxygen equipment installed in the aircraft.

Should the crew determine that the passengers are familiar with the contents of this briefing, it may be omitted.

3.25 Severe Weather

Weather hazards affect the safety of all flights. The following guidelines have been established to assist crews in the avoidance of hazardous weather.

Thunderstorms

Thunderstorms are particularly dangerous and should be avoided if possible. Thunderstorm avoidance can be accomplished by use of airborne weather radar, ATC and/or visual means. If necessary, request a change of route and altitude. When flying in an area of thunderstorms, all pilots should adhere to the following procedures:

1. When the temperature at flight level is 0°C or warmer, avoid all echoes exhibiting sharp shear by 5 miles.

2. When the temperature at flight level is less than 0°C, avoid all echoes exhibiting sharp shear by 10 miles.

   NOTE: Increase the above distances by 50% for echoes that are changing shape rapidly or exhibiting hooks, fingers or scalloped edges.

3. When flying above 23,000 feet avoid all echoes by 20 miles, even though no sharp shear is indicated.

4. Even though "in the clear" on top, use these distances if the flight is clearing the tops by less than 5,000 feet.

   NOTE: As a guideline, sharp shear shall mean a distance less than 3 miles across the remaining portion of the echo after it is contoured.

3.25 Severe Weather (Cont'd)

Icing

No aircraft may take off with frost, snow or ice adhering to any rotor blade, windshield, or powerplant installation, or to an airspeed, altimeter, rate of climb, or flight attitude instrument system, or to the wings, stabilizers, or control surfaces.

Unless an aircraft has ice protection provisions that meet the requirements in Section 34 of Special Federal Aviation Regulation No. 23, or those for transport category airplane type certification, no pilot may fly:

1. Under IFR into known or forecast moderate icing condition; or

2. Under VFR into known light or moderate icing conditions unless the aircraft has functioning de-icing or anti-icing equipment protecting each rotor blade, windshield, wing, stabilizing or control surface, and each airspeed, altimeter, rate of climb, or flight altitude instrument system; or

3. Into known or forecast severe icing conditions.

If current weather reports indicate that the forecast icing conditions that would otherwise prohibit the flight, will not be encountered during the flight because of changing conditions since the forecast, the restrictions based on the forecast conditions would not apply.

3.26 Helicopter Operating Criteria

Whenever possible Category A procedures (as defined in FAR part 29) which state that a helicopter shall be able to safely land or continue flight during all flight regimes will be used. It is recognized that there will be times when full Category A performance cannot be obtained. During these times, crews will utilize as much of the Category A procedure as possible. All operations will limit exposure to other than Category A operations.

Appendix H **341**

3.27 Take-Off Weather Minimums

The takeoff minimums prescribed in the Take-off and IFR Departure Procedure section of the Jeppesen approach charts will be used for takeoff on all runways. With centerline lighting (CL) and runway centerline marking (RCLM) the visibility requirement may be as low as 600 ft. RVR. Reduced visibility runway takeoff's are limited to specific runway use and you should refer to the chart before take-off.

When operating from an uncontrolled airport, the crew must determine that the runway of intended use is free of obstructions and that runway marking or lighting will provide adequate visual reference to continuously identify the take-off surface and maintain directional control throughout the take-off roll. The standard take-off visibility is ¼ mile unless a higher visibility requirement is stated in the Take-off and IFR Departure Procedure section of the Jeppesen approach chart.

Special attention should be given to the Take-off and IFR Departure Procedure for obstacle avoidance and climb gradient requirements at all airports.

If weather conditions at the time of takeoff are below the approved landing minimums, a takeoff alternate airport within the distances listed below must be specified by including the takeoff alternate in the remarks section of your IFR flight plan.

Before takeoff the crew shall determine from weather reports, forecasts and NOTAMS that the takeoff alternate is at or above landing minimums and is expected to remain so for the time period that such takeoff alternate may be required. It must also be determined that the airport conditions at the takeoff alternate are suitable for landing.

1. Rotorcraft: Approximately one half hour from the departure airport at normal cruising speed in still air with one engine inoperative.
2. Fixed Wing: Approximately one hour from the departure airport at normal cruising speed in still air with one engine inoperative.

When weather conditions at the time of take off are equal to or above published landing minimums for the airport of departure, a takeoff alternate is not required.

3.39 Emergency Equipment

Emergency equipment will be carried as permanent aircraft equipment and as per FAR Part 91.513. Each item will be noted as being in place and usable on preflight checks. The operation and capabilities will be known by all crew members.

3.41 Equipment for Over-water Operations

UTC crews will comply with FAR 91.509, survival equipment, and 91.511, radio equipment for over-water flights.

3.43 Oxygen Requirements

Before take off, each flight deck crew member shall personally preflight their oxygen equipment to insure that the oxygen mask is functioning, fitting properly, connected to appropriate supply terminals and that the oxygen supply and pressure are adequate for use.

If the oxygen system has been utilized and the requirement for oxygen no longer exists, the pilot must check the controls on their regulator to assure that the remaining oxygen supply is not accidentally being depleted.

3.45 Use of Exterior Lights

The rotating beacon will be turned on prior to the starting of any engine (except the APU) and shall remain on until all engines have been shut down.

Strobe lights will be turned on any time the aircraft is on the active runway and will remain on throughout the flight until the aircraft has exited the active runway. This paragraph does not prevent pilots from extinguishing the strobes in flight when there is adverse reflections from clouds.

The aircraft landing or recognition lights will be turned on when the takeoff clearance is received or when starting the takeoff roll and will remain on until the aircraft has reached 10,000 feet AGL.

During descent, when the aircraft passes through 10,000 feet AGL, the landing or recognition lights must be on and should remain on until the aircraft has exited the landing runway or made a 180 degree turn on the runway. The recognition lights may be turned on during the Landing Preliminary Checklist.

Landing or recognition lights need not be turned on below 10,000 feet if weather or specific aircraft operational characteristics preclude their use.

3.47 Aircraft Security

The following procedures have been established concerning aircraft security at home base as well as transient bases.

1. All entrance doors and all lockable hatches of the aircraft shall be closed and locked at any time the crew or maintenance personnel are not in attendance.

2. Diplomatically request passengers who are not readily recognizable as company personnel or bona fide guests for proper identification prior to boarding the aircraft.

3. Non-company personnel who are not on the flight manifest cannot fly on aircraft without the approval of the principal passenger or Director, Aviation Services.

4. The manifest shall be protected and not revealed to any unauthorized person.

5. Flight personnel shall not disclose to anyone who does not have the "need to know" the destination of any company flight, identity of its passengers, or the purpose of the flight.

6. No unauthorized person shall be allowed aboard, or near, the aircraft. Any person loitering around or near the aircraft shall be deemed suspicious and be challenged. Airport authorities should be immediately notified if necessary.

7. Unaccompanied baggage or packages are not allowed on the aircraft unless opened and inspected by the crew.

8. Before securing the aircraft on an RON, all interior compartments of the aircraft should be thoroughly checked to determine that all smoking materials are extinguished and the aircraft is otherwise secure.

9. If possible, overnight parking away from home base shall be in a well lighted and patrolled area. The fixed base operator or handler should be questioned as to what precautions are taken to provide for aircraft and personnel security. Additional security precautions may be taken at the discretion of the Captain.

3.67 Use of Minimum Equipment Lists

operates all its aircraft in accordance with FAR 91.213 - Inoperative Instruments and Equipment. Minimum Equipment Lists (MEL's) have been prepared under the provisions of FAR 91.213 for all aircraft. These MEL's together with the Letter of Authorization signed by the Director, Aviation Services, constitute a supplemental type certificate for the aircraft.

There are two official copies of the MEL. One copy is provided to the Maintenance Department. The other copy is located aboard the aircraft in a white binder which is conspicuously marked. If a mechanical problem occurs, or should an employee of the Federal Aviation Agency request to see the MEL, one of the two official copies should be referred to.

If an item or component malfunctions during a trip, the crew should consult the MEL to determine if the item is addressed there. If so, the crew will follow the procedures outlined in the MEL. Any items which are not addressed in the MEL must be operational for departure.

If the crew returns to home base with an equipment malfunction, they should discuss these items with the Maintenance Department prior to making log book entries. This will help assure accuracy of the write-up and a thorough understanding on the part of the Maintenance Department.

As a part of the normal pre-flight activities, the Pilot-in-Command is required to review the Maintenance Logbook for discrepancies. If, during this review, an open MEL item is noted, the pilot should confirm that any required Operational and Maintenance (O & M) Procedures have been completed. The pilot should also evaluate any operational restrictions and ensure that they do not adversely affect the safety of flight.

Appendix H **343**

FLIGHT OPERATIONS MANUAL

STANDARD OPERATING PROCEDURES

SECTION 4.00

4.01 Use of Checklists...4-1
4.03 Preparation for Flight..4-2
4.05 Before Starting Engines Checklists:....................4-3
4.07 After Starting Engines Checklists:......................4-3
4.09 Taxi..4-4
4.11 Before Take-Off...4-5
4.13 Take-Off..4-5
4.15 After Take-Off..4-7
4.17 Landing Preliminary..4-8
4.19 Landing Final...4-9
4.21 After Landing..4-12
4.23 Secure Cockpit and Post Flight...........................4-12

4.03 Preparation for Flight

All flight crew members will report to operations a minimum of one hour before the scheduled departure to ensure adequate time for flight planning and preflight inspection of the aircraft. This time may be varied dependent upon the complexity of the aircraft and the nature of the flight. An earlier report time may be necessary.

Upon arriving for a flight, the PIC will review the Aircraft Maintenance Log, Passenger Manifest, and those items required in FAR 91.103, Preflight Actions. The PIC is responsible to file the flight plan, if appropriate. Fixed wing operations are expected to be conducted under IFR except in those few cases where it may be more practical to go VFR (refer to section 3.05). Crews electing to operate under VFR flight rules must make use of ATC radar services or, if those services are unavailable, the crew will operate under a VFR flight plan.

The SIC (and Flight Attendant when assigned) will make contact with the PIC upon arrival for a flight, and then perform the preflight checks. It is imperative that these preflight checks be performed in sufficient time to enable maintenance to address discrepancies prior to the scheduled departure.

The crew will meet, following completion of their individual preflight activities and review the manifest, trip plan, and aircraft status. A good review technique, is to use the acronym, AWARE.

A = Aircraft status-Previous write-up's, deferred items, MEL items, etc.
W = Weather-Departure enroute, destination, hazards to flight, etc.
A = Airport Information-Runway length, NOTAM, FBO, etc.
R = Route to be flown.
E = Extra-Manifest notes, catering, passenger requirements, fuel load, etc.

For passenger flights, the aircraft should be ready for departure, with crews standing by, thirty minutes prior to the scheduled departure time.

Passengers will be greeted by a crew member who will confirm the trip details including passenger names, destinations and itinerary. At least one crew member will supervise the loading of passengers and baggage. Before starting any engine or turning rotor blades, a crew member must ensure that all exterior doors are secured, that all pins and chocks are removed and that the area around the aircraft is free from any hazard. At those locations where a lineman is stationed in front of the aircraft for the start, the chocks may be left in.

The crew should provide the passengers with information concerning destination weather, time enroute, any unusual flight conditions such as turbulence, and information regarding catering or special passenger arrangements.

344 *Appendix H*

4.03 Preparation for Flight (Cont'd)

In addition to this, as required by FAR Part 91.519, passengers will be orally briefed on:

1. Smoking.
2. Use of safety belts.
3. Location and use of entry doors and emergency exits.
4. Location of survival equipment.
5. Ditching procedures and the use of flotation equipment for overwater flights.
6. The normal and emergency use of oxygen equipment installed in the aircraft.

Should the crew determine that the passengers are familiar with the contents of this briefing, it may be abbreviated.

4.05 Before Starting Engines Checklists:

It is not necessary for both crew member to be in the cockpit prior to the engine start sequence, however in all cases the BEFORE STARTING ENGINES CHECKLIST will be accomplished by the PF prior to the start.

When ground crew personnel are available, the flight deck crew should establish communication with them prior to starting any engines. When ground personnel are not available, visually clear the area around the aircraft, and under the rotor in the case of rotorcraft, prior to starting any engines.

4.07 After Starting Engines Checklists:

The AFTER STARTING ENGINES CHECKLIST will be accomplished prior to releasing the brakes for taxiing to the take-off position. The PF will delay calling for the checklist until the PNF is seated and ready to participate in the checklist challenge and response process.

FLIGHT OPERATIONS MANUAL

STANDARD CALLOUTS

SECTION 5.00

5.01 Callouts, Fixed and Rotary Wing 5-1
5.03 Callouts, Fixed Wing Only 5-4
5.05 Callouts, Rotary Wing Only 5-5

5.01 Callouts, Fixed and Rotary Wing

BUG SETTINGS

Right Seat Pilot - "BUGS SET"
Left Seat Pilot - "BUGS SET AND CROSS-CHECKED"

CALL FOR _____ CHECKLIST

Example:
PF - "AFTER TAKEOFF CHECKLIST"
PNF - "AFTER TAKEOFF CHECKLIST"

WHEN SETTING TAKE-OFF POWER

PF - "CHECK POWER"
PNF - The PNF will ensure that the power is set at the proper value, adjust if necessary, that all engine parameters are indicating normally, that no abnormalities exist, then state:
"POWER CHECKED"

Note: The PNF on rotary wing aircraft does not adjust the power prior to calling "POWER CHECKED" and stating % torque.

DECISION SPEED

PNF - "V1"
PF - (Fixed Wing) Places both hands on control yoke.
PF - (Rotary Wing) Rotates to achieve V2.

ALTIMETER SETTINGS

Anytime a new altimeter setting is received from ATC the PF will state the new setting, check that the PNF has changed their altimeter and state the phrase "Set and Cross Checked".

PF - 31.13 Set and Cross Checked

ONE THOUSAND FEET PRIOR TO ALL LEVEL-OFF'S *

Example - When leaving 7,000 for 8,000 feet:

PNF - "7 for 8"
PF - "7 for 8"

Appendix H 345

5.01 Callouts, Fixed and Rotary Wing (Cont'd)

THREE HUNDRED FEET PRIOR TO ALL LEVEL-OFF *

PNF - "300 FEET"
PF - "300 FEET"

* Not applicable for VFR helicopter operations.

INSTRUMENT APPROACHES (when CDI and/or glide slope indicators come alive)

PNF - "COURSE ALIVE"
PF - "COURSE ALIVE"

PNF - "GLIDE SLOPE ALIVE"
PF - "GLIDE SLOPE ALIVE"

GEAR AND FLAP CALL-OUTS

PF - "GEAR UP" After positive-rate call by PNF
PNF - "ROGER, GEAR UP"
PNF - "GEAR IS UP" After confirming that the gear is up.

PF - "GEAR DOWN"
PNF - "ROGER, GEAR DOWN"

NOTE: Refer to the respective ASM for which crew member moves the gear selector level. If the PF moves the gear lever, he should still state the phrase "GEAR DOWN".

Challenge during the landing final checklist:
PNF - "GEAR"
PF - "GEAR DOWN AND 3 GREEN"

Flap movement:
PF - "FLAPS 15"
PNF - "ROGER, FLAPS 15"
PNF - "FLAPS ARE 15" After confirming flaps are at 15.

5.01 Callouts, Fixed and Rotary Wing (Cont'd)

OUTER MARKER/FINAL APPROACH FIX

Each pilot will call the name of the fix and their indicated altitude crossing the fix.

Example:
PF - "DUSTY 1,700"
PNF - "DUSTY 1,700"

100 FEET ABOVE MINIMUMS

PNF - "APPROACHING MINIMUMS"
PF - "ROGER"

AT MINIMUMS (DH or MDA)

PF - "MINIMUMS"

AS RUNWAY ENVIRONMENT COMES INTO VIEW

Example:

PNF - "STROBES TWELVE O'CLOCK"
       "RUNWAY TWELVE O'CLOCK"
PF - Transitions to outside visual cues and states:
     "I HAVE THE RUNWAY"

IF RUNWAY ENVIRONMENT NOT IN SIGHT

PNF - "NO RUNWAY"
PF - "MISSED APPROACH"
     Advances throttles, rotates aircraft to stop the sink rate, and calls:
     "CHECK POWER"
PNF - Adjusts throttles to computed go-around power if necessary, checks for a positive rate of climb, and states:
      "POWER CHECKED"
PF - "Flaps ___" (As required by specific Aircraft Standards Manual)
PNF - "Positive Rate"
PF - "Gear Up"

At the completion of the missed approach, the crew must conduct an After Take-off checklist.

346 *Appendix H*

5.03 Callouts, Fixed Wing Only

AIRSPEED INDICATOR CHECK

As each pilots airspeed indicator indicates 80 knots:

PNF - "80 KNOTS"
PF - "80 KNOTS"

V1 Speed

PNF - "V1"
PF - Removes their hand from the power levers

ROTATION SPEED

PNF - "Vr"
PF - Smoothly rotates aircraft.

EMERGENCY (MAXIMUM) POWER

Should a situation occur where maximum available thrust is necessary:

PF - Advances power levers full forward and commands: "MAX POWER"
PNF - Ensures that power levers are full forward and states: "MAX POWER"

CLIMBING THROUGH 17,500 FEET OR TRANSITION LEVEL

PF - "29.92 SET, OXYGEN CHECKS, PRESSURIZATION CHECKS"
PNF - "29.92 SET AND CROSS-CHECKED, OXYGEN CHECKS

DESCENDING THROUGH FL 180 OR TRANSITION ALTITUDE

PNF - "30.13 SET"
PF - "30.13 SET AND CROSS-CHECKED"

500 FEET ABOVE THE ALTITUDE SELECTED ON THE ALTIMETER BUG

The 500 foot call-out states the airspeed reference to the computed V_{ref}, the sink rate and the landing gear position. This call-out is made by the PNF at 500 feet above the altitude selected on the altimeter bug.

PNF - "500, V_{ref} PLUS 10, SINK 700, GEAR DOWN AND 3 GREEN "
PF- "ROGER"

5.05 Callouts, Rotary Wing Only

TAKEOFF SAFETY SPEED (V2)

PNF - "V2"
PF - Rotates to maintain V2 until clear of obstacles - "EAPS OFF, FLOATS OFF" (When Applicable).

VBROC (75 KIAS)

PNF - "75 KNOTS"
PF - "GEAR UP" or "ROGER"
PNF - "GEAR UP" - Monitors position indicators and states: "GEAR IS UP"

APPROACHING LDP

PNF - "APPROACHING LDP, ___ KTS"
PF - "ROGER"

500 FEET ABOVE ALTITUDE SELECTED ON THE ALTIMETER BUG

PNF - "AIRSPEED ___ KTS, SINK ___, GEAR DOWN AND 3 GREEN"
PF - "ROGER"

The 500 foot call may be delayed during VFR conditions for rotor wing operations but shall be made at an altitude no lower than 100 feet above the altitude selected on the altimeter bug.

Appendix H **347**

FLIGHT OPERATIONS MANUAL

EMERGENCY PROCEDURES

SECTION 6.00

6.01 General 6-1
6.02 Engine Failure 6-1
6.03 Movement of Critical Switches 6-1
6.04 Pilot Incapacitation 6-2
6.05 Ditching 6-2
6.07 Ground Proximity Warning Systems .. 6-2
6.09 Aircraft Accidents and Incidents 6-4

6.01 General

The emergency procedures in each manufacturer's aircraft flight manual represents the best known available facts about the subject. Flight crews should follow the recommended procedures as they fit the emergency. However, it is recognized that all conceivable situations cannot be anticipated, therefore, the leadership of the Captain and the judgment of the crew will be utilized to successfully cope with the particular circumstances at the time.

Memory items will be accomplished without delay. These procedures would normally be accomplished by the pilot flying (PF); however, some situations could dictate that the Captain designate another crew member to accomplish these steps.

As in normal procedures, challenge and response is the key to the successful completion of a checklist procedure. The PF will initiate the checklist and the PNF will read aloud the items on the checklist including the memory items already accomplished. A response will be given for each checklist item, normally by the PF. However, circumstances may dictate that the PNF accomplish and respond to the checklist items while the other pilot flies the airplane. Remember, someone must fly the airplane. The Captain will ensure that a pilot has been designated to FLY THE AIRPLANE at all times.

6.02 Engine Failure

In the event of an engine failure condition, crew members should not relate the problem engine to a particular engine because the pilot's attention must be directed on controlling the aircraft at this time. The crew member recognizing the problem should state "ENGINE FAILURE". Generally, loss of thrust or mechanical failure will be indicated by two or more instruments so that individual instrument failures can usually be isolated from engine failures.

6.03 Movement of Critical Switches

If a power lever, fire-pull handle, fuel control switch, or any other switch or lever that is critical to the safe operation of the flight is to be repositioned, a second flight deck crew member will confirm and visually verify that the correct switch or lever is being moved to the correct position.

6.04 Pilot Incapacitation

As previously stated, the Corporate Flight Department crew coordination philosophy is based on challenge and response.

If the pilot flying the aircraft fails to respond to any of the required callouts or fails to act in a positive manner with the aircraft controls, the pilot not flying will promptly challenge the pilot a second time. If the pilot fails to respond a second time, the pilot not flying must immediately ascertain the pilot's mental/physical condition. If the pilot flying has become incapacitated, the pilot not flying will take control and state, "I HAVE THE CONTROLS".

6.05 Ditching

The minimum equipment requirements as stated in the Federal Air Regulations Manual are necessary for successful survival at sea. [1] However, this specified equipment will be of relatively little value unless aircrews and passengers are thoroughly familiar with the location of the equipment, how to use it and the ditching procedures for the aircraft in use.

It would be difficult to outline specific procedures in this manual for each aircraft; therefore, it is imperative that each crew member know and periodically review the ditching procedures as outline in each aircraft's flight manual or checklist and those outlined in the Airman's Information Manual or J-Aid.

6.07 Ground Proximity Warning Systems

The Controlled Flight Into Terrain (CFIT) accident represents the single greatest threat to aircraft, flight crews and passengers according to the Flight Safety Foundation (FSF). A FSF task force finding indicated that in CFIT accidents where the aircraft were equipped with a GPWS, an alarming number of flight crews did not follow the recommended pull-up procedure or their actions were too slow in responding to GPWS warnings and the aircraft crashed into the terrain.

The FSF has launched an ambitious international project to reduce by 50% the number of CFIT accidents during the next five years. CAD management supports this effort by the FSF and recognizes the need for a formal GPWS policy. Effective immediately, the following policy and procedure is implemented into our Flight Operations Manual and expected to be adhered in all CAD flight operations.

348 *Appendix H*

6.07 Ground Proximity Warning Systems (Cont'd)

When the Ground Proximity Warning System (GPWS) "Terrain" warning occurs, the PF should immediately and without hesitating to evaluate the warning, execute a pull-up similar to that of a wind sheer encounter. The climb should be continued until the warning stops and the crew determines that terrain clearance is assured. Attempt to contact ATC as soon as practical to inform them of your intentions. This procedure should be followed except in clear daylight visual meteorological conditions when the flight crew can immediately and unequivocally confirm a false GPWS warning.

A pull-up similar to a wind sheer encounter means;

- Apply maximum power
- Aggressively pitch the nose up to an angle of attack that could activate the stick shaker.
- Lower the nose slightly and maintain that attitude until clear of the obstacle and the warning stops.

During your preflight planning, use all available aids to heighten your awareness of the terrain surrounding your destination. Jeppesen approach charts and area charts offer limited terrain avoidance information. In the US, VFR sectional charts are the best resource for terrain awareness. When planning a trip to a destination out of the country, World Aeronautical Charts (WAC) should be used.

The quickest way to get any of these charts is through Sporty's Pilot Shop. calling (800   Sporty's has world wide charts available and offers same day shipping.

Brief the terrain during the landing preliminary checklist so that all crew members are thoroughly aware of the terrain at the destination. In anything other than clear daylight VMC, the safest approach to an airport in a high terrain environment will be to remain on a published route to the initial approach fix followed by a precision approach if available. Avoid the temptation to go direct if offered by ATC. Request to fly the complete published approach.

CFIT accidents have happened on departure too. Review the approach plate to see if there is an IFR departure procedure for the runway you will be using. Crews flying FMS equipped aircraft should be particularly cautious about going "direct to" after takeoff in a rapidly rising terrain area like Laconia, Aspen or Seattle.

FLIGHT OPERATIONS MANUAL

INTERNATIONAL PROCEDURES

SECTION 7.00

7.01 International Flights 7-1
7.03 ICAO Flight Plans - BDL Departures 7-2
7.05 ICAO Flight Plans - Other Departures 7-2
7.07 Flight Following 7-2
7.09 Airport Slots 7-2
7.11 Ground Transportation 7-2
7.13 Hotels 7-2
7.15 Trip Responsibility 7-3

7.01 International Flights

When traveling outside the United States, the flight crew must be prepared to deal with problems and situations unique to the countries to be visited.

Flight crews must acquaint themselves with the following areas in preparation for international flights.

1. Personal Documentation - requirements for most countries likely to be visited are clearly spelled out in the International Flight Information Manual (IFIM).

2. Aircraft Documentation - appropriate to the geographic area of operation and type of flight.

3. Landing and Overfly Permits - aircraft entry requirements are delineated in the IFIM. In most cases, prior permission to land in a country or to overfly a country, must be obtained directly from civil aviation authorities or through the American Embassy in the country in question. It is important to understand that receipt of overflight and landing permits may require anywhere from four hours to several weeks depending upon the destination or country to be overflown and that the requirements vary from country to country.

4. Private Source Services - the NBAA Associate Membership Directory has the names, addresses and telephone numbers of companies that will provide a wide range of services.

5. Military Airfields Overseas - "Hold Harmless" agreements and certification of insurance are prerequisites to obtaining permission to use military fields. Follow procedures outlined in the IFIM and International NOTAMS.

6. Foreign User Charges and Other Fees - charges assessed for ground handling and other services rendered are generally based on the aircraft size and/or weight. Schedules of such charges can only be obtained in writing directly to the handling agent at the airport. Most of the charges paid to government agencies are delineated in the ICAO Manual of Airport and Air Navigation Facility Tariffs.

7. Entry to the United States - procedures to be followed upon return to the U.S. are in the IFIM and Customs Guide for Private Flyers.

Appendix H **349**

FLIGHT OPERATIONS MANUAL

ADMINISTRATIVE

SECTION 8.00

8.01 Manifest Departure Delay Codes......................... 8-1
8.03 Crew Log System Activity Codes........................ 8-2
8.05 Fuel Credit Cards.. 8-4

8.01 Manifest Departure Delay Codes

We now have the ability to extract data concerning departure delays from the PFM computer system. For this data to be meaningful, each base must be using the same guidelines for recording delays.

Please enter the appropriate code into your aircraft flight log whenever you block out <u>more than 15 minutes</u> beyond your proposed departure time.

(A) **ATC Delay** Use this code any time you receive a delay from ATC. This could occur due to traffic saturation, a disabled aircraft on a runway, weather problems which cause ATC to issue a delay, etc.

(C) **Crew Delay** This could occur if a crew member must be replaced due to illness or, perhaps, is involved in a minor auto accident which delays their arrival at work. This code would also be used if the crew finds it necessary to delay a departure due to the 12-hour minimum crew rest period requirement.

(M) **Mechanical Delay** Use this code whenever an aircraft malfunction occurs. You might discover this during taxi out and have to return to the gate. Even though you initially blocked out on time, this would still be a delay because your block out was not followed by a takeoff.

(P) **Passenger Delay** Enter a "P" code when the delay is caused by a passenger or group of passengers. If you are advised of a revised departure time, you should use the new time for your delay calculation <u>unless</u> that revision causes you to be late for a subsequent group of passengers. In that case you should enter a "P" code on the trip segment that initially caused the delay.

(W) **Weather Delay** Use this code any time the crew feels it would be prudent to delay the departure due to weather.

The primary purpose of the crew duty log is to accurately account for the activities performed by aircrew personnel. This data is used for payroll allocation, crew scheduling purposes, and manpower justification. It is therefore important that this data be logged in an accurate and timely manner.

7.03 ICAO Flight Plans - BDL Departures
(Reserved)

7.05 ICAO Flight Plans - Other Departures

When utilizing an international handling agent i.e., Universal or Air Routing:

- Captain will complete flight plan and Wx request for each leg of flight during trip brief.
- Dispatch will coordinate with handling agent.
- Handling agent will provide all plans and Wx at international locations.

When   Flight Ops handles the trip:

- Captain is responsible for handling flight plan and Wx arrangements with local aircraft handling agent or   agent   in London.

7.07 Flight Following

When utilizing an international handling agent i.e., Universal or Air Routing:

- Captain will flight follow through the handling agent.

When   Flight Ops handles the trip:

- Captain will flight follow directly with Dispatch by FAX

7.09 Airport Slots

- Flight Ops will coordinate all slots prior to departure. Crews will be advised at Int'l trip brief.
- The Captain is responsible for all ATC slots and airport slots (if changes have occurred) after the trip has started. The Captain is to use whatever means that are available to him to coordinate slot changes.

7.11 Ground Transportation

- The Captain is responsible for coordinating ground transportation for the crew to the hotel. Considerations should be: cost of transportation, number of crews, amount of baggage/cargo, etc.

7.13 Hotels

- Flight Ops will arrange crew hotels at international locations. Considerations will be: Cost of hotel, location, availability, and principal passenger considerations. If The Captain has a specific location request i.e., near airport or downtown, he should make that known early in the planning stages of trip. <u>Once reservations are made, they are final</u>.

7.15 Trip Responsibility

- The Captain will be identified as early as possible so as to be jointly involved with UTC Flight Ops in the planning stages of the trip.
- Once the trip has started, the Captain has full responsibility for the trip.

350 *Appendix H*

8.03 Crew Log System Activity Codes

Flight Related Activity Codes

These activities are directly related to the primary function of flying aircraft.

(F) - Flight Activity. The "F" code is used when flight activities have been performed in support of executive, utility, or maintenance flights. Days traveling to support these flight functions should also be logged as "F" days.

The flight duty period typically begins when reporting to work for the purpose of flying and ends after post flight activities are completed at the end of the day.

(T) - Aircraft Training. Log "T" any time aircraft related training is given or received, including simulator training. Log actual hours in training. Travel days associated with the aircraft related training function should be logged as "T" days.

(Y) - Standby. Log "Y" when assigned to crew an aircraft for the purpose of flying. Normal standby hours are between the hours of 0800 and 1800 local time (10.0 hrs). If assigned to standby for a different period of time, log the appropriate times and number of hours actually standing by on the aircraft.

Flight Support Activity Codes

Flight support activities are indirectly related to the flight function.

(G) - General Training. Log "G" when participating in training that is not aircraft specific, i.e., management seminars, leadership courses, CRM, CPR, etc. Days used for traveling to and from the training session should be logged as "G" days. Log actual hours worked.

(A) - Administrative Duty. When actual administrative duty is performed, log the "A" code. Log only actual hours worked.

8.03 Crew Log System Activity Codes (Cont'd)

Non-Support Activity Codes

The following activities use up man days; therefore they must be accounted for in manpower justification.

(H) - Holidays. Log "H" on all holidays if not assigned to work. If a crew member is assigned to work a holiday, the appropriate work related duty code should be logged. Corporate holidays do not fall on weekends. The "H" code should be logged on either the Monday or Friday, closest to the weekend holiday.

(V) - Vacation Days. Log "V" for vacation days. One week of vacation equates to five days Monday through Friday. Therefore, weekends on either side of a Monday through Friday vacation period should be logged as "X" (off - not scheduled).

(M) - Military Duty. The "M" code should be logged whenever a crew member is performing their annual (15 days) military duty commitment. Weekend days between two weeks of military drill should be logged as "X" (off - not scheduled). Weekend drill should be logged as "P" (personal days off).

(I) - Illness. "I" should be logged on days when a crew member was sick and not available for flight duty.

(R) - Crew Rest. If you cannot be scheduled for flight duty the next day because of a late night arrival, log the "R" code for the next day's activity. For example: You landed at 2200 hours on Monday and the aircraft was scheduled out again at 0800 on Tuesday. You could not be scheduled for the Tuesday trip because of the 12-hour crew rest rule; therefore, you should log the "R" code on Tuesday. Do not log an "X" when the "R" code is appropriate.

In this example, if you come to work on Tuesday and perform administrative or training activities, you should log the appropriate code: "T", "G", "A", etc.

Appendix H 351

8.03 Crew Log System (Cont'd)

<u>Non-Work Activity Codes</u>

These activity codes are not work related but are tracked for crew scheduling and manpower justification purposes.

(P) - <u>Off, Personal Time</u>. Personal days off can be requested in advance to allow you to plan time off for special occasions with the family or for other personal reasons.

(X) - <u>Off, Not Scheduled</u>. There are no start or stop times associated with the "X" code.

NOTE: Log start and stop times for Flight, Aircraft, Training, Standby, General Training, and Administrative days. Do NOT log hours against the Non-Support and Non-Work days.

8.05 Fuel Credit Cards

When purchasing fuel at an FBO, use the credit card that is indicated in the FBO section of the flight manifest. If the indicated credit card is not acceptable or no credit card is listed, use the credit cards in the following order:

1. UVAir
2. AVCard
3. Multi-Service
4. Individual Oil Company Credit Cards

FLIGHT OPERATIONS NOTICE

Number: 97-01
Effective Date: January 1, 1997
Expiration Date: Permanent

Application: GENERAL

Subject: 2.03 Flight and Duty Time Limits

Flight is capable of operating its aircraft 24 hours-a-day, worldwide to meet Corporation's business demands. The growth in global long haul and domestic operations will continue to increase these around the clock requirements. Flight Aircrew must be available to support 24 hour-a-day operational demands.

The threat to safely operating our aircraft during these operational demands is Aircrew fatigue. The FAA, NTSB and NASA has recently cited fatigue as a contributing factor in several recent aircraft accidents. They have also indicated the necessity for responsible management of maximum On Duty and minimum Off Duty times to ensure adequate levels of safety in air transportation.

At Flight, we have a responsibility to manage our flight operations in a manner supporting the latest industry recommendations in order to ensure our customers of at least the same level of safety found elsewhere in the industry.

The policies and guidelines established here are based on the latest scientific knowledge on the subject. They are based on studies and reports from the NASA Fatigue Countermeasures Program, The Flight Safety Foundation's Special Task Force Report on Principles and Guidelines For Scheduling Duty and Rest Scheduling in Corporate and Business Aviation and Flight's Fatigue Countermeasures Team.

The policies for maximum and minimum times have taken into account the unique operational aspects of Flight's corporate aircraft activities including fixed wing and rotary wing aircraft. This policy was designed with domestic and international operations, varying number of take-off's and landings day and night scheduling and the disruption of normal circadian rhythms in mind.

It is expected that these policies will be respected by Aircrew, Aircrew Schedulers, Flight Management and the passengers and customers of United Technologies Corporation. A scheduled Off Duty Period is important for the Aircrew to properly manage their relaxation, exercise, nourishment and sleep in order to be well prepared to safely conduct their next On Duty Period. It is important to protect this period from outside disruptions that may impede sleep, including phone calls.

Flight Aircrew have a responsibility too. They are expected to be rested and physically fit for their next On Duty Period. Aircrew are reminded that they may not act as a required crewmember while they have a known medical deficiency that would make them unable to meet the requirements of their current medical certificate. In the NASA study, alcohol was identified as the most widely used sleep aid. Flight Aircrew are reminded that they shall not consume alcoholic beverages while On Duty or within 10 hours prior to acting as a crewmember or during a Stand By period.

The Fatigue Countermeasures Policy was designed as a "work in progress" policy. A number of other approaches to managing fatigue are in different stages of development. Research continues and may provide further findings on fatigue and personal strategies that will help us manage this problem in aviation operations.

Number 97-01
Page 3: of 6

Application: GENERAL

Flight Aircrew Fatigue Countermeasures Policy

Standard Duty Period per 24 hours	All Aircraft
On Duty Period	14 Hours
Off Duty Period	10 Hours
Off Duty Period for Flights Beyond North America	12 Hours

Extended Duty Period per 24 hours |
--- | ---
On Duty Period | 16 Hours
Off Duty Period | 12 Hours
Augmented Crew | 18 Hours
Off Duty Period | 12 Hours
On Duty Rest Period Provided | 18 Hours
Off Duty Period | 12 Hours

If either the Standard Duty Period or the Extended Duty Period penetrates a circadian low time or if the number of take-off and landings exceeds six, then the On Duty Period must be reduced by 2 hours.

Aircrew may not be scheduled to fly more than six consecutive days in a seven day period.

Definitions & Policies

Augmented Crew An additional Captain qualified Aircrew member must be onboard the aircraft during at least the extended duty portion of the trip. During Augmented Crew flights, a reclining seat in the passenger cabin must be available for the resting crew member. This is a condition for an Augmented Crew flight and must be understood in advance with the passengers, as the resting crew member should not be confined to the cockpit. The Captain must prepare and use a schedule that will ensure adequate rest for all crew members including cabin crew members.

Appendix H **353**

FLIGHT OPERATIONS NOTICE
Number 97-01
Application: GENERAL   Page: 5 of 6

Deviations to the Policy and Required Reports
Planned Deviations

Occasionally there may be circumstances that require an exception to this policy. The exception process must not be used to routinely deviate from the intent of this Fatigue Countermeasures policy. Whenever a scheduled deviation to this policy is necessary, a written request shall be made by the requester. The request must be approved by the Aircrew involved <u>and</u> any two Flight personnel in the following succession order:

Chief Pilot (local base)
Manager Safety, Standards & Training
Chief Pilot (of another base)
VP & General Manager, UTFlight
Manager, Flight Operations Support
Supervisor of Flight

The request must be documented and state the rationale for the deviation. Before approval can be made, there must be identifiable and planned fatigue countermeasure strategies to be undertaken to mitigate the consequences of the deviation. A copy of the written request must be forwarded to the Manager SST upon approval and retained in the managers office for two years from the time of occurrence. A post flight debriefing with the original requester, Chief Pilot, Manger SST or VP & General Manager Flight and Aircrew to discuss the effectiveness and consequences of the occurrence must take place following the trip and filed with the request.

FLIGHT OPERATIONS NOTICE
Number: 97-01
Application: GENERAL   Page: 6 of 6

Unplanned Deviations

There may be unforeseen circumstances such as a passenger delay, mechanical delay, weather or other operational delay that cause a well planned trip to exceed these guidelines. However, the trip should not just go on endlessly oblivious to this policy and these guidelines. A reasonable limit to an unplanned deviation is considered 2 hours beyond the allowable Duty Period. This policy further mandates that if these unplanned limits are met, or will be, that the aircraft not be allowed to depart or that it stop while enroute and a 12 hours rest period obtained. Anytime an unplanned deviation occurs, or has the potential to occur, the Captain must contact the Chief Pilot or SOF. At the Captain's next opportunity, the Captain must submit a verbal or written NOR to the first available person in the preceding UTFlight personnel succession list. A written report must be made by the person receiving the reported deviation and a follow-up report made to the Manager SST by the base Chief Pilot.

FLIGHT OPERATIONS NOTICE
Number 97-01
Application: GENERAL   Page 4 of 6

Circadian Low Time The period of time from 0200 to 0559 calculated from the time zone where the On Duty Period begins.

Fatigue Countermeasure A proactive undertaking to lessen the consequences of being fatigued while operating in an environment that requires alertness. In simple terms it means taking steps to assure the Aircrew adequate rest to do their job safely.

North America Alaska, Bermuda, Canada, Caribbean Islands, Mexico and the USA. The day prior to a trip beyond North America, crews will be assigned an Administrative day for trip preparation and rest. Upon return to home base from a trip beyond North America, crews will be scheduled for a 48 hour Off Duty Period for rest.

Off Duty Period The Period when the Aircrew is relieved of all duties for UTFlight.

On Duty Period The period beginning when the Aircrew first performs any duties for UTFlight and ending when the Aircrew last performs any duties for UTFlight. This includes, but is not limited to, flight duty, administrative work, management duties, training and deadheading. For flights within North America, the on Duty Period begins one hour before scheduled take-off. For flights beyond North America, the On Duty period begins two hours before scheduled take-off. The On Duty period ends one hour after landing.

On Duty Rest Period At least six hours of uninterrupted rest during a minimum eight hour block of time between flights. A hotel or other suitable rest facility should be used.

Standby An Aircrew status whereby the Aircrew has completed a normal Off Duty Period and is readily available for call to an On Duty Period. The Aircrew is expected to be able to have the aircraft airborne within two hours of notification. The combination of Standby time and On Duty time cannot exceed allowable On Duty limits. Unless otherwise informed, the standard Standby period is from 0800 to 1800 local time.

FLIGHT OPERATIONS NOTICE

354   *Appendix H*

Appendix I

Case Studies

Several brief case studies are provided for individual or group analysis and reporting. For each short scenario, you should be able to determine whether business aircraft can be used to increase management's productivity. Since complete data are not available, make all assumptions necessary, but each should be reasonable and practical. The steps necessary to complete the case analysis include:

1. Critically analyze the situation.
2. Define the problem.
3. Identify assumptions and limitations.
4. Develop alternative plans.
5. Recommend a solution.
6. Prepare a formal written response (may be an oral briefing).
7. Include necessary appendices.

Formal Response

In a formal written response, prepare a **cover letter** to the company's CEO or president. Write this letter as the aviation consultant or company employee asked to review the situation and make a recommendation. The purpose of this letter is to make a specific recommendation; therefore, clearly indicate this and provide any data necessary as **Appendices.** Important in this cover letter is your approach and writing style. Be positive, eliminate wordiness, and confirm appointment/dates. It is suggested that the following supporting documents be attached, when appropriate:

1. *Aircraft matrix.* Use a maximum of three aircraft to compare.
2. *Aircraft specifications/performance data.* For the recommended aircraft, provide current data.
3. *Listing of tangible/intangible benefits.* Present in a table.
4. *Time/cost comparisons for trip requirements.* For origin and destinations for charter, commercial, and/or business aircraft.
5. *Summary of costs.* Costs of aircraft acquisition and possession. Show full ownership, fractional ownership, charter, lease, or other costs.
6. *Cash flow analysis.* Show at least a five-year costing of the aircraft recommendation.

Assumptions: Acting as the aviation consultant or knowledgeable employee, make the necessary assumptions on the number of aircraft engines (safety and operating concerns), expansion of aviation operations, budgetary limits, equipment requirements, and other pertinent issues. Be sure to identify any assumption made. Use the current literature; for example,

Business & Commercial Aviation's Planning & Purchasing Handbook, Professional Pilot, NBAA's Travel$ense program, FAA publications, or other pertinent data. Be sure to cite all sources. Alexander T. Wells' and Franklin D. Richey's text, *Commuter Airlines,* published by Krieger Publishing Company (Florida), 1996, is an excellent source for cost/cash flow analysis. For all citations, provide a reference listing. Finally, ensure that grammar, punctuation, and writing style result in a quality product.

Group Presentation: The documents identified above should be presented using overhead transparencies or charts. The major difference between the group and individual presentation should be the oral briefing that accompanies the data. Additionally, responses during a question and answer session should allow for additional assessment.

Case I -1

KenKov Associates Incorporated is a firm that specializes in sales and electronics repairs for high-technology equipment used by major businesses. The chief executive, Sally Katris, has asked for an evaluation of employee travel and the need for using a business aircraft. She has been receiving brochures about the benefits of using business aircraft. Fractional ownership has interested her, and she wonders if this would be better than buying outright. Mainly, she wonders if a business aircraft is needed.

KenKov is located at 1385 Pearl Road, Cleveland, Ohio. Since its startup in 1990, the company has increased its number of employees to 200 people. Many of these people are electronics technicians and high-tech salespersons who need to travel to various customer locations, often at short notice. So far, all travel has been on commercial carriers.

Being proactive and young, Sally and her top managers have determined that any reasonable aircraft expense for their growing company can be considered because excellent financial terms are available. For the last three years, KenKov employees have traveled to many locations in the United States, but the major destinations have been as follows:

Destination	Passengers per Trip	Average Monthly Roundtrips
Atlanta, Georgia	3	8
New York, New York	3	10
Cambridge, Ohio	6	10
Flagstaff, Arizona	3	4
Orlando, Florida	4	8
Denver, Colorado	4	4

Case I - 2

StayRight Incorporated is a progressive company involved in the repair of large production machinery and is located at 13585 Simonsonl Street, St. Louis, Missouri. Its chairman, Al Mitchell, is considering whether the company's employees would be more productive and useful if the firm used business aircraft. At present, long-range travel is accomplished using the commercial carriers. Plans are being made to expand the company operations to the U.S. eastcoast within the next three years.

Since the company's inception in 1992, StayRight personnel have traveled to various destinations as indicated below. Al is wondering about buying or chartering aircraft for the company to support the employees who serve major contracted firms. The contracts earn StayRight substantial revenues.

Destination	Passengers/Cargo per Trip	Average Monthly Trips
Denver, Colorado	4/200 pounds	8
Chicago, Illinois	2/300 pounds	12
Charlotte, N. Carolina	3/200 pounds	8
Roanoke, Virginia	2/100 pounds	8
Charleston, S. Carolina	4/200 pounds	10

Case I - 3

Joan Neal, president of Expando Incorporated, 2525 Stecklow Street, Columbus, Ohio, decided that she would like to purchase a small company plane so that she can make business trips throughout the United States. She presently drives and uses the commercial carriers; however, she does have her private pilot's license. For some of her journeys, Joan takes two mid-level managers on the business trips. Besides her trips, other mid-level managers visit sales regions, usually traveling in pairs for the short two-hour presentations. Last-minute report presentation preparations are routine.

Joan has decided to seek outside advice in this matter and has asked you to provide a recommendation. As an aviation consultant, would you recommend a business aircraft? If recommended, which type and model? Other recommendations?

	Monthly Roundtrips	
Destination	Joan Neal	Others
Harrisburg, Pennsylvania	2	4
Columbia, S. Carolina	1*	3
Atlanta, Georgia	3*	4
Montgomery, Alabama	2	2
Frankfort, Kentucky	2	2
Nashville, Tennessee	2	4
Richmond, Virginia	2*	2
Cleveland, Ohio	2	6

* - Two mid-level managers accompany Joan on these trips.

Case I - 4

Terry Costar is the senior vice-president of operations for Sunny Oil Corporation. The firm is located at 3131 West 111th Street, Dallas, Texas. Terry is preparing a recommendation for the Board of Trustees to begin helicopter operations for the firm's oil rigs in the Gulf and throughout Oklahoma and Texas, approximately 54 total sites. Presently, managers and technicians use contracted airlift services, but the costs seem too high to Terry.

He is seeking outside advice about a company-owned helicopter operation. Costs are not the main worry, although they are to be considered extremely important. Flexibility and support are the main concerns for the routine patrols and team visits made to the sites. The latter involve an eight-person team visiting each site at least once a year, with 600 pounds of equipment. The average load factor for the routine patrols is two people with spare parts and equipment (approximately 400 pounds). A patrol visit lasts about two hours, and at least 30 routine site visits are made per month.

Sunny Oil operates under a six day-a-week schedule. Emergency response is considered vital to the company managers. Terry has asked you to recommend the model and number of helicopters needed, along with the costs to organize and operate an aviation department. After his review, he will make a final recommendation to the board.

Case I - 5

Global Enterprises is a firm that sells pharmaceuticals throughout the United States, Europe, and the Far East. Its home base is in Atlanta, Georgia. Terry Swift is senior vice-president of marketing and has been considering that the firm obtain business aircraft after discussing this with other executives during a convention he recently attended. Before he makes a serious recommendation to the CEO, he contracts you to review the need and make a recommendation.

Terry provides you basic data relating to executive/mid-level managers' travels. Six senior company representatives visit major cities in Europe and four senior representatives visit the Far East at least once each month. The visits last two weeks for each person. Complaints have been received about the long time away from home, although personnel schedules are fairly even across management. Some managers are beginning to look for other work.

Additionally, once an executive visits the major cities, he/she must travel around various local sites. Commercial transportation has been the only method used by all personnel. Usually employees travel in pairs.

Terry noted that the overall productivity of the managers was low and morale was poor. Global Enterprises made almost $500 million in profit last year, and the firm has excellent credit ratings. What is your recommendation to Terry?

Case I - 6

Logistics Express is a firm located in Cleveland, Ohio, and provides various managerial support services to companies throughout the United States and Europe. The firm owns and operates two Falcon 50 aircraft, both purchased new in 1988 for 70 percent of today's Falcon 50 value. Joe Moveitt, CEO, has been asked by his aviation department manager, Ken Stable, to consider replacing the Falcon 50s with newer aircraft having greater range (at least 3,400 sm) and capacity, with the possibility of adding another aircraft. The flight department falls under the CEO.

Ken told Joe that the flight department cannot provide the required corporate airlift support because there are more company users than ever and the aircraft require more maintenance. The situation looks like it will continue because of the company's expansion plans to Europe. The load factor has been averaging 90 percent per plane per trip, and some European scheduled flights have been canceled at the last minute because of maintenance. The aircraft have been averaging about 800 flight hours each per year.

Presently, four captains head the flight crew of six total pilots. Four maintenance workers (one is chief of maintenance), two dispatchers/schedulers, and one administrative employee, besides Ken, make up the flight department. Personnel from other company departments support the flight operations as needed.

Joe contracted you as an outside aviation consultant to evaluate the situation and determine what aircraft are needed and how many. He knows something has to be done fairly soon. How many aircraft and aviation specialists should be added and how much will it all cost the company?

Joe knows that company expansion plans at its locations in Paris, Frankfurt, London, and Berlin will necessitate that more senior executives travel. Costs will be important, but not overly decisive.

References

Chapter 1

Assistant Administrator for System Safety. (1996). *Aviation system indicators: 1995 Annual report.* Washington, DC: U.S. Government Printing Office.

Baran, R. T. (1990, March 16). *Anatomy of a flight department.* Washington, DC: National Business Aircraft Association.

Castro, R. (1995). *Corporate aviation management.* Carbondale: Southern Illinois University Press.

Federal Aviation Administration. (1995, March). *FAA Aviation forecasts—fiscal years 1995–2006.* Washington, DC: Office of Aviation Policy, Plans, and Management Analysis.

General Aviation Manufacturers Association. (1996). *General aviation statistical databook.* Washington, DC: author.

Gilbert, G. A. (Ed.). (1993, March). GAMA 1992 deliveries slip below 1,000. *Business and Commercial Aviation,* 31.

Hotson, F. (1995, Spring). Goodyear's DC-3. *Leading edge,* 2–4.

Hotson, F.W. (1991). *Business wings.* Ottawa, Ontario: Canadian Business Aircraft Association, Inc.

National Business Aircraft Association. (1997). *NBAA Business aviation fact book 1997.* Washington, DC: author.

North, D. M. (1994, October 10). Corporate flying activities accelerates. *Aviation Week and Space Technology,* 39.

Office of Aviation Policy and Plans. (1994). *General aviation and air taxi activity survey.* Washington, DC: U.S. Government Printing Office.

Office of Aviation Policy and Plans. (1996). *6th Annual FAA General aviation forecast conference proceedings.* Washington, DC: U.S. Government Printing Office.

Office of Federal Register. (1994). *Code of federal regulations: Aeronautics and space—Parts 200 to 1199.* Washington, DC: U.S. Government Printing Office.

Statistics and Forecast Branch. (1996, March). *FAA aviation forecasts: Fiscal years 1996–2007.* Springfield: National Technical Information Service.

Story of Gulfstream, The. (n.d.). Gulfstream news release. Savannah: Gulfstream.

Wartenberg, S. (1989, May). Portrait of a pioneer: Igor Sikorsky. *Rotor & Wing International,* 38–42, 92–93.

Wells, A. T. & Chadbourne, B. D. (1987). *General aviation marketing.* Malabar: Robert E. Krieger Publishing Company.

Whempner, R. J. (1982). *Corporate Aviation.* New York: McGraw-Hill.

Wings at Work. (n.d.). [Film]. Produced by Lockheed-Georgia Company, Division of Lockheed Aircraft Corporation.

Chapter 2

Canadian Business Aircraft Association. (n.d.). *Business Aviation in Canada.* [Brochure]. Montreal, Canada: Author.

European Business Aviation Association. (n.d.). *Expanding the Horizon of Business Aviation.* [Brochure]. Zaventem, Belgium: Author.

General Aviation Manufacturers Association. (n.d.). *1997 Statistical databook.* Washington, DC: Author.

Gilbert, G. A. (1996, December). U.S. Business jet deliveries continue to decline. *Business & Commercial Aviation, 17.*

International Business Aviation Council. (1988, January). *What is IBAC?* [Brochure]. Wallingford, England: Author.

Jackson, K. S. & Brennan, J. T. (1995). *Federal aviation regulations explained, Parts 1, 61, 91, 141, and NTSB 830.* (Available from Jeppeson, 55 Inverness Drive East, Englewood, CO 80112-5498, phone 303-799-9090).

National Business Aircraft Association. (1997). *1997 Business aviation fact book.* Washington, DC: Author.

Office of Federal Register. (1994, January 1). *Code of federal regulations: Parts 60 to 139.* Washington, DC: U.S. Government Printing Office.

Phillips, E. H. (1994, August 15). Clinton to sign product liability bill. *Aviation Week & Space Technology, 143,* 34.

Rossier, R. N. (1994, October). NBAA: Shaping the future of general aviation. *Flight Training,* 62-65.

Truitt, L. J. & Tarry, S. E. (1995, Summer). The rise and fall of general aviation: Product liability, market structure, and technological innovation. *Transportation Journal, 34,* 52–70.

Woolsey, J.P. (1996, November). Good business. *Air Transport World,* 107–108, 111.

Chapter 3

Canadian Business Aircraft Association. (n.d.). *Business aviation in Canada.* [Brochure]. Montreal: Author.

Kovach, K. J. (1993). *Corporate and business aviation independent study guide.* Daytona Beach, FL: Embry-Riddle Aeronautical University.

National Business Aviation Association. (1997a). *NBAA Business aviation fact book 1997.* Washington, DC: Author.

National Business Aviation Association. (1997b). *Travel$ense.* [On-line], Available: http://www.nbaa/T$.htm. Accessed 16 May 1997.

National Business Aircraft Association & General Aviation Manufacturers Asociation. (1997). [Brochure]. *Face to Face.* Washington, DC: Author.

PRC Aviation. (1995). *Business aircraft operations: Financial benefits and intangible advantages.* Study for GAMA and NBAA. Tucson, Arizona: PRC Aviation, Inc.

Chapter 4

General Aviation Manufacturers Association. (n.d.). *Government aircraft: an essential tool.* Washington, DC: Author.

Gormley, M. (1992, January). Justifying business aircraft: the analysis. *Business and Commercial Aviation,* 54–61.

McLaren, G. (1996). Travolta's flight department. *Professional Pilot,* 66.

National Business Aircraft Association. (1982). *NBAA Recommended standards manual.* Washington, DC: Author.

National Business Aircraft Association. (1997a). *Business aviation fact book 1997.* Washington, DC: Author.

National Business Aircraft Association. (1997b). *Travel$ense Program.* [On-line], Available: http://www.nbaa/T$.htm. Accessed November 6, 1997.

National Business Aircraft Association & General Aviation Manufacturers Association. (n.d.). No plane—no gain. Washington, DC: National Business Aircraft Association.

PRC Aviation. (1995). *Business aircraft operations: financial benefits and intangible advantages.* Tucson, AZ: Author.

What, me overpaid? CEOs fight back. (1992, May 4). *Business Week,* 60.

Chapter 5

Aastad, A. (1996–97, Winter). After the first three quarters—a recap. *Rotor,* 44/46.

Boeing. (1995, December). 737-600/-700/-800 for Corporate Operations. [Brochure]. Seattle, WA: Author.

Business Jet Solutions. (n.d.). [Brochure]. Dallas, TX: Author.

Center for Professional Development. (n.d.). Aircraft selection seminar. Daytona Beach, FL: Embry-Riddle Aeronautical University.

Executive Jet Aviation. (1996). Information packet. Montvale, NJ: Author.

Federal Aviation Administration. (1997). *FAR 91.501.* [On-line], Available: http://www.faa.gov/avr. Accessed May 17 1997.

General Aviation Manufacturers Association. (1997a). *1996 General Aviation Statistical Databook.* Washington, DC: Author.

General Aviation Manufacturers Association. (1997b). *Annual industry review.* Washington, DC; Author.

National Business Aircraft Association. (1997a). *Business aviation fact book 1997.* Washington, DC: Author.

National Business Aircraft Association. (1997b). *NBAA Management Guide,* [On-line], Available: http://www.nbaa.org. Accessed November 15, 1997.

Office of Aviation Policy and Plans. (1996, March). *FAA Aviation forecasts fiscal years 1996–2007.* Washington, DC: Federal Aviation Administration.

Chapter 6

Baty, P. (1996). Women in aviation—An untapped market segment. *6th Annual FAA General Aviation Forecast Conference Proceedings.* Washington, DC: Federal Aviation Administration.
Cannon, J. R. (1988, March). A time to rest. *Business & Commercial Aviation, 67–70.*
Federal Aviation Administration. (1996, March). *FAA Aviation Forecasts.* Washington, DC: Author.
Federal Aviation Administration. (1997). *Federal aviation regulations.* [On-line], Available: http://www.faa.gov/avr. Accessed May 16, 1997.
Gormley, M. (1997, April). Taking the corner office. *Business & Commercial Aviation, 48–53.*
Higdon, D. (1996). Corporate pilot shortage? *Business Aviation Management, 3,* 61–65.
Jackson, K. S. & Brennan, J. T. (1995). *Federal Aviation Regulations Explained, Parts 1, 61, 91, 141, and NTSB 830.* Englewood, CA: Jeppesen Sanderson, Inc.
National Business Aircraft Association. (1997). *NBAA Management guide.* [On-line], Available: http://www.nbaa.org. Accessed November 18, 1997.
Olcott, J. W. (1996). Leadership. *Business Aviation Management, 3,* 1.
Quilty, S. M. (1996a). Meeting company expectations. *Business Aviation Management, 3,* 30–33.
Quilty, S. M. (1996b). What does it take to be chief? *Business Aviation Management, 3,* 26–29.
Sheehan, J. J. (1996). Talking the talk. *Business Aviation Management, 3,* 6–10.
Whempner, R. J. (1982). *Corporate aviation.* New York: McGraw-Hill.

Chapter 7

Bradley, P. (1995, December). Separation anxiety. *Business & Commercial Aviation, 86–92.*
Cannon, J. R. (1988, March). A time to rest. *Business & Commercial Aviation, 67–70.*
Federal Aviation Administration. (1996, March). *FAA Aviation forecasts.* Washington, DC: Author.
Federal Aviation Administration. (1997). *FAR Part 65.* [On-line], Available: http:www.faa.gov/avr. Accessed May 16, 1997.
Jackson, K. S. & Brennan, J. T. (1995). *Federal aviation regulations explained, Parts 1, 61, 91, 141, and NTSB 830.* Englewood, CA: Jeppesen Sanderson, Inc.
National Business Aircraft Association. (1994). *NBAA Management guide.* Washington, DC: Author.
Office of Federal Register. (1994, January 1). *Code of federal regulations: Parts 60 to 139.* Washington, DC: U.S. Government Printing Office.
Parke, R. B. (1990, September). International handling: The brave new world of business. *Business & Commercial Aviation,* 84–87.
Van Cleave, F. (1996, March 22). The development of a cockpit resource management (CRM) program in a corporate or business aviation flight department. Unpublished graduate report. Daytona Beach, FL: Embry-Riddle Aeronautical University.
Wrobel, T. B. (1995, December). Analysis of the key factors affecting international corporate aviation operations. Unpublished graduate paper. Daytona Beach, FL: Embry-Riddle Aeronautical University.

Chapter 8

Federal Aviation Administration. (1991). *AC No. 91-67. Minimum equipment requirements for general aviation operations under FAR Part 91.* Washington, DC: Author.

Federal Aviation Administration. (1997a). *FAR Part 43.* [On-line], Available: http://www.faa.gov/avr. Accessed May 17, 1997.

Federal Aviation Administration. (1997b). *FAR Part 91.* [On-line], Available: http://www.faa.gov/avr. Accessed May 17, 1997.

National Business Aircraft Association. (1997). *NBAA Management guide.* [On-line], Available: http://www.nbaa.org. Accessed May 17, 1997.

Chapter 9

Bradley, P. (1997, February). Helicopter safety statistics. *Business & Commercial Aviation,* 42–46.

Federal Aviation Administration. (1996). *Aviation system indicators.* Washington, DC: U.S. Government Printing Office.

Federal Aviation Administration. (1997). *ASRS Program overview.* [On-line], Available: http://www.faa.gov/overview.html. Accessed November 15, 1997.

Flight Safety Foundation. (1996, May 8). *FSF Safety seminar focuses on globalization of corporate flight operations.* Alexandria, VA: Author.

Griffin, J. E. (1992). Corporate participation in accident investigation. *ISASI Forum proceedings 1992.*

Kane, R. M. (1996). *Air transportation.* Dubuque, Iowa: Kendall/Hunt Publishing Company.

National Business Aircraft Association. (1994). *NBAA Management guide.* Washington, DC: Author.

National Business Aircraft Association. (1996). *1996 NBAA Business aviation fact book.* Washington, DC: Author.

National Business Aircraft Association. (1997). *NBAA Business aviation safety.* [On-line], Available: http://www.nbaa.org/culture.htm. Assessed November 14, 1997.

National Business Aircraft Association & Mescon Group. (1989). *Flight operations security and safety.* Unpublished course leader's guide.

National Transportation Safety Board. (1997). *Office of Accident Investigation.* [On-line], Available: http://www.ntsb.gov/iirform.htm. Accessed November 15, 1997.

Turbine Business Aircraft Accidents. (1997, November). *Business & Commercial Aviation,* 44.

Chapter 10

Federal Aviation Administration. (1996). Growth of pilot population. *6th Annual FAA General Aviation Forecast Conference Proceedings.* Washington, DC: Office of Policy and Plans, 175–180.

Index

A

Accident, aircraft, 257, 263-265
Accident investigations, 268-269
Accurate costing, 187
Acquisition costs, 109
Advantages of using business aircraft, 5, 49-60, 63-64, 70
Aerial application, 5
Aerial observation, 5
Affiliate members of NBAA, 37
Air carriers (air operators), 3, 28
Air crew certificates and ratings, 191, 208
Air crew expenses, 191, 207-208
Air crew members, 217, 261, 281
Air Incorporated, 181
Air taxi, 4, 27, 29, 58
Air traffic control, 15-16
Aircraft costs, 16-17, 109, 279
Aircraft fueling, 14, 208-210
Aircraft maintenance, 174, 233-254
Aircraft management companies, 106, 167
Aircraft manufacturers, 6, 11, 13, 15, 97, 104-105
Aircraft Operating Costs Form, 111
Aircraft security, 260-261
Aircraft selection, 106-113, 278-279
Aircraft selection matrix, 110-113
Aircraft types, 33, 95-101
Airline Deregulation Act, 14, 22, 28-30
Airmail Act of 1925, 22
Airmail Act of 1934, 10, 22
Airmail Acts, 9
Airmail service, 8
Airport and airway access, 53-58, 101, 281
Airport and Airway Development Act of 1970, 22
Airport and Airway Trust Fund, 17, 22
Alcohol and drugs, 28, 258
Annual aircraft costs, 109
AOPA (Aircraft Owners and Pilots Association), 3, 4, 16, 31

ASRS (Aviation Safety Reporting System), 266-267
Associate members of NBAA, 37
ATA System codes, 238, 239, 240
Autogyro, 9
Aviation categories, 3
Aviation Data Services, 64
Aviation department manager (ADM), 14, 24, 113, 114, 167-173, 174, 175, 181, 186, 192, 202, 204, 207, 208, 258, 274, 277
Aviation department organizational structure, 174-177
Aviation flight department, 14, 113, 161-194, 167
Aviation law, source/foundation, 21-22
Aviation maintenance department, 235-236
Aviation personnel, 177-180

B

BAUA (Business Aircraft Users Association), 45
Beachey, Lincoln, 6
Beech, Walter, 7, 12
Bell, 9, 13, 15, 279
Bellanca, 8
Benefits of business aircraft, 5, 49-60, 63-64
Boeing's *Winds on World Air Routes,* 214
Budget, flexible, 186
Budgeting, 186-187, 277
Business aviation (defined), 1-2
Business aviation associations, 33-46
Business members of NBAA, 37
Business strategies, 17-18, 87-88
Business tool, 16, 116
Business transportation, 2, 5
Buying aircraft, 113-114

C

CAMP Systems Incorporated, 203, 243-248
Canadair, 15
Capital lease, 114

CBAA (Canadian Business Aircraft Association), 35, 39-43, 58
Certificate of public convenience and necessity (Section 401), 4
Cessna, Clyde, 6
Chartering, 114-115
Chief pilot, 172, 179
City-pair service, 28
Civil Aeronautics Board, 3, 14, 22
Cockpit resource management (CRM), 174, 215-216
Code of Federal Regulations, 25
Combination aircraft maintenance, 242-249
Commerce clause, 21
Commercial air carriers, 3, 5, 14-15, 16, 28-29
Commercial operators, 4, 27
Common carrier, 72
Communications, 212
Commuters, 4, 14, 29
Comparative analysis, 70, 71, 85
Consideration, in decision making, 70
Contract aircraft maintenance, 242
Contract carrier, 72
Copilot, 179
Corporate aviation (defined), 1-3
Corporate members of NBAA, 36-37
Cost, 14, 16, 18, 52, 59-60, 70-71, 72, 187
Costs of acquisition, 109
Costs of possession, 109
Crew duty limitations, 203-204
Crew expenses, 191, 207-208
Crew rest, 204
Curtis, Glenn H., 6

D

Dassault, 15
Data collection, 78, 84
Department of Transportation, 14, 22, 30
Depression years, 7
Disadvantages of business aircraft, 61-62, 70, 71
DLAND (Development of Landing Areas for National Defense) Act, 8
Documentation, 212
Douglas, Donald, 7

Drugs and weapons, 207
Dry lease, 114
Duty-time, 174, 204

E

EBAA (European Business Aviation Association), 44
Economic rules, 25
Embry-Riddle Aeronautical University, 61, 106, 107, 172, 181, 280
Emergency communications, 260
Emergency equipment, 261
Emergency procedures, 190
Enroute operations, 214
Environment, 14, 27, 209
Environmentalists, 14
Equal-time point (ETP), 213-214
Executive transportation, 2
Executive/Corporate transportation, 5
Extended range operations (ETOPS), 216, 279-280

F

FAA (Federal Aviation Administration), 2, 3, 5, 14, 18, 22, 92, 189, 207, 211, 235, 262, 264
FAA regional boundaries, 23-24
Fairchild, 7, 8
Falcon, 15
FAR (Federal Aviation Regulation), 2-4, 22, 24-28, 61, 72, 115, 188, 189, 191, 200, 207, 208, 211, 234-236, 267
Federal Aviation Act of 1958, 22
Federal regulators, 21-22, 27
Financing aircraft, 17, 28
Fixed-base operator (FBO), 162-166, 209, 242
Fixed costs of an aircraft, 109
Fixed expenses, 187
Flexibility as a benefit of business aircraft, 5, 50, 51, 52, 53
Flight attendants, 177, 179, 186
Flight crew members, 217, 261, 281
Flight department purpose, 167
Flight handlers, 212

Flight operations, 174, 193, 197-231, 277, 281-282
Flight personnel other than pilots and mechanics, 185-186
Flight planning software, 219-227, 282
Flight preparations, 210-214
Flight Standards District Office (FSDO), 234, 236
Flight-time, 174, 204
FlightPak (software), 223, 228
Flying machine, 6
Fokker, 7, 8
Fortune 500 Study, 64-65
Fractional ownership, 17, 115-117, 275
Fringe benefit rule, 30
Fuel, 14, 15, 208-210
Fuel crises, 14, 16

G

GAMA (General Aviation Manufacturer's Association), 3, 4, 5, 16, 17, 30, 31, 52, 65, 93, 95
GBAA (German Business Aircraft Association), 36, 45
General aviation, 4-5, 92-101
General aviation aircraft types and uses, 5
General Aviation Revitalization Act (GARA), 16, 17, 30-33, 280
General Aviation Team 2000, 181, 277
Global competition (marketplace), 15, 16, 70
Government carrier, 72
Ground transportation elements, 56
Grumman, 12, 13
Gulfstream, 12, 15, 279

H

Health of the air crew, 212
Helicopters, 9, 13, 15, 96
Hiring model, 182
Honeywell, 10
Hub-and-spoke, 14-15, 28-29, 281
Hubbing, 28
Human resources, 167, 172, 192, 193

I

IATA, 34
IBAC (International Business Aviation Council), 33-34, 35, 39, 176, 219
IBAC members, 34
ICAO (International Civil Aviation Organization), 24, 34, 176, 217
In-house maintenance, 241-242
Incident, aircraft, 257
Indirect carrier, 72
Industrial Revolution, 6
Inspection programs, aircraft, 234-235
Instructional aviation, 5
Instrument rating, 191, 208
Insurance, 31
Intangible benefits, 51, 53, 59, 70
Interchange, 115
International airmail service, 8
International Flight Information Manual (IFIM), 211, 217, 218
International flight planning, 210-211, 213
International flights, 15, 198, 199, 202, 205, 217
International operations checklist, 218
Intoxicant or drug use, 207
Investment tax credit, 17, 18

J

Jet Star, 12, 13, 15
Job descriptions, 177-180
Joint ownership, 115

K

King Air, 12
Knox, Charles B., 6
Knox Gelatin, 6
Korean War, 11, 13

L

Labor, 27, 28
Large certificated air carriers, 3
Leadership of an ADM, 168-170

Index **369**

Lear, William, 11
Learjet, 12-13, 279
Learn to Fly program, 181, 278
Leasing aircraft, 17, 114
Lindbergh, Charles, 8
Lockheed (Loughead), 6-7, 8, 12, 15
Luxury tax, 17

M

Maintenance, combination aircraft, 242-249
Maintenance, contract aircraft, 242
Maintenance department, 174, 235-236
Maintenance, in-house, 241-242
Maintenance issues, 236-238
Maintenance methods, 238-249
Maintenance personnel, 181
Major airlines, 3
Management, 11, 14, 15, 39, 58, 66-67, 167-173, 277-278
Management accountability factor, 76
Martin, Glenn L., 7
Master Mechanic award, 234
Mechanics, number of, 183-184
MEL (Minimum equipment list), 236-238
Millennium factors, 92
MMEL (Master minimum equipment list), 236-237
MNPS (Minimum Navigation Performance Specifications), 211
Modes of transportation, 3, 6

N

NASA (National Aeronautics and Space Administration), 22, 256
NATA (National Air Transportation Association), 31
National airlines, 3
National Airspace System Plan, 16
NBAA (National Business Aviation Association), 1, 2, 4, 5, 17, 30, 31, 34-39, 50, 52, 58-59, 65, 66, 67, 72, 85, 87, 106, 172, 183, 185, 189, 191, 208, 210, 218, 235-236, 262, 267

NBAA fleet, 98-101
NBAA, members of, 36-37
NBAA publications, 39, 40
Necessary and proper clause, 21
No Plane-No Gain, 17, 66, 87
Noise, 14, 16, 26, 28
Noise Control Act, 14, 22
Non-economic rules, 25
Northrop, John K., 7
Notices to airmen (NOTAMs), 212
NTSB (National Transportation Safety Board), 22, 27, 256-257, 262
NTSB safety programs, 267-269

O

Operations manual, 61, 174, 177, 188-190, 193, 206, 207
Optimizing costs, 70-71
Organizational structure of an aviation department, 174-177

P

PAMA (Professional Aviation Maintenance Association), 234
PANS (Procedures for Air Navigation), 24, 218
Passenger briefing, 206, 207, 214
Personal appearance and conduct, 205-206
Personal/Recreation, 5
Personnel, aviation, 177-180
Pilot certification, 15, 18, 104, 191, 278
Pilot decision sequence, 237-238
Pilot recruitment, 180-181
Pilot-in-command (PIC), 191, 207, 208, 217
Pilot's Choice (software), 223, 229
Pilots, number of, 183-184
Pilots, qualified, 101, 103-104
Placement of aviation department, 177
Planning times, 58
Postflight actions, 216-217, 261
PPS (Pre-Flight Planning System), 223, 224-227
PRC valuation methods, 53-57, 73-75
Preflight planning, 210-211
Private carrier, 72

Private pilot, 18, 101, 191, 208
Product liability, 16, 17, 30-31
Productivity, 5, 9, 14, 16, 58-60, 66-67
Project management, 106-109
Project Pilot, 181, 278
Pseudo lease, 114

R

Recession, 14
Recruitment sources, 180-181
Reference checks, 182
Regional airlines, 3
Regulations, 21-33, 279-280
Rest-time, 174, 204, 265
Rockwell, 12
RON (Remain overnight), 204
Royal barge, 2, 14
RVSM (Reduced Vertical Separation Minimum), 211

S

Safety, aviation, 39, 73, 193, 261-270
Safety programs, 265-268
Safety and security, 255-271, 276
Safety statistics, 262-263
Salaries, 77, 184-185
Salary multipliers, 78
SARPS (Standards and Recommended Practices), 24, 217-218
SaSIMS System (Safety by Simplicity Maintenance Management System), 248-253
Scheduling/dispatching, 200-203, 260, 282
Second-in-command, 179, 208
Security, 257-261
Security, information/physical, 258-259
Selection process, 181-183
SIFL (Standard Industry Fare Level), 30
Sikorsky, Igor I., 8, 9, 13
Small regionals, 4
SMART OPS, 275-282
Spreadsheet, 79
State control, 28

Subjective assessment, 112-113
Sun Flight program, 181

T

Tangible benefits, 51, 52-53, 59, 70
Tax considerations, 17
Tax Reform Act, 16, 17
Taxes, 17, 28, 64
Team concept, 106-107, 112-113, 243
Technological advances, 9, 17, 18, 280
Theft, internal/external, 258-259
Three C's, 70
Time savings, 52, 55, 59-60, 72-73, 76
Time sharing, 115
Transportation carriers, 41
Transportation modes, 3, 6, 22, 72
Transportation request, 200
Transportation Safety Act of 1974, 22
Travel Air, 7, 8
Travel analysis, 67, 78-85
Travel pattern graph, 81-85
Travel pattern map, 80-81
Travel request form, 201, 202
Travel$ense® program, 52, 56, 58, 59, 64, 66-67, 85-86, 187
Trimotor, 7, 10
Turbojet, 9, 17
Turboprop, 9, 12, 17, 31, 58, 95, 100

U

U.S. civil aviation categories, 3-4
U.S. Joint-use airports, 102
U.S. Post Office, 8, 22, 27

V

Value of business aircraft, 49-60, 63-64
Valuing a person's worth, 73-78
Variable costs of an aircraft, 109
Variable expenses, 187
Vega, 8

W

Wet footprint, 213
Wet lease, 114
Whempner, Robert, 6, 7, 10
Wings at Work, 8, 13, 65
Women in Aviation, International, 181, 278
World War I, 6, 7
World War II, 8, 9-11, 13, 28
Wright brothers, 6

Y

Young Eagles program, 181, 277